Focus on Exercise Physiology and Sports Performance

Guest Editor
Laikang Yu
Department of Strength and
Conditioning Assessment and
Monitoring
Beijing Sport University
Beijing
China

Editorial Office
MDPI AG
Grosspeteranlage 5
4052 Basel, Switzerland

This is a reprint of the Special Issue, published open access by the journal *Life* (ISSN 2075-1729), freely accessible at: https://www.mdpi.com/journal/life/special_issues/T0OIL0J01Y.

For citation purposes, cite each article independently as indicated on the article page online and as indicated below:

Lastname, A.A.; Lastname, B.B. Article Title. *Journal Name* **Year**, *Volume Number*, Page Range.

ISBN 978-3-7258-3721-2 (Hbk)
ISBN 978-3-7258-3722-9 (PDF)
https://doi.org/10.3390/books978-3-7258-3722-9

Contents

About the Editor

Laikang Yu

Laikang Yu is an Associate Professor at Beijing Sport University. He earned his PhD in Sports Science from Beijing Sport University and spent 16 months as an exchange scholar in the Department of Pharmacology and Toxicology at the Medical College of Wisconsin (MCW). His research focuses on exercise and neuroplasticity, as well as exercise and health promotion. He has authored over 35 SCI-indexed scientific papers, either as the first or corresponding author, and has led or participated in 22 research projects. Currently, he serves as an Academic Editor for 6 SCI journals and as a reviewer for 34 SCI journals.

life

MDPI

Editorial

Focus on Exercise Physiology and Sports Performance

Laikang Yu [1,2]

[1] Beijing Key Laboratory of Sports Performance and Skill Assessment, Beijing Sport University, Beijing 100084, China; yulaikang@126.com
[2] Department of Strength and Conditioning Assessment and Monitoring, Beijing Sport University, Beijing 100084, China

1. Introduction

Exercise physiology is a crucial scientific discipline that explores the complex manner in which physical activity influences the physiological responses and adaptations of the human body [1,2]. With the development of exercise science, more emphasis has been placed on understanding the underlying physiological processes that can augment sports performance [3]. This is essential because performance hinges not only on technical skill and psychological factors but also on an individual's physiological condition. Recent studies have demonstrated the complex interaction between various physiological systems during exercise, especially the cardiovascular, pulmonary, and metabolic responses, which are crucial for optimizing both performance and health outcomes [4–6].

Aerobic exercise, for instance, primarily demands energy through oxidative adenosine triphosphate (ATP) synthesis. The delicate balance between oxygen consumption and carbon dioxide production within muscle cells is closely related to maximal oxygen uptake (VO_2max), a key indicator of aerobic capacity and a predictor of sports performance across diverse populations [7]. This correlation highlights the critical role of the cardiovascular and pulmonary systems in sustaining physical activity. Moreover, the physiological changes that occur with regular exercise, such as enhanced mitochondrial function and improved oxygen delivery, are pivotal in enhancing performance and overall health [8].

Contemporary research has further emphasized the importance of understanding the physiological reactions to various exercise modalities, including high-intensity interval training (HIIT) [9,10]. These forms of exercise elicit distinct metabolic responses and adaptations, influencing factors such as muscle hypertrophy, strength development, and metabolic flexibility [11,12]. Additionally, the significance of exercise in managing chronic diseases and neurodegenerative diseases has gained increasing recognition, making it a fundamental part of preventive healthcare [13,14].

In short, exercise physiology forms the basis for understanding how physical activity affects human health and performance. By exploring the physiological mechanisms involved, researchers and practitioners can develop more effective training programs and interventions to enhance sports performance and promote overall wellbeing.

Received: 2 January 2025
Accepted: 10 January 2025
Published: 12 January 2025

Citation: Yu, L. Focus on Exercise Physiology and Sports Performance. *Life* **2025**, *15*, 84. https://doi.org/10.3390/life15010084

2. An Overview of the Published Articles

The goal of the Special Issue, "Focus on Exercise Physiology and Sports Performance", is to bring together the latest research that examines how the body adapts to exercise and provides insights into how training can enhance performance and optimize an athlete's capabilities. Thirty-two manuscripts were submitted for consideration for the Special Issue, and thirteen papers were finally accepted for publication and inclusion. The key contents and findings of each paper are as follows.

Zhu et al. (contribution 1) explored how birth season and gender influence the gross and fine motor skills development of 2-year-olds. Their results indicated no gender effect on overall motor skills but a notable influence of birth season on fine motor quotient and total motor quotient. Notably, girls born during winter months demonstrated superior fine motor skills in comparison to those born in summer, hinting at crucial seasonal environmental impacts, especially for girls. These factors should be taken into account when designing early childhood programs aimed at advancing sports performance.

In a meta-analysis by Hu et al. (contribution 2), the effectiveness of isoinertial flywheel training (FWT) versus traditional resistance training (TRT) in augmenting muscle strength and power among healthy participants was examined. Their results confirmed that FWT is superior to TRT in enhancing muscle power, particularly when performed with squat and lunge exercises and within a specific session range, in both healthy untrained and well-trained individuals. Consequently, they advised coaches to integrate FWT into training programs for varied stimuli and improved power, suggesting a 6-week implementation with 2–3 sessions weekly, allowing at least a 48 h gap between sessions.

The study by Wang et al. (contribution 3) evaluated the immediate impact of various breath-hold (BH) scenarios on the aerobic fitness of 18 male elite rugby athletes. They found that among the various BH conditions, dynamic dry BH warm-up significantly improved aerobic fitness indicators such as peak oxygen uptake (VO_2peak) and peak stroke volume (SVpeak) in elite rugby players, with these improvements strongly correlated with changes in red blood cells and hematocrit, suggesting that dynamic dry BH warm-up optimizes subsequent aerobic performance.

Ye et al. (contribution 4) designed and validated a comprehensive assessment framework for physical fitness in elite male badminton singles players, employing the Delphi method and analytic hierarchy process. This framework, comprising three primary, nine secondary, and twenty-one tertiary indicators, proved highly feasible and valid in assessing athletes' fitness, providing practical guidance for enhancing competitive performance. This aids coaches in formulating targeted training plans and optimizing outcomes.

The study by Ieong et al. (contribution 5) investigated the relationship between musculoarticular stiffness and pedaling rate in sprint cycling, finding that participants with higher musculoarticular stiffness exhibited higher pedaling rates, peak crank force, and rate of crank force development, leading to higher power output in a 6 s sprint cycling test. The results suggest that optimizing cycling resistance or gear ratio to enhance these factors may be crucial for improving sprint cycling performance.

A meta-analysis conducted by Zhou et al. (contribution 6) focused on the impact of exercise on cancer-related fatigue in people with breast cancer, aiming to establish an optimal exercise prescription. They found that exercise interventions, particularly combining aerobic and resistance exercise conducted ≥ 3 times weekly for more than 60 min each session, totaling 180 min weekly, significantly improve cancer-related fatigue in breast cancer patients, with middle-aged patients benefiting the most. The optimal exercise prescription for breast cancer patients should incorporate combined exercise as the principal intervention.

In Hu et al.'s study (contribution 7), the effectiveness of inertial flywheel training (FWT) versus accentuated eccentric loading training (AELT) in augmenting neuromuscular performance among well-trained male college sprinters was examined. Their results showed that both training methods significantly improved lower-body strength, power, and speed. Notably, FWT demonstrated a superior ability to enhance the elastic energy storage and stretch-shortening cycle (SSC), as evidenced by greater improvements in countermovement jump (CMJ) and eccentric utilization ratio (EUR) compared to AELT.

The study by Zhou et al. (contribution 8) explored how transcranial direct current stimulation (tDCS) combined with resistance training influenced jump performance and brain activity using electroencephalography. Their results showed that the combination significantly enhanced vertical jump height and altered α-wave and β-wave power in frontal and temporal lobes, compared to either intervention alone.

The study by Ni et al. (contribution 9) analyzed changes in the electromyogram properties of agonist and antagonist muscles during a fatiguing heel-raise task. They found that during the initial stage of fatigue, both muscles exhibited increased activity and decreased mean frequency, with later differentiation in control strategies, where agonist muscle parameters stabilized, while antagonist muscle activity decreased. The findings suggest a differentiated control strategy by the central nervous system, enhancing the understanding of neuromuscular adaptations during fatigue and potential strategies for fatigue assessment and management.

Li et al. (contribution 10) investigated how stroboscopic visual conditions affected the key performance aspects of elite curling athletes. Their results showed that stroboscopic conditions significantly increased errors in these areas, suggesting that such conditions could be used as a training tool to enhance elite curling performance by challenging athletes to adapt and improve under limited visual information. Therefore, stroboscopic training has the potential to improve elite curling performance by enhancing the visual processing speed, reaction time, and motor skill control.

Bagchi et al. (contribution 11) found that the K-Deltas force platform exhibited high reliability and strong correlations with the other two tools, although small but consistent measurement differences were observed. Despite these discrepancies, the study highlights the K-Deltas force platform as a feasible choice for assessing CMJ, suggesting its potential use in monitoring training, predicting injury risk, assessing neuromuscular fatigue, and informing decision making in practical settings.

The study by Zhang et al. (contribution 12) examined the correlations between sprint force–velocity (Fv) characteristics and change of direction (COD), along with their impact on asymmetries in COD speed performance in volleyball and basketball players. Their results showed that the velocity (V_0) and maximal ratio of force (RFmax) were crucial for improving COD performance, including linear sprints. Meanwhile, the force application technique (D_{RF}), the force (F_0), the ratio between F_0 and V_0 (Fv_{slope}), and the maximal power (Pmax) collectively influenced the 180° COD performance. Additionally, the DRF and Fv_{slope} were important factors for asymmetries in COD speed performance.

Han et al. (contribution 13) conducted a meta-analysis to evaluate the impact of inspiratory muscle training (IMT) on chronic obstructive pulmonary disease (COPD) patients. By reviewing data from sixteen studies, they found that IMT significantly improved all three outcomes, and subgroup analysis revealed that IMT at $< 60\%$ maximal inspiratory muscle pressure (PImax), conducted for ≤ 20 min and more than 3 times per week, had the greatest benefits. Therefore, they recommended that COPD patients engage in IMT with these specific parameters to enhance their inspiratory muscle strength, reduce dyspnea, and improve their quality of life.

3. Conclusions

Exercise physiology involves many complex factors that greatly affects sports performance. Current research indicates that understanding key aspects such as energy systems, oxygen uptake, cardiovascular adaptations, muscle physiology, and effective recovery strategies is crucial for developing scientifically based training programs. These factors do not work independently; instead, they interact dynamically to shape an athlete's overall abilities. Given these findings, a one-size-fits-all training method is not enough.

Individual physiological responses to exercise can vary widely, highlighting the need for customized training regimens that account for an athlete's unique characteristics. This requires ongoing research that not only explores the individual parts of exercise physiology but also introduces ideas from other fields like nutrition, psychology, and biomechanics.

Future studies should focus on personalized training methods, using new technology and data analysis to better understand how to best adapt training for different athletes. Additionally, using wearable devices and real-time monitoring tools can help with more precise recovery and training strategies, enhancing performance outcomes. As we move forward, it is crucial to consider all the different ideas and findings in the field. While some studies might focus more on certain physiological factors, having a complete view that recognizes how these systems all connect provides a more comprehensive understanding of sports performance. By working together across different fields and encouraging creative research, we can make large strides that not only benefit elite athletes but also improve health and fitness among the general population.

Funding: This research was funded by the Humanities and Social Science Fund of Ministry of Education of China, grant number 24YJC890065.

Conflicts of Interest: The author declares no conflicts of interest.

List of Contributions:

1. Zhu, Y.; Wang, S.; Qian, Y.; Hu, J.; Zhou, H.; Korivi, M.; Ye, W.; Zhu, R. The Impact of Birth Season and Sex on Motor Skills in 2-Year-Old Children: A Study in Jinhua, Eastern China. *Life* **2024**, *14*, 836. https://doi.org/10.3390/life14070836.
2. Hu, Z.; Liu, Y.; Huang, K.; Huang, H.; Li, F.; Yuan, X. Comparing the Effect of Isoinertial Flywheel Training and Traditional Resistance Training on Maximal Strength and Muscle Power in Healthy People: A Systematic Review and Meta-Analysis. *Life* **2024**, *14*, 908. https://doi.org/10.3390/life14070908.
3. Wang, W.; Wu, D.; Wang, H.; Zhang, Z.; Jiang, X.; Li, S.; Shi, Y.; Gao, X. Acute Effects of Breath-Hold Conditions on Aerobic Fitness in Elite Rugby Players. *Life* **2024**, *14*, 917. https://doi.org/10.3390/life14080917.
4. Ye, B.; Zhu, H.; Yang, Z.; He, Z.; Liu, G.; Pan, H.; Guo, H. Construction and Analysis of the Physical Fitness Evaluation Index System for Elite Male Singles Badminton Players: Based on Delphi and AHP Methods. *Life* **2024**, *14*, 944. https://doi.org/10.3390/life14080944.
5. Ieong, L.; Ni, X.; Xie, H.; Liu, Y. Impact of Stiffness of Quadriceps on the Pedaling Rate of Maximal Cycling. *Life* **2024**, *14*, 956. https://doi.org/10.3390/life14080956.
6. Zhou, R.; Chen, Z.; Zhang, S.; Wang, Y.; Zhang, C.; Lv, Y.; Yu, L. Effects of Exercise on Cancer-Related Fatigue in Breast Cancer Patients: A Systematic Review and Meta-Analysis of Randomized Controlled Trials. *Life* **2024**, *14*, 1011. https://doi.org/10.3390/life14081011.
7. Hu, Z.; Liu, Y.; Huang, K.; Huang, H.; Zhang, Y.; Yuan, X. Effects of Inertial Flywheel Training vs. Accentuated Eccentric Loading Training on Strength, Power, and Speed in Well-Trained Male College Sprinters. *Life* **2024**, *14*, 1081. https://doi.org/10.3390/life14091081.
8. Zhou, Y.; Zhai, H.; Wei, H. Acute Effects of Transcranial Direct Current Stimulation Combined with High-Load Resistance Exercises on Repetitive Vertical Jump Performance and EEG Characteristics in Healthy Men. *Life* **2024**, *14*, 1106. https://doi.org/10.3390/life14091106.
9. Ni, X.; Ieong, L.; Xiang, M.; Liu, Y. Analysis of Wavelet Coherence in Calf Agonist-Antagonist Muscles during Dynamic Fatigue. *Life* **2024**, *14*, 1137. https://doi.org/10.3390/life14091137.
10. Li, T.; Zhang, C.; Wang, X.; Zhang, X.; Wu, Z.; Liang, Y. The Impact of Stroboscopic Visual Conditions on the Performance of Elite Curling Athletes. *Life* **2024**, *14*, 1184. https://doi.org/10.3390/life14091184.
11. Bagchi, A.; Raizada, S.; Thapa, R.K.; Stefanica, V.; Ceylan, H.İ. Reliability and Accuracy of Portable Devices for Measuring Countermovement Jump Height in Physically Active Adults: A Comparison of Force Platforms, Contact Mats, and Video-Based Software. *Life* **2024**, *14*, 1394. https://doi.org/10.3390/life14111394.

12. Zhang, Z.; Jiang, M.; Jing, Y.; Li, M.; Li, Y.; Yang, X. Associations Between Sprint Mechanical Properties and Change of Direction Ability and Asymmetries in COD Speed Performance in Basketball and Volleyball Players. *Life* **2024**, *14*, 1434. https://doi.org/10.3390/life14111434.
13. Han, B.; Chen, Z.; Ruan, B.; Chen, Y.; Lv, Y.; Li, C.; Yu, L. Effects of Inspiratory Muscle Training in People with Chronic Obstructive Pulmonary Disease: A Systematic Review and Meta-Analysis. *Life* **2024**, *14*, 1470. https://doi.org/10.3390/life14111470.

References

1. Joyner, M.J.; Dempsey, J.A. Physiological Redundancy and the Integrative Responses to Exercise. *Cold Spring Harb. Perspect. Med.* **2018**, *8*, a029660. [CrossRef] [PubMed]
2. Heinonen, I.; Kalliokoski, K.K.; Hannukainen, J.C.; Duncker, D.J.; Nuutila, P.; Knuuti, J. Organ-specific physiological responses to acute physical exercise and long-term training in humans. *Physiology* **2014**, *29*, 421–436. [CrossRef] [PubMed]
3. Uchida, M.; Fujie, S.; Yano, H.; Iemitsu, M. Aerobic exercise training-induced alteration of gut microbiota composition affects endurance capacity. *J. Physiol.* **2023**, *601*, 2329–2344. [CrossRef] [PubMed]
4. Ashcroft, S.P.; Stocks, B.; Egan, B.; Zierath, J.R. Exercise induces tissue-specific adaptations to enhance cardiometabolic health. *Cell Metab.* **2024**, *36*, 278–300. [CrossRef] [PubMed]
5. Thyfault, J.P.; Bergouignan, A. Exercise and metabolic health: Beyond skeletal muscle. *Diabetologia* **2020**, *63*, 1464–1474. [CrossRef] [PubMed]
6. Iellamo, F. Acute responses and chronic adaptations to exercise in humans: A look from the autonomic nervous system window. *J. Sports Med. Phys. Fitness* **2024**, *64*, 137–150. [CrossRef] [PubMed]
7. Sietsema, K.E.; Rossiter, H.B. Exercise Physiology and Cardiopulmonary Exercise Testing. *Semin. Respir. Crit. Care Med.* **2023**, *44*, 661–680. [CrossRef] [PubMed]
8. Tanaka, T.; Nishimura, A.; Nishiyama, K.; Goto, T.; Numaga-Tomita, T.; Nishida, M. Mitochondrial dynamics in exercise physiology. *Pflugers Arch.* **2020**, *472*, 137–153. [CrossRef]
9. Van der Stede, T.; Van de Loock, A.; Lievens, E.; Yigit, N.; Anckaert, J.; Van Thienen, R.; Weyns, A.; Mestdagh, P.; Vandesompele, J.; Derave, W. Transcriptomic signatures of human single skeletal muscle fibers in response to high-intensity interval exercise. *Am. J. Physiol. Cell Physiol.* **2024**, *327*, C1249–C1262. [CrossRef] [PubMed]
10. Moncion, K.; Rodrigues, L.; De Las Heras, B.; Noguchi, K.S.; Wiley, E.; Eng, J.J.; MacKay-Lyons, M.; Sweet, S.N.; Thiel, A.; Fung, J.; et al. Cardiorespiratory Fitness Benefits of High-Intensity Interval Training After Stroke: A Randomized Controlled Trial. *Stroke* **2024**, *55*, 2202–2211. [CrossRef] [PubMed]
11. Thomas, A.C.Q.; Stead, C.A.; Burniston, J.G.; Phillips, S.M. Exercise-specific adaptations in human skeletal muscle: Molecular mechanisms of making muscles fit and mighty. *Free Radic. Biol. Med.* **2024**, *223*, 341–356. [CrossRef]
12. Hernández-Lepe, M.A.; Hernández-Ontiveros, D.A.; Chávez-Guevara, I.A.; Ramos-Jiménez, A.; Hernández-Torres, R.P.; López-Fregoso, R.J.; Ramos-Lopez, O.; Amaro-Gahete, F.J.; Muñiz-Salazar, R.; Olivas-Aguirre, F.J. Impact of Exercise Training at Maximal Fat Oxidation Intensity on Metabolic and Epigenetic Parameters in Patients with Overweight and Obesity: Study Protocol of a Randomized Controlled Trial. *J. Funct. Morphol. Kinesiol.* **2024**, *9*, 214. [CrossRef] [PubMed]
13. Du, L.; Xi, H.; Zhang, S.; Zhou, Y.; Tao, X.; Lv, Y.; Hou, X.; Yu, L. Effects of exercise in people with multiple sclerosis: A systematic review and meta-analysis. *Front. Public Health* **2024**, *12*, 1387658. [CrossRef] [PubMed]
14. Qiu, B.; Zhou, Y.; Tao, X.; Hou, X.; Du, L.; Lv, Y.; Yu, L. The effect of exercise on flow-mediated dilation in people with type 2 diabetes mellitus: A systematic review and meta-analysis of randomized controlled trials. *Front. Endocrinol.* **2024**, *15*, 1347399. [CrossRef] [PubMed]

life

MDPI

Article

Reliability and Accuracy of Portable Devices for Measuring Countermovement Jump Height in Physically Active Adults: A Comparison of Force Platforms, Contact Mats, and Video-Based Software

Amritashish Bagchi [1], Shiny Raizada [1], Rohit K. Thapa [1], Valentina Stefanica [2,*] and Halil İbrahim Ceylan [3,*]

[1] Symbiosis School of Sports Sciences, Symbiosis International (Deemed University), Pune 412115, India; amritashish.bagchi@ssss.edu.in (A.B.); shiny.raizada@ssss.edu.in (S.R.); rohit.thapa@ssss.edu.in (R.K.T.)

[2] Department of Physical Education and Sport, Faculty of Sciences, Physical Education and Informatics, National University of Science and Technology Politehnica Bucharest, 110040 Pitesti, Romania

[3] Department of Physical Education of Sports Teaching, Faculty of Sports Sciences, Atatürk University, Erzurum 25240, Türkiye

* Correspondence: valentina.stefanica@upb.ro (V.S.); halil.ceylan@atauni.edu.tr (H.İ.C.)

Abstract: Measuring countermovement jump (CMJ) height accurately is essential for evaluating lower-body explosive power in athletes and other active populations. With technological advancements, various portable tools have been developed for this purpose, including force platforms, contact mats, and video-based software. This study aimed to (a) investigate the test–retest reliability of the KINVENT K-Deltas force platform for CMJ height measurement and (b) compare its accuracy with a contact mat (Chronojump, Spain) and a video-based software (My Jump app, version 3). Twenty-two physically active collegiate athletes (mean age of 19.7 ± 1.2 years) from various sports backgrounds completed five CMJ trials with simultaneous height measurements using all three tools. Intra-class correlation coefficients (ICC), Cronbach's alpha, and coefficient of variation (CV) were calculated to assess reliability. In contrast, Pearson correlations and Bland–Altman plots were used to compare device results. The K-Deltas force platform exhibited high test–retest reliability (ICC = 0.981), closely matching the contact mat (ICC = 0.987) and the My Jump app (ICC = 0.986). Correlations between the instruments were strong (force platform vs. contact mat: r = 0.987; force platform vs. My Jump: r = 0.987; contact mat vs. My Jump: r = 0.996), with no between-instrument differences (*t*-test p = 0.203–0.935, effect size ≤ 0.01–0.16), demonstrating the interchangeability of these tools for practical purposes. However, Bland–Altman analysis revealed limits of agreement between the devices, indicating small but consistent measurement differences. While all instruments were reliable, discrepancies in the absolute values suggest practitioners should consider device-specific variations when comparing CMJ data. These findings highlight the reliability of the K-Deltas force platform as a viable alternative for measuring CMJ height, though differences between devices should be accounted for in applied settings. Therefore, the portable force plates can monitor training, predict injury risk, assess neuromuscular fatigue, and lead to informed decision-making.

Keywords: plyometric exercise; athletic performance; physical fitness; validity; testing battery; motor activity

Citation: Bagchi, A.; Raizada, S.; Thapa, R.K.; Stefanica, V.; Ceylan, H.İ. Reliability and Accuracy of Portable Devices for Measuring Countermovement Jump Height in Physically Active Adults: A Comparison of Force Platforms, Contact Mats, and Video-Based Software. *Life* **2024**, *14*, 1394. https://doi.org/10.3390/life14111394

Academic Editor: Nicola Smania

Received: 24 September 2024
Revised: 23 October 2024
Accepted: 28 October 2024
Published: 29 October 2024

1. Introduction

The countermovement jump (CMJ) assesses the stretch-shortening cycle (i.e., eccentric, amortization, and concentric phases) function of the lower body and is amongst the most frequently used tests to measure lower body strength and power capabilities due to the simplicity of the testing procedure [1]. Indeed, previous studies have reported that CMJ requires the least number of familiarization sessions compared to other vertical jump assessments [2–4]. Due to the test being simple to learn and perform, CMJ is used by practitioners for multiple reasons. For example, CMJ is used to evaluate the effectiveness of training [5], monitor neuromuscular fatigue [6], measure inter-limb asymmetry [7], predict injury risk [8], and guide the return to sports decision-making process [9]. Similarly, researchers have conducted studies in various sports where the CMJ test was used to assess athletes' explosive qualities and their relationship with other variables [10,11]. Of note, CMJ has also been reported to be associated with other physical fitness outcomes such as sprinting, jumping, and change in direction [5]. In addition, the CMJ tests have been suggested to differentiate between elite versus sub-elite athletes [12,13] and professional versus amateur athletes [14,15].

With the advancement of science, several technologies (e.g., force platforms) have emerged to measure the CMJ performance of individuals, with the laboratory-based force platform considered the gold standard [16]. However, many portable technologies have emerged as we transition from laboratory-based testing (i.e., athletes visiting the laboratory) to creating a mobile laboratory (i.e., bringing the laboratory to the athlete). Some examples of such portable technologies are video-based mobile software (e.g., My Jump Lab [17]), contact platforms (e.g., Chronojump [18]), and wireless force platforms. Although new technologies are frequently developing and older technologies are becoming redundant, assessing the reliability and validity of the new technologies is important [19,20]. In simple words, the technology should provide the same measurement if conducted multiple times (i.e., reliability), and it should measure the outcome that it intends to measure (i.e., validity) [21]. Moreover, without knowing the validity and reliability of technology, practitioners would have very little confidence in the outcomes generated by the technology.

Therefore, assessing the validity and reliability of new technologies is important. For example, the video-based mobile software has been found reliable when compared against force platforms (r = 0.98, $p \leq 0.01$, interclass correlation coefficient [ICC] = 0.997) and contact mats (r = 0.99, $p \leq 0.01$, ICC = 0.948) [17]. Similarly, portable contact mats have also been found reliable (r = 0.99, $p \leq 0.01$, ICC = 0.99) against a proprietary jump mat (Globus Ergo Tester) [18]. In line with new emerging technologies, portable force platforms are also gaining popularity with a significant price decrease compared to the very high cost of embedded force platforms. This makes portable force platforms accessible to a larger population. However, with the popularity of portable force platform technology, it is also very important to assess the validity and reliability of those instruments.

One of the new portable force platform technologies (K-Deltas) on the market is manufactured by Kinvent Physio (Montpellier, France). The K-Deltas measure the CMJs with a high sampling rate of 1000 Hz and have a user-friendly interface. It is equipped with a unidirectional strain gauge (vertical axis; records vertical force data) that is easily connected to a mobile device (Android or iOS) via Bluetooth and can be a good alternative for traditional in-ground force platforms [22,23]. As discussed previously, ensuring the validity and reliability of such new technologies is important in maintaining scientific integrity and enhancing the confidence of practitioners in using those technologies [24]. For instance, Mylonas, Chalitsios, and Nikodelis [23] have established the validity of the K-Deltas using the gold standard Bertec Force platform during bipedal stance, measuring the ground reaction force and center of pressure. Considering the information and studies mentioned above, CMJ is crucial for assessing lower-body explosive power, but the tools used for measurements, such as force platforms, contact mats, and video-based software, vary in accessibility and cost. While force platforms are the gold standard, they are often expensive and less portable. This study addresses the gap in the literature by comparing the test-retest reliability of the K-Deltas force platform with a contact mat and the My Jump app, aiming to determine if more affordable, portable tools can reliably measure CMJ height. By providing comparative reliability data, the study results may offer valuable insights for practitioners, potentially expanding the use of CMJ testing tools in various settings and making accurate assessments more accessible across different contexts. To the best of the author's knowledge, no studies have analyzed the reliability of K-Deltas force platforms for measuring CMJ height. Therefore, this study aimed to assess the test-retest reliability of the portable K-Deltas force platforms to measure CMJ height. In addition, the study also aimed to compare the force platform-derived jump height to the Contact mat (Chronojump, Spain) and My Jump (with an iPad) app. It was hypothesized that K-Deltas force platforms would have high test-retest reliability in comparison with Contact Mat and My Jump app.

2. Materials and Methods

2.1. Participants

A total of twenty-two healthy physically active participants (6 females and 16 males; age: 19.68 ± 1.21 years; body mass: 63.32 ± 12.77 kg; height: 171.65 ± 9.23 cm; body mass index: 21.35 ± 2.95 kg/m^2) volunteered (via convenience sampling) to be a part of the study. Each participant was instructed to perform five CMJ (total 110 jumps). G-Power software 3.1.9.7. (University of Dusseldorf, Dusseldorf, Germany) was used to determine the sample size (110 jumps recorded by three distinct instruments resulted in a total of 330 CMJ data points), assuming power = 0.95, alpha error < 0.05, and effect size f = 0.22. The inclusion criteria for the study required participants to be currently enrolled as students in a Sports and Exercise Science program, with no recent injury or any other medical conditions that could interfere with performance and safety. Additionally, participants required at least two years of training experience (minimum 6 h of training per week) and must have actively participated in collegiate-level sports such as football (soccer), tennis, gymnastics, volleyball, wrestling, swimming, or table tennis. Participants with a history of surgeries or severe lower extremity injuries requiring medical attention and follow-up in the last 12 months were excluded. Those currently involved in another study that might interfere with their participation were also excluded. Participants with incorrect CMJ techniques during the end of the familiarization session were excluded from the study to maintain data integrity and participant safety. The performance of CMJ was visually inspected by an accredited strength and conditioning coach and a sports scientist. The testing protocols, the study's objective, and the potential risks and benefits of the study were explained to the participants before the study was conducted. After that, informed consent forms were signed by the participants. The study was approved by the Institute Research Committee of Symbiosis School of Sports Science, Symbiosis International (Deemed University), and conducted according to the Declaration of Helsinki's guidelines.

2.2. Experimental Procedure

2.2.1. Contact Mat

The Contact Mat (Chronojump Boscosystem, Barcelona, Spain) was placed and fixed over the Kinvent K-Deltas portable force platforms with adhesive tape (Supplementary Figure S1). Monitoring of the setup was performed regularly, i.e., on the completion of the fifth trial of every participant (i.e., before the trials of the next participants). To maintain consistency across trials and participants, the placement of the contact mat on K-Deltas was marked using a marker. The contact mat (size: DIN-A2 420 × 590 mm) was connected to a microcontroller using an RCB cable, and a USB cable from the same microcontroller was connected to the laptop through a USB port. An open-source software (Chronojump, version 2.3.0-79) was used to record the data on jump height from the contact mat.

2.2.2. Force Platform

The KINVENT Physio software (version 2.11.2) installed on an iPad (Apple Inc., Cupertino, CA, USA) was used to record data from the portable force platform. The portable force platforms were connected via Bluetooth (BLE 5.1) to the iPad. The force platforms recorded the data with a sampling frequency of 1000 Hz.

2.2.3. Video-Based Mobile Software

Simultaneously, an iPad Air (Apple Inc., Cupertino, CA, USA) device was used to record the participants' CMJ trials in the sagittal plane. The camera was placed approximately three meters away and one meter above the ground (Supplementary Figure S2). The same camera setting was used across two weeks of data collection. The videos were recorded at a frequency of 240 frames per second. Thereafter, the CMJ performance (jump height in centimeters) was extracted from the videos via the My Jump app installed on the same iPad Air device. Two independent assessors conducted the data extraction for all the CMJ jumps. One of the assessors was not the author of this study. The calculation of jump height using the application and contact mat was conducted using protocols set in previous studies [25,26].

The experimental setup for data collection was conducted in an indoor laboratory environment, ensuring optimal lighting and adequate temperature. The force platform was positioned on a concrete surface to ensure stability and accuracy in the jump height measurement during the assessment. The data collection was completed in two weeks.

2.3. Data Collection

2.3.1. Familiarization Sessions

One week before the start of the data collection, two familiarization sessions, 60 min each, were conducted for the participants. The session included a demonstration of technique with instruction (15 min), a standard warm-up (10 min), low- to moderate-intensity jumps with sufficient recovery, instructor feedback (30 min), and cool-down (5 min). Detailed instructions followed by demonstrations were given by an accredited strength and conditioning coach and sports scientist. The first session focused on the correct jumping and landing techniques. During the second familiarization session, minor corrections in the jump technique were made if required. Both sessions ensured the athletes' preparedness to perform CMJ with proper technique. The final selection of participants eligible for the study was conducted during this stage (i.e., assessing the correct CMJ technique by participants).

2.3.2. Warm-Up

On the testing day, the participants performed a standard (RAMP) 10 min of warm-up before the performance of CMJ. The warm-up consists of 4 min of low to moderate-intensity on-the-spot jogging with one minute of high knees, two minutes of body-weight squats, three minutes of leg swings, and walking lunges with twist and arm swings. After that, the participants were allowed to perform low-moderate effort CMJs for two minutes.

2.3.3. Testing Procedure

Before the testing day, participants were instructed not to be involved in any strenuous physical activity 24 h before the testing that might affect their participation. They were asked to wear comfortable clothing to allow ease of movement and proper athletic footwear. For the testing, the participants were instructed to stand on the force platform (with the contact mat securely placed over it). The participants were asked to stand straight, stable, with their hands on their hips. Once the participant was ready, all the administrators recorded the CMJ performance for each jump using the three instruments (i.e., force platforms, contact mats, and My Jump). Participants were allowed to perform sub-maximal trials before performing the actual jumps. Each participant performed a total of five CMJ trials with hands-on-hips, with each trial separated by one minute of recovery [27]. The participants were instructed to perform CMJ with moderate to high intensity, focusing on the correct technique. This was requested to record data across the jump height spectrum (as jump intensity would not affect the study's outcome).

2.4. Statistical Analysis

Descriptive statistics with mean and standard deviation values were used to characterize the central tendency and scatteredness of the dataset. The Kolmogorov–Smirnov test was used to assess the data's normality. Intraclass Correlation Coefficient (ICC; two-way random single measures [absolute agreement]), Pearson Product Moment Correlation Coefficient, one-way ANOVA, and independent sample *t*-test were employed to analyze the three instruments' results. Levene's Test was used to measure the homoscedasticity among the groups wherever applicable. A Bland–Atman plot was used to study the bias between the mean difference in the two instruments and to describe an agreement between these instruments. In addition, ICC, Cronbach Alpha, and Coefficient of variation (CV) were used to analyze the test-retest reliability of the collected data. ICC values were categorized based on the lower bound of the 95% confidence interval (CI) into poor (<0.50), moderate (0.50–0.75), good (0.75–0.90), and excellent (>0.90) reliability [28]. A correlation value of r = 0.10 meant a low correlation, r = 0.30 meant a medium correlation, and r = 0.50 meant a higher correlation [29]. Effect sizes (ES) were calculated as Hedge's *g* and were interpreted as trivial (<0.2), small (0.2–0.6), moderate (>0.6–1.2), or large (>1.2–2.0) [30]. The CV, indicating the typical error of measurements as a percentage of dispersion around the mean, was acceptable if below 10% [31]. Statistical software (SPSS version 24) was used for analysis, and the significance level was set at 5%.

3. Results

Table 1 shows the results of descriptive statistics, comparative analysis, ICC, and correlation values from the jump height assessment during CMJ. Individual participants' data across the three measurement devices are presented in Supplementary Figure S3. No significant differences ($p = 0.368$) were reported between the force platform, contact mat, and My Jump app in the jump height of the participants. A nearly perfect correlation coefficient (r = 0.987; $p < 0.001$) with excellent reliability levels (ICC = 0.993) was found between the force platform and contact mat (non-significant differences $p = 0.935$, ES < 0.01 [trivial]). Similarly, there was no significant difference ($p = 0.241$, ES = 0.16 [trivial]) reported between the force platform and My Jump, with a nearly perfect correlation coefficient (r = 0.987; $p < 0.001$) and excellent reliability levels (ICC = 0.987). Additionally, an almost perfect correlation coefficient (r = 0.996; $p < 0.001$) with excellent reliability levels (ICC = 0.990) was found between the contact mat and My Jump (non-significant differences $p = 0.203$, ES = 0.16 [trivial]).

Table 1. Validity statistics.

	Force Platform	Contact Mat	My Jump	ANOVA		Force Platform vs. Contact Mat	Force Platform vs. My Jump	Contact Mat vs. My Jump
	Mean ± Standard Deviation			*p*-Value				
CMJ (cm)	31.9 ± 8.4	31.9 ± 8.3	30.5 ± 8.5	0.368	MD ± SE	−0.091 ± 1.123	1.336 ± 1.137	1.428 ± 1.128
					ES	<0.01	0.16	0.16
					ICC (95% CI)	0.993 (0.990–0.995)	0.987 (0.903–0.996)	0.990 (0.578–0.998)
					r	0.987	0.987	0.996

Note: CMJ—Counter Movement Jump. ES—effect size. MD—Mean Difference. SE—Standard Error. ICC—interclass correlation coefficient. CI—confidence interval. *r*—Pearson correlation coefficient.

The test-retest reliability statistics are presented in Table 2. The average measure values were used for reporting ICC. All the instruments reported excellent reliability levels with high Cronbach alpha values (>0.90) and acceptable CV values (<10%) (Force platform: ICC = 0.981, $\alpha = 0.983$, CV% = 7.266; Contact mat: ICC = 0.987, $\alpha = 0.988$, CV% = 6.057; My Jump: ICC = 0.986, $\alpha = 0.987$, CV% = 6.625).

The scatter plot graph displays an almost perfect correlation between the instruments in Figure 1. The Bland–Altman plot (Figure 2) depicts a large limit of agreement between the instruments (i.e., force platform vs. contact mat: LoA = −2.744, 2.561; Bias = −0.091, force platform vs. My Jump: LoA = −1.329, 4.001; Bias = 1.336, and contact mat vs. My Jump: LoA = −0.144, 3.000; Bias = 1.428).

Table 2. Test–retest reliability.

	Force Platform			Contact Mat			My Jump		
	ICC (95% CI)	α	CV (%)	ICC (95% CI)	α	CV (%)	ICC (95% CI)	α	CV (%)
CMJ (cm)	0.981 (0.964–0.991)	0.983	7.266	0.987 (0.975–0.994)	0.988	6.057	0.986 (0.974–0.994)	0.987	6.625

Note: CMJ—Counter movement jump. ICC (95% CI)—Interclass correlation coefficient with 95% confidence interval. α—Cronbach's alpha. CV—coefficient of variation.

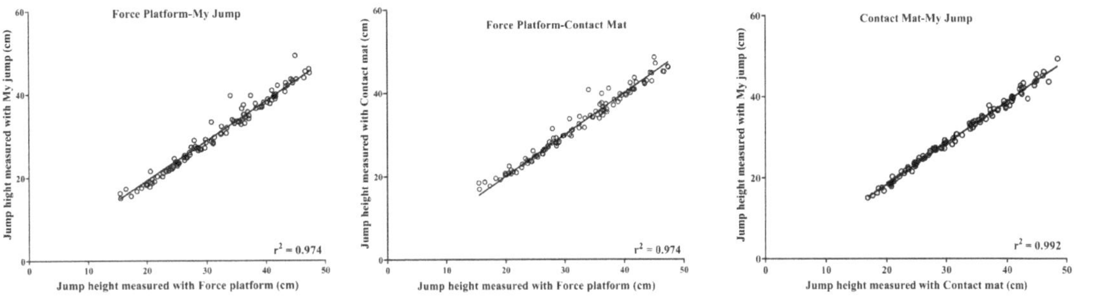

Figure 1. A scatter plot depicting the relationship between all the methods (i.e., force platform-contact mat, force platform-My Jump, contact mat-My Jump), along with the coefficient of determination R^2.

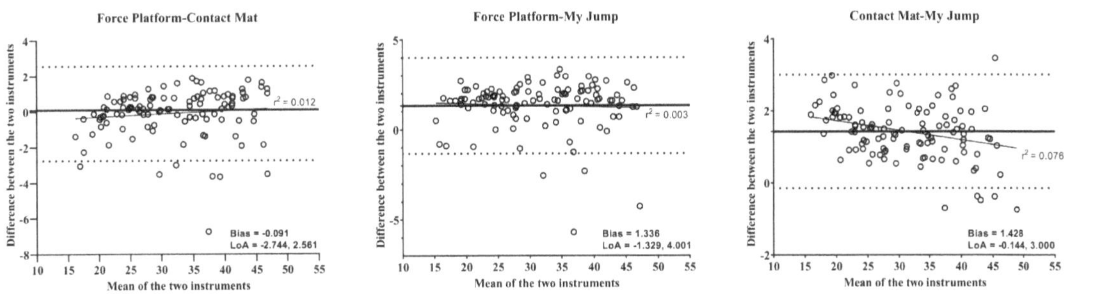

Figure 2. Bland–Altman Plots representing CMJ height data for all the methods (i.e., force platform-contact mat, force platform-My Jump, contact mat-My Jump). The thick central line represents the mean difference between the two methods, while the dotted lines represent ± level of agreement (±1.96 standard deviations), and the thin line represents the coefficient of determination R^2.

4. Discussion

Reliability is an important factor that is considered while implementing the use of testing instruments. No previous study examined the reliability of portable K-Deltas. Therefore, the present study aimed to assess the test–retest reliability of the portable K-Deltas force platform to measure CMJ height and compare the force platform-derived jump height to the contact mat and My Jump app. The K-Deltas force platform was highly reliable, with an acceptable coefficient of variation acquired for both the contact mat and My Jump for measuring CMJ height. Additionally, the jump height measured from all three instruments (i.e., force platform, contact mat, and My Jump) was similar (non-significant difference; $p > 0.05$). The Bland–Atman plots also depicted a large limit of agreement between the instruments. The mean differences were relatively large but were within the 95% agreement limits (i.e., mean difference ± 1.96 SD of the difference). The spread of each score in all plots was around the bias line, and there were no cases of proportional bias [32].

Though there were no significant differences between the instruments, on average, CMJ height was found to be slightly greater in the contact mat (31.9 ± 8.3), followed by the force platform (31.9 ± 8.4), and then My Jump app (30.5 ± 8.5). Studies have reported that CMJ height is generally higher when measured with a contact mat than with a force platform [33–35]. Additionally, studies have also reported greater limits of agreement for CMJ height between the contact mat vs. force platform [35,36] and My Jump vs. force platform [37]. In this study, an excellent intra-class correlation coefficient with an almost perfect correlation was obtained using all three instruments (force platform vs. contact mat: ICC = 0.993, r = 0.987, force platform vs. My Jump: ICC = 0.987 r = 0.987, contact mat vs. My Jump: ICC = 0.990, r = 0.996) for measuring CMJ height. Additionally, an excellent between-trial reliability (force platform: ICC = 0.981, contact mat: ICC = 0.987, My Jump: ICC = 0.986) was reported along with an acceptable CV (7.266%, 6.057%, 6.625%).

To the author's knowledge, this is the first study comparing the CMJ height measures of the Kinvent K-Deltas portable force platform with the Chronojump contact mat and My Jump app. Several studies were conducted previously to test the reliability and validity of the My Jump app [37–39] and contact mat [26,36,40] with a force platform for measuring CMJ. My Jump studies reported excellent ICC and an almost perfect correlation with the scores obtained from the force platform in CMJ. Similarly, studies on the contact mat showed good ICC [36] and a strong correlation coefficient [40] with the force platform. Similar results were obtained by Plakoutsis et al. [41], where the authors established the reliability and validity of the k-force plate with the My Jump app. The findings indicated excellent reliability (ICC = 0.999, 1.000) and validity (r = 0.999) of the K-force plate in CMJ height. The low difference (or high association) obtained between the portable K deltas and contact mat and My Jump application may be potentially due to the high frequency at which the instrument records the data. Moreover, this low difference also suggests that the portable force platforms are accurate in identifying the take-off and landing, which is required to calculate the jump height using the flight time data.

Our findings suggest that the K-Deltas portable force platform has high accuracy and could be used for field-based or laboratory-based assessments. The K-Deltas force platform is less expensive, lightweight, easy to carry, and can provide accurate feedback to practitioners. The underlying mechanisms that potentially contribute to greater neuromuscular adaptations observed during the use of the K-Deltas force platform in the countermovement jump (CMJ) test can be linked to the enhanced functionality of the stretch-shortening cycle (SSC), which is especially relevant for athletes [42]. This cycle, which is crucial in optimizing muscle performance during dynamic activities such as jumping, involves complex neuromuscular processes, including the stretch reflex and the H-reflex, as highlighted by Taube et al. [43]. These mechanisms are particularly pertinent for athletes, as they directly influence the ability to generate powerful, explosive movements.

Greater jumping performance, as recorded by the K-Deltas force platform, may indicate a more effective engagement of the SSC. This is characterized by increased muscle activation, particularly in key muscle groups such as the medial gastrocnemius, biceps femoris, and rectus femoris, which are essential for optimal performance in youth athletes [42,44]. Enhancing muscle activation contributes to the efficient storage and release of elastic energy during the eccentric and concentric phases of muscle action, respectively. This energy release is crucial for producing the forceful contractions needed to achieve higher jumps [45–47].

Moreover, the pre-activation of leg muscles before the jump, which is facilitated by the SSC, plays a critical role in increasing muscle stiffness. This stiffness prepares the agonist muscles to better resist the high-impact loads encountered during the vertical jump, thereby improving overall jump performance [43]. The ability of the K-Deltas portable force platform to capture these nuances in muscle activity and performance makes it an invaluable tool for coaches and physical trainers who seek to assess and enhance an athlete's vertical jump capabilities. This device, by providing detailed insights into muscle power performance, can help tailor training programs to maximize athletic output, particularly in sports where explosive leg power is a key determinant of success.

This study has some limitations that need to be acknowledged. The participants were from different sports, including volleyball, soccer, gymnastics, tennis, table tennis, and swimming. Both male and female participants were also included, making the participants more heterogeneous. However, it did not affect the results, as the study aimed to assess the test-retest reliability of and compare the CMJ height between different instruments. Indeed, heterogeneous data represents the reliability of the instruments across a wide range of jump heights. Of note, future studies should aim to include more homogeneous data, possibly including elite athletes, to assess the sensitivity of the instrument. The test-retest reliability could have been slightly better if the participants had performed the CMJ with maximum effort (all five trials) instead of the instructed moderate- to high-intensity effort. The lack of a gold standard measure (ground-embedded force platforms) may have affected the study results. In addition, the My Jump app may have individual errors, as the data from the pre-recorded video was manually analyzed with the My Jump app by two assessors separately. However, an excellent ICC (0.997) was reported between the assessors' data. Additionally, future studies could investigate the use of the K-Deltas platform for measuring additional performance metrics beyond CMJ height, such as rate of force development, peak power, or reactive strength index. These parameters are critical for understanding neuromuscular adaptations and performance in various dynamic movements, and their measurement could further validate the platform's versatility. These studies could focus on sports-specific parameters, involving elite or amateur athletes to ensure a homogeneous sample. Additionally, the authors suggest assessing the reliability and validity of K-Deltas in comparison with the gold standard laboratory-based in-ground force platform for better implication. Lastly, long-term studies examining the impact of using portable force platforms like K-Deltas on injury prevention, training adaptation, and performance enhancement would provide valuable insights. These studies could track athletes over multiple seasons to determine whether regular monitoring of jump performance with portable platforms leads to improved outcomes in terms of performance gains and injury reduction.

The practical implications of this study's findings are highly relevant for coaches and practitioners who seek efficient and accurate methods for assessing athletic performance. The portability, affordability, and reliability of the K-Deltas force platform allow it to be a practical alternative to traditional, lab-based in-ground force platforms. These features enable practitioners to conduct frequent assessments (pre-, mid-, and post-training) in field-based settings without needing expensive or immobile equipment. For coaches, this means they can more easily integrate jump performance testing into regular training sessions, offering real-time feedback to athletes. This capability can help tailor training programs to individual athletes' needs, monitor progress, and adjust training loads to optimize performance. Additionally, by regularly tracking neuromuscular fatigue and injury risk, practitioners can reduce the likelihood of overtraining or injury, improving athlete longevity and performance. The ability to make quick and accurate assessments using the K-Deltas platform also enables coaches and trainers to implement data-driven decisions in competitive environments, where timely insights can lead to performance enhancements and strategic advantages. The platform's ease of use, combined with its accuracy, offers a powerful tool for both routine athlete monitoring and high-performance decision-making.

5. Conclusions

The present study aimed to (1) assess the test-retest reliability of the portable K-Deltas force platform for measuring countermovement jump (CMJ) height and (2) compare its accuracy with a contact mat and video-based software. Both objectives were successfully met. The findings demonstrated that the K-Deltas force platform exhibited excellent test-retest reliability, comparable to the contact mat and the My Jump app, with no significant differences in CMJ height measurements across the devices. Additionally, the high correlation between the instruments confirmed the K-Deltas force platform as a reliable and valid tool for field-based assessments. These results suggest that the K-Deltas force platform is a practical alternative to traditional laboratory-based equipment, offering practitioners a cost-effective and portable solution without sacrificing accuracy, and can quickly connect to a mobile device via Bluetooth. However, future research should explore its application across different athletic populations and performance metrics, as well as its long-term impact on athlete development and injury prevention.

Supplementary Materials: The following supporting information can be downloaded at: https://www.mdpi.com/article/10.3390/life14111394/s1, Supplementary Figure S1: Contact mat placement on the k-delta force platform with adhesive tape; Supplementary Figure S2: Measurement setup during the data collection; Supplementary Figure S3: Illustrations of individual participant data across the three instruments.

Author Contributions: Conceptualization, A.B. and R.K.T.; methodology, R.K.T.; software, A.B., S.R., and R.K.T.; formal analysis, A.B. and R.K.T.; investigation, A.B.; resources, A.B.; data curation, A.B., S.R., and R.K.T.; writing—original draft preparation, A.B. and H.İ.C.; writing—review and editing, S.R., R.K.T., H.İ.C. and V.S.; visualization, A.B., and R.K.T.; supervision, R.K.T. and H.İ.C.; project administration, A.B., S.R. and R.K.T. All authors have read and agreed to the published version of the manuscript.

Funding: This research received no external funding.

Institutional Review Board Statement: The study was conducted in accordance with the Declaration of Helsinki and approved by the Institute Research Committee (IRC) of Symbiosis School of Sports Sciences (protocol code IRC-SSSS/2024-25/197 and date of approval: 11 March 2024).

Informed Consent Statement: Informed consent was obtained from all subjects involved in the study.

Data Availability Statement: The datasets presented in this study can be found in online repositories. The names of the repository/repositories and accession number(s) can be found at: https://figshare.com/ (accessed on 21 October 2024), https://doi.org/10.6084/m9.figshare.26936401.v1 (accessed on 21 October 2024).

Acknowledgments: The author would like to thank the students who volunteered to participate in this study.

Conflicts of Interest: The authors declare no conflicts of interest.

References

1. Anicic, Z.; Janicijevic, D.; Knezevic, O.M.; Garcia-Ramos, A.; Petrovic, M.R.; Cabarkapa, D.; Mirkov, D.M. Assessment of countermovement jump: What should we report? *Life* **2023**, *13*, 190. [CrossRef] [PubMed]
2. Nibali, M.L.; Tombleson, T.; Brady, P.H.; Wagner, P. Influence of familiarization and competitive level on the reliability of countermovement vertical jump kinetic and kinematic variables. *J. Strength Cond. Res.* **2015**, *29*, 2827–2835. [CrossRef] [PubMed]
3. Moir, G.; Button, C.; Glaister, M.; Stone, M.H. Influence of familiarization on the reliability of vertical jump and acceleration sprinting performance in physically active men. *J. Strength Cond. Res.* **2004**, *18*, 276–280.
4. Moir, G.; Sanders, R.; Button, C.; Glaister, M. The influence of familiarization on the reliability of force variables measured during unloaded and loaded vertical jumps. *J. Strength Cond. Res.* **2005**, *19*, 140–145.
5. Pleša, J.; Kozinc, Ž.; Šarabon, N. A Brief Review of Selected Biomechanical Variables for Sport Performance Monitoring and Training Optimization. *Appl. Mech.* **2022**, *3*, 144–159. [CrossRef]
6. Claudino, J.G.; Cronin, J.; Mezêncio, B.; McMaster, D.T.; McGuigan, M.; Tricoli, V.; Amadio, A.C.; Serrão, J.C. The countermovement jump to monitor neuromuscular status: A meta-analysis. *J. Sci. Med. Sport* **2017**, *20*, 397–402. [CrossRef]
7. Bishop, C.; Read, P.; McCubbine, J.; Turner, A. Vertical and horizontal asymmetries are related to slower sprinting and jump performance in elite youth female soccer players. *J. Strength Cond. Res.* **2021**, *35*, 56–63. [CrossRef]
8. Stevenson, J.H.; Beattie, C.S.; Schwartz, J.B.; Busconi, B.D. Assessing the effectiveness of neuromuscular training programs in reducing the incidence of anterior cruciate ligament injuries in female athletes: A systematic review. *Am. J. Sports Med.* **2015**, *43*, 482–490. [CrossRef]
9. Bishop, C.; Jordan, M.; Torres-Ronda, L.; Loturco, I.; Harry, J.; Virgile, A.; Mundy, P.; Turner, A.; Comfort, P. Selecting Metrics That Matter: Comparing the Use of the Countermovement Jump for Performance Profiling, Neuromuscular Fatigue Monitoring, and Injury Rehabilitation Testing. *Strength Cond. J.* **2023**, *45*, 545–553. [CrossRef]
10. Nuzzo, J.L.; McBride, J.M.; Cormie, P.; McCaulley, G.O. Relationship between countermovement jump performance and multijoint isometric and dynamic tests of strength. *J. Strength Cond. Res.* **2008**, *22*, 699–707. [CrossRef]

11. González-Badillo, J.J.; Marques, M.C. Relationship between kinematic factors and countermovement jump height in trained track and field athletes. *J. Strength Cond. Res.* **2010**, *24*, 3443–3447. [CrossRef] [PubMed]
12. Beattie, K.; Tawiah-Dodoo, J.; Graham-Smith, P. Countermovement jump characteristics of world-class elite and sub-elite male sprinters. *Sport Perf. Sci. Rep.* **2020**, *1*, 1–4.
13. Trecroci, A.; Longo, S.; Perri, E.; Iaia, F.M.; Alberti, G. Field-based physical performance of elite and sub-elite middle-adolescent soccer players. *Res. Sports Med.* **2019**, *27*, 60–71. [CrossRef]
14. Cerrah, A.; Gungor, E.O.; Soylu, A.; Ertan, H. Muscular activation differences between professional and amateur soccer players during countermovement jump. *Turk. J. Sports Med.* **2014**, *16*, 51–58. [CrossRef]
15. Papaevangelou, E.; Metaxas, T.; Riganas, C.; Mandroukas, A.; Vamvakoudis, E. Evaluation of soccer performance in professional, semi-professional and amateur players of the same club. *J. Phys. Educ. Sport* **2012**, *12*, 362.
16. Lake, J.; Mundy, P.; Comfort, P.; McMahon, J.J.; Suchomel, T.J.; Carden, P. Concurrent validity of a portable force plate using vertical jump force–time characteristics. *J. Appl. Biomech.* **2018**, *34*, 410–413. [CrossRef]
17. Gençoğlu, C.; Ulupınar, S.; Özbay, S.; Turan, M.; Savaş, B.Ç.; Asan, S.; İnce, İ. Validity and reliability of "My Jump app" to assess vertical jump performance: A meta-analytic review. *Sci. Rep.* **2023**, *13*, 20137. [CrossRef]
18. Pueo, B.; Penichet-Tomas, A.; Jimenez-Olmedo, J. Reliability and validity of the Chronojump open-source jump mat system. *Biol. Sport* **2020**, *37*, 255–259. [CrossRef]
19. Bunn, J.A.; Navalta, J.W.; Fountaine, C.J.; Reece, J.D. Current state of commercial wearable technology in physical activity monitoring 2015–2017. *Int. J. Exerc. Sci.* **2018**, *11*, 503. [CrossRef]
20. Knowles, B.; Smith-Renner, A.; Poursabzi-Sangdeh, F.; Lu, D.; Alabi, H. Uncertainty in current and future health wearables. *Commun. ACM* **2018**, *61*, 62–67. [CrossRef]
21. Currell, K.; Jeukendrup, A.E. Validity, reliability and sensitivity of measures of sporting performance. *Sports Med.* **2008**, *38*, 297–316. [CrossRef] [PubMed]
22. Meras Serrano, H.; Mottet, D.; Caillaud, K. Validity and Reliability of Kinvent Plates for Assessing Single Leg Static and Dynamic Balance in the Field. *Sensors* **2023**, *23*, 2354. [CrossRef] [PubMed]
23. Mylonas, V.; Chalitsios, C.; Nikodelis, T. Validation of a Portable Wireless Force Platform System to Measure Ground Reaction Forces During Various Tasks. *Int. J. Sports Phys. Ther.* **2023**, *18*, 1283. [CrossRef] [PubMed]
24. Mahar, M.T.; Rowe, D.A. Practical guidelines for valid and reliable youth fitness testing. *Meas. Phys. Educ. Exerc. Sci.* **2008**, *12*, 126–145. [CrossRef]
25. Gallardo-Fuentes, F.; Gallardo-Fuentes, J.; Ramírez-Campillo, R.; Balsalobre-Fernández, C.; Martínez, C.; Caniuqueo, A.; Cañas, R.; Banzer, W.; Loturco, I.; Nakamura, F.Y. Intersession and intrasession reliability and validity of the My Jump app for measuring different jump actions in trained male and female athletes. *J. Strength Cond. Res.* **2016**, *30*, 2049–2056. [CrossRef]
26. Stanton, R.; Doering, T.M.; Macgregor, C.; Borges, N.; Delvecchio, L. Validity of a contact mat and accelerometric system to assess countermovement jump from flight time. *Meas. Phys. Educ. Exerc. Sci.* **2019**, *23*, 39–46. [CrossRef]
27. Gathercole, R.J.; Stellingwerff, T.; Sporer, B.C. Effect of acute fatigue and training adaptation on countermovement jump performance in elite snowboard cross athletes. *J. Strength Cond. Res.* **2015**, *29*, 37–46. [CrossRef]
28. Koo, T.K.; Li, M.Y. A guideline of selecting and reporting intraclass correlation coefficients for reliability research. *J. Chiropr. Med.* **2016**, *15*, 155–163. [CrossRef]
29. Cohen, J. *Statistical Power Analysis for the Behavioral Sciences*; Routledge: London, UK, 2013.
30. Hopkins, W.G.; Marshall, S.W.; Batterham, A.M.; Hanin, J. Progressive statistics for studies in sports medicine and exercise science. *Med. Sci. Sports Exerc.* **2009**, *41*, 3–13. [CrossRef]
31. Cormack, S.J.; Newton, R.U.; McGuigan, M.R.; Doyle, T.L. Reliability of measures obtained during single and repeated countermovement jumps. *Int. J. Sports Physiol. Perform.* **2008**, *3*, 131–144. [CrossRef]
32. Bland, J.M.; Altman, D. Statistical methods for assessing agreement between two methods of clinical measurement. *Lancet* **1986**, *327*, 307–310. [CrossRef]
33. Reeve, T.C.; Tyler, C.J. The validity of the SmartJump contact mat. *J. Strength Cond. Res.* **2013**, *27*, 1597–1601. [CrossRef] [PubMed]
34. Whitmer, T.D.; Fry, A.C.; Forsythe, C.M.; Andre, M.J.; Lane, M.T.; Hudy, A.; Honnold, D.E. Accuracy of a vertical jump contact mat for determining jump height and flight time. *J. Strength Cond. Res.* **2015**, *29*, 877–881. [CrossRef]
35. Dobbin, N.; Hunwicks, R.; Highton, J.; Twist, C. Validity of a jump mat for assessing countermovement jump performance in elite rugby players. *Int. J. Sports Med.* **2017**, *38*, 99–104. [CrossRef] [PubMed]
36. Ruf, L.; Altmann, S.; Müller, K.; Rehborn, A.; Schindler, F.; Woll, A.; Härtel, S. Concurrent validity of countermovement and squat jump height assessed with a contact mat and force platform in professional soccer players. *Front. Sports Act. Living* **2024**, *6*, 1437230. [CrossRef] [PubMed]
37. Balsalobre-Fernández, C.; Glaister, M.; Lockey, R.A. The validity and reliability of an iPhone app for measuring vertical jump performance. *J. Sports Sci.* **2015**, *33*, 1574–1579. [CrossRef]
38. Bishop, C.; Jarvis, P.; Turner, A.; Balsalobre-Fernandez, C. Validity and reliability of strategy metrics to assess countermovement jump performance using the newly developed smartphone application. *J. Hum. Kinet.* **2022**, *83*, 185–195. [CrossRef]
39. Carlos-Vivas, J.; Martin-Martinez, J.P.; Hernandez-Mocholi, M.A.; Perez-Gomez, J. Validation of the iPhone app using the force platform to estimate vertical jump height. *J. Sports Med. Phys. Fitness* **2016**, *58*, 227–232. [CrossRef]

40. Pojskić, H.; Papa, E.V.; Wu, S.S.X.; Pagaduan, J. Validity, reliability, and usefulness of jump performance from a low-cost contact mat. *J. Hum. Sport Exerc.* **2022**, *17*, 261–271. [CrossRef]
41. Plakoutsis, G.; Zapantis, D.; Panagiotopoulou, E.-M.; Paraskevopoulos, E.; Moutzouri, M.; Koumantakis, G.A.; Papandreou, M. Reliability and Validity of the Portable KForce Plates for Measuring Countermovement Jump (CMJ). *Appl. Sci.* **2023**, *13*, 11200. [CrossRef]
42. Andrade, D.C.; Manzo, O.; Beltrán, A.R.; Álvarez, C.; Del Rio, R.; Toledo, C.; Moran, J.; Ramirez-Campillo, R. Kinematic and neuromuscular measures of intensity during plyometric jumps. *J. Strength Cond. Res.* **2020**, *34*, 3395–3402. [CrossRef] [PubMed]
43. Taube, W.; Leukel, C.; Lauber, B.; Gollhofer, A. The drop height determines neuromuscular adaptations and changes in jump performance in stretch-shortening cycle training. *Scand. J. Med. Sci. Sports* **2012**, *22*, 671–683. [CrossRef] [PubMed]
44. Radnor, J.M.; Oliver, J.L.; Waugh, C.M.; Myer, G.D.; Lloyd, R.S. The influence of maturity status on muscle architecture in school-aged boys. *Pediatr. Exerc. Sci.* **2020**, *32*, 89–96. [CrossRef] [PubMed]
45. Seiberl, W.; Hahn, D.; Power, G.A.; Fletcher, J.R.; Siebert, T. Editorial: The Stretch-Shortening Cycle of Active Muscle and Muscle-Tendon Complex: What, Why and How It Increases Muscle Performance? *Front. Physiol.* **2021**, *12*, 693141. [CrossRef]
46. Radnor, J.M.; Oliver, J.L.; Waugh, C.M.; Myer, G.D.; Moore, I.S.; Lloyd, R.S. The influence of growth and maturation on stretch-shortening cycle function in youth. *Sports Med.* **2018**, *48*, 57–71. [CrossRef]
47. Su, E.Y.-S.; Carroll, T.J.; Farris, D.J.; Lichtwark, G. Increased force and elastic energy storage are not the mechanisms that improve jump performance with accentuated eccentric loading during a constrained vertical jump. *PLoS ONE* **2024**, *19*, e0308226. [CrossRef]

life

MDPI

Article

Analysis of Wavelet Coherence in Calf Agonist-Antagonist Muscles during Dynamic Fatigue

Xindi Ni, Loi Ieong, Mai Xiang and Ye Liu *

School of Sport Science, Beijing Sport University, Beijing 100084, China; nxd_bsu@foxmail.com (X.N.); loi_326@hotmail.com (L.I.); michael9789@126.com (M.X.)
* Correspondence: liuye_tg@163.com

Abstract: Dynamic muscle fatigue during repetitive movements can lead to changes in communication between the central nervous system and peripheral muscles. This study investigated these changes by examining electromyogram (EMG) characteristics from agonist and antagonist muscles during a fatiguing task. Twenty-two healthy male university students (age: 22.92 ± 2.19 years) performed heel raises until fatigue. EMG signals from lateral gastrocnemius (GL) and tibialis anterior (TA) muscles were processed using synchrosqueezed wavelet transform (SST). Root mean square (RMS), mean frequency (MF), power across frequency ranges, wavelet coherence, and co-activation ratio were computed. During the initial 80% of the task, RMS and EMG power increased for both muscles, while MF declined. In the final 20%, GL parameters stabilized, but TA showed significant decreases. Beta and gamma intermuscular coherence increased upon reaching 60% of the task. Alpha coherence and co-activation ratio remained constant. Results suggest that the central nervous system adopts a differentiated control strategy for agonist and antagonist muscles during fatigue progression. Initially, a coordinated "common drive" mechanism enhances both muscle groups' activity. Later, despite continued increases in muscle activity, neural-muscular coupling remains stable. This asynchronous, differentiated control mechanism enhances our understanding of neuromuscular adaptations during fatigue, potentially contributing to the development of more targeted fatigue assessment and management strategies.

Keywords: muscle fatigue; wavelet coherence; antagonistic muscles; EMG

Citation: Ni, X.; Ieong, L.; Xiang, M.; Liu, Y. Analysis of Wavelet Coherence in Calf Agonist-Antagonist Muscles during Dynamic Fatigue. *Life* **2024**, *14*, 1137. https://doi.org/10.3390/life14091137

Academic Editor: Nicola Smania

Received: 14 August 2024
Revised: 6 September 2024
Accepted: 6 September 2024
Published: 9 September 2024

1. Introduction

During human motion, the central nervous system regulates the active and antagonist muscles' cooperation through feed-forward adjustments via descending signals and feedback adjustments from ascending signals, enabling the execution of various complex movements [1]. The modulatory role of descending signals commonly manifests as oscillations in the beta and gamma frequency bands in EMG signals, often interpreted using De Luca's "common drive" concept [2,3]. Meanwhile, ascending feedback regulatory information typically appears as oscillations in the alpha frequency band [4,5]. The weakening of bidirectional signal communication between the cerebral cortex and muscles is thought to be the primary neurophysiological mechanism driving muscle fatigue [3]. However, this perspective largely stems from studies comparing non-fatigued and fatigued states [3,6], potentially overlooking the nuances of fatigue evolution [7].

Fatigue resulting from prolonged physical activity is a progressive process [7]. During fatigue onset, an increase in the discharge rate of motor unit action potentials (MUAP) is observed, followed by the accumulation of fatigue [8]. Concurrently, a decline in local muscle conduction velocity during fatigue leads to a reduction in muscle EMG amplitude [9]. Moreover, fatigue can also affect antagonist muscle EMG activity and co-activation levels [6,10–12]. Research has observed the H-reflex of the antagonist-prime mover pair in the calf during isometric fatigue processes, identifying a phenomenon of bidirectional spinal reflex excitability regulation in the antagonist muscle during fatigue [13,14]. This regulation of antagonist muscles seems to be modulated by structures of the spine, aiming to maintain a specific functional level [12]. However, when this modulation occurs, and what its relationship with prime mover and antagonist muscle activity is, remains unclear. Hence, further understanding of the neuromuscular system's changes during the progression of fatigue is crucial for comprehending fatigue [7,15].

The central nervous system drives muscle movement by regulating the discharge frequency of muscles. These different frequency muscle electrical activities are often considered manifestations of functional activities in different brain circuits [16]. Such manifestations are commonly studied through EEG-EMG coherence or EMG-EMG coherence [11,17–20]. Research has suggested that EMG alpha band coherence may originate from the reticular spinal cord [18], possibly reflecting ascending or feedback interactions and being influenced by Renshaw cells in the spinal cord [5]. The beta and gamma bands are primarily driven by the motor cortex [21]. Previous studies have found strong beta frequency band coherence during postural tasks, while gamma frequency band (31–60 Hz) coherence is associated with dynamic force output [22]. These results indicate that different tasks induce changes in cortico-muscular coherence of different frequency bands. Coherence research has mainly focused on long-term recordings of basic, quasi-static physiological activities [1,16], consistently observing an increase in EMG beta coherence [5,11]. Dynamic movements have been observed to shift beta band coherence to the gamma band [1]. As most human dynamic movements involve both isometric and isotonic contractions, studying the changes in coherence and co-activation of different frequency bands during the dynamic fatigue process can more comprehensively reflect the central nervous system's regulation of muscles.

Although intermuscular coherence analysis has been widely used to reflect the neuromuscular control mechanism of movement, most related research uses the Fourier coherence method. This power spectrum estimation method, based on periodograms, has limitations. For instance, spectral leakage after windowing and large variance may affect the reliability of results [23]. Furthermore, in most movements and some pathological conditions, the actions are continuous, making the Fourier coherence method potentially unsuitable [16]. This necessitates the use of non-stationary analysis methods, such as wavelet methods. Using wavelet methods for coherence analysis can not only improve the accuracy of coherence but also accommodate non-stationary signals [1]. Wavelet transform is a common and mature method that can address the non-stationarity, randomness, and multi-component properties of sEMG signals. It relies on the mother wavelet to represent signals in the frequency space and offers higher accuracy than traditional methods [24]. The Synchrosqueezing Wavelet Transform (SST), developed from the Continuous Wavelet Transform (CWT), provides better frequency resolution. Moreover, SST has shorter computation time and better stability against errors [25]. Therefore, using SST, we can perform coherence analysis with high time-frequency resolution.

The aim of this study is to use the SST method, which is more suitable for dynamic signals, to investigate the changes in intermuscular coherence and co-activation in the alpha, beta, and gamma frequency bands of the prime mover and antagonist muscles during the fatigue process. As calf muscles play a crucial role in both daily activities and sports performance [26], the gastrocnemius muscle's high proportion of Type I fibers [27] leads to a gradual fatigue process, offering a better opportunity to study neuromuscular changes. By using the calf as an example, we aim to understand the changes in neuromuscular control during the dynamic fatigue process in calf muscles.

2. Materials and Methods

2.1. Subjects

A total of 22 healthy participants were recruited from the student population of Beijing Sport University (all males; age: 22.92 ± 2.19 years, average height: 1.79 ± 0.06 m, average weight: 76.08 ± 9.01 kg; all participants were right-leg dominant). Participants were recruited through convenience sampling from the student population of Beijing Sport University. The study was approved by the Ethics Committee of Beijing Sport University, and all participants received an informed consent form that they signed prior to the experiment. The EMG of the gastrocnemius and tibialis anterior of each participant's dominant leg was analyzed. Inclusion criteria: males aged between 18 and 25 years old, in good health (defined as having no history of cardiovascular, respiratory, musculoskeletal, or neurological disorders), no history of lower limb injuries in the past one year, and not systematically trained athletes or sports students. Exclusion criteria: fear of the experiment; allergy to any part of the instruments used in the measurement; presence of infections or signs of fatigue on the assessment day; consumption of medications, drugs (including antidepressants and pain medications), or caffeine on the assessment day that may influence neuromuscular function during the fatigue test.

2.2. Experimental Protocol

All measurements were conducted in the Biomechanics Laboratory of Beijing Sport University. The recordings were carried out on a 30 cm high jump box. The EMG of the GL and TA of the dominant leg of the participants was measured. The fatigue protocol included performing heel raises at a rhythm of 75 BPM. Participants were allowed to put their fingers on a wall to maintain balance and were instructed to keep their knees straight and try to achieve the maximum amplitude during heel raise. EMG signals were continuously recorded throughout the experiment, and the Rating of Perceived Exertion (RPE) was recorded every 30 s. The experiment was terminated when the participant's RPE $\geq$ 18 and they could not follow the specified rhythm. The RPE used the Borg Scale [28], ranging from 6 (very, very light) to 20 (very, very hard). The employed calf fatigue protocol is a commonly used test for assessing calf muscle function under clinical conditions. It can test the endurance of the calf muscles, and in this test, both the RMS amplitude and MF of the EMG signal change progressively during repeated heel raises [29]. Subjects were verbally motivated to give their best effort.

The Delsys wireless EMG system was used to record muscle activity. It is an amplifier with a gain of 1000. It was set to record at a sampling frequency of 1926 Hz. The EMG system included four 5×1 mm^2 Ag electrodes with an inter-electrode spacing of 10 mm, arranged in a 2×2 configuration. Skin preparation included shaving and wiping the skin with alcohol. Electrode placements followed the recommendations of the ISEK tutorials (Merletti, R., and G. L. Cerone, 2020) [30]. Electrodes were placed in the middle of the belly of the GL and TA muscles, parallel to the direction of the muscle fibers. The EMG acquisition module was bound with an elastic band to minimize the occurrence of motion artifacts.

2.3. Data Processing

Data were analyzed off-line using custom-written programs with MATLAB 2022b (The MathWorks Inc., Natick, MA, USA) to apply all necessary processing to obtain spectral representations of the EMG signals. First, each participant's signal was normalized using the maximum value of the signal during the entire fatigue process. This was followed by preprocessing with a 4th order Butterworth band-pass filter (5–500 Hz) to reduce motion artifacts and high-frequency noise. Research has shown [7] that dividing signals into 10 segments during wavelet coherence analysis can reveal details of fatigue effectively. Thus, in this study, each participant's signal was divided into 10 equal segments according to its duration for subsequent analysis.

2.4. Synchrosqueezed Wavelet Transform

In order to extract the power of EMG signals at different frequency bands, we first used the SST method to calculate the time-frequency matrix of the segmented EMG signals. SST, based on the Continuous Wavelet Transform (CWT), reassigns the signal to improve the time-frequency (TF) representation.

In this study, we chose the Morlet wavelet as the mother wavelet due to its excellent frequency localization characteristics. The Morlet wavelet is defined as follows:

$$\psi(t) = \frac{1}{\sqrt[4]{\pi}}(e^{i\omega_0 t} - e^{\frac{-\omega_0^2}{2}})e^{-\frac{t^2}{2}} \tag{1}$$

where *ψ(t)* represents the Morlet wavelet, *t* is the non-dimensional time, *i* is the imaginary unit, ω_0 is the non-dimensional frequency of the Morlet wavelet, and *e* is the base of the natural logarithm. The surface electromyographic signal *e(t)* is first decomposed into a continuous wavelet representation *We(a, b)*, defined as follows:

$$W_e(a,b) = \frac{1}{\sqrt{a}}\int_{-\infty}^{\infty} e(t)\psi^*(\frac{t-b}{a})dt \tag{2}$$

Here, *a* represents scale, and *b* represents time shift. When choosing the mother wavelet *ψ*, two conditions need to be satisfied: (i) the Fourier transform of the wavelet has strictly positive support; (ii) the wavelet satisfies the standard admissibility condition, as shown in the following formula:

$$\int_0^{\infty} z^{-1}\hat{\psi}(z)dz < \infty \tag{3}$$

Next, we calculate the derivative of the complex phase of *We(a, b)* to obtain the phase transform. For any *(a, b)* values satisfying *We(a, b)* $\neq 0$, the instantaneous frequency *ωe(a, b)* of the TF spectrum is calculated as follows:

$$\omega_e(a,b) = -i(W_e(a,b))^{-1}\frac{\partial}{\partial b}W_e(a,b) \tag{4}$$

Afterwards, we map the existing time-scale plane to the TF plane, transforming from *(b, a)* to (b, $\omega_e(a, b)$). On the TF plane, we estimate the SST, denoted as TS(*ω*, *b*), only at the center of the continuous interval ω_1, as shown in the following formula:

$$[\omega_l - \frac{1}{2}\Delta\omega, \omega_l + \frac{1}{2}\Delta\omega] \ with\ \omega_l - \omega_{l-1} = \Delta\omega \tag{5}$$

$$T_s(\omega_l,b) = (\Delta\omega)^{-1}\sum\nolimits_{a_k:|\omega(a_k,b)-\omega_l|\leq\Delta\omega/2} W_e(a_k,b)a_k^{\frac{-3}{2}}(\Delta a)k \tag{6}$$

Based on the obtained time-frequency distribution matrix TS, the time-domain integration is performed on different frequency bands in the electromyographic signal to calculate their average power. We will focus on the following frequency bands: alpha, beta, gamma, and high frequency bands.

2.5. EMG-EMG Coherence

To obtain the EMG-EMG coherence in different frequency bands, we first calculate the coherence matrix. The coherence matrix is computed using the analytic Morlet wavelet. The wavelet coherence of two time series *x* and *y* is defined as follows:

$$WCoh = \frac{|S(T_x^*(a,b)T_y(a,b))|^2}{S(|T_x(a,b)|_2) \times S(|T_y(a,b)|_2)} \tag{7}$$

Here, *Tx (a, b)* and *Ty(a, b)* denote the SST of *x* and *y* at scales a and positions *b*. The superscript * indicates the complex conjugate and *S* is a smoothing operator in time and scale [31].

2.6. RMS, MF and Coactivation Ratio

Root Mean Square (RMS):

$$RMS = \sqrt{\frac{1}{N}\sum_{i=1}^{N} x_i^2} \tag{8}$$

Median frequency (MF):

$$MF = f_K \tag{9}$$

$$\sum_{m=1}^{K} \Phi(f_m) = \sum_{m=K+1}^{M} \Phi(f_m) \tag{10}$$

The co-activation index is calculated using the following formula:

$$CI = 2 * \frac{EMG_{ANT}}{EMG_{AG} + EMG_{ANT}} * 100 \tag{11}$$

where EMG_{ANT} represents the EMG of the antagonist muscle, and EMG_{AG} denotes the EMG of the agonist muscle.

2.7. Statistical Analysis

In this study, statistical analysis was carried out using IBM SPSS Statistics 25 software. Data for each individual, divided into ten parts based on time, was subjected to a normality test for the extracted indicators, including different frequency band EMG power, co-activation ratio, first principal component, and intermuscular coherence in different frequency bands. For data not adhering to a normal distribution, a log10 transformation was applied. Subsequently, a one-way repeated measures analysis of variance (ANOVA) was performed.

Prior to conducting the ANOVA, the Mauchly's sphericity test was initially conducted to assess if the data satisfied the assumption of sphericity. If the assumption was violated, we employed the Greenhouse-Geisser correction method to handle the data, ensuring the accuracy of the results.

Upon identifying statistical significance, post hoc testing was conducted using the LSD method, further comparing differences between time points. The level of significance was set at $p < 0.05$, implying statistical significance when the p-value was less than 0.05.

In reporting the experimental results, the data in the text was presented as mean ± standard deviation, while data in graphs was presented as mean ± standard error.

3. Results

The experimental process is illustrated in Figure 1. During the dynamic contraction phase, the subjects exhibited an average duration of (110.76 ± 58.74) s.

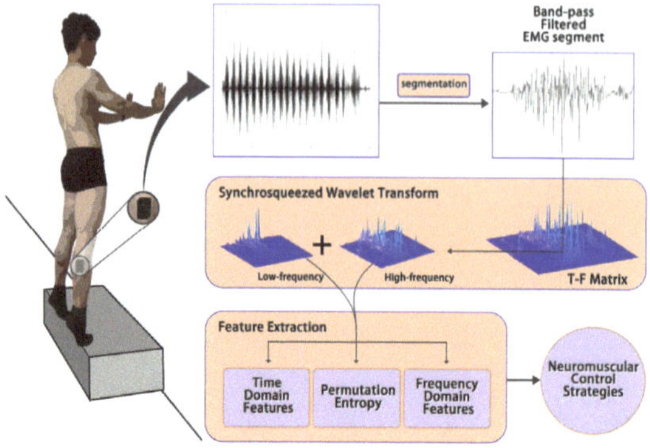

Figure 1. Workflow.

3.1. *Gastrocnemius Lateralis (GL)*

In the GL analysis, as demonstrated in Figures 2 and 3, clear alterations were observed in RMS (Partial Eta Squared = 0.605), MF (Partial Eta Squared = 0.719), the power within the 8–60 Hz frequency band (Partial Eta Squared = 0.691), and the power within the 100–400 Hz frequency band (Partial Eta Squared = 0.321), all of which corresponded with the level of fatigue ($p < 0.05$). Upon post hoc analysis, we discerned the following patterns:

(a) RMS: Compared to the onset of the motion, a significant increase occurred at time points T3 through T10 ($p < 0.05$). However, there were no significant changes noted when comparing T9 and T10 to T8.

(b) MF: There was a significant decline at all points from T2 to T10 when compared to the beginning of the action ($p < 0.05$). In comparison to T8, T9 and T10 also exhibited a significant decrease ($p < 0.05$).

(c) Low frequency band (8–60 Hz): Power significantly increased at time points T4 to T10, compared to the beginning of the action ($p < 0.05$). No significant power changes were observed in T9 and T10 relative to T8 ($p < 0.05$).

(d) Frequency band above 100 Hz: Compared to the onset of the motion, power significantly increased at time points T5 through T10 ($p < 0.05$), while it significantly decreased in T9 and T10 when compared to T8 ($p < 0.05$).

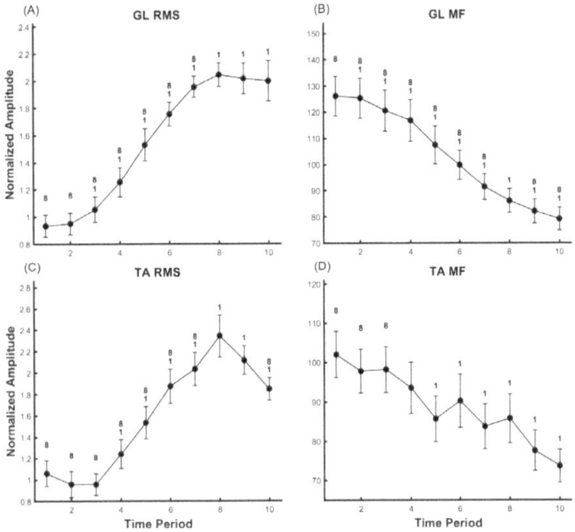

Figure 2. Representing the RMS (**A**,**C**) and MF (**B**,**D**) of EMG for the subjects' GL and TA muscles. Data are presented as mean ± standard error (SE). '1' indicates significance compared to time point 1 ($p < 0.05$), '8' indicates significance compared to time point 8 ($p < 0.05$).

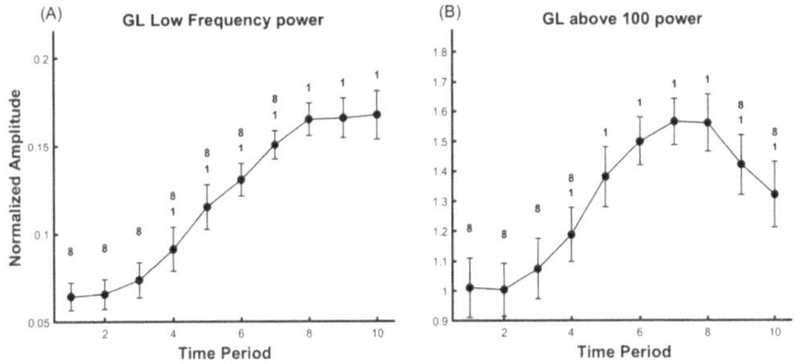

Figure 3. *Cont.*

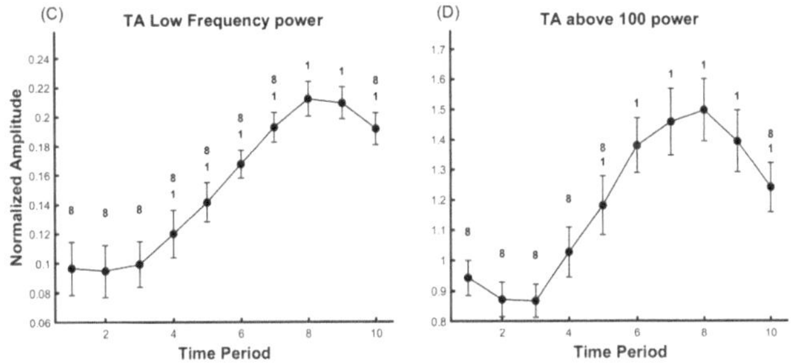

Figure 3. Representing the power changes in low frequency (**A**,**C**) and high frequency (**B**,**D**) of the EMG for GL and TA. Data are presented as mean ± standard error (SE). '1' indicates significance compared to time point 1 ($p < 0.05$), '8' indicates significance compared to time point 8 ($p < 0.05$).

3.2. Tibialis Anterior (TA)

In the TA analysis, as shown in Figures 2 and 3, significant changes were observed in RMS (Partial Eta Squared = 0.532), MF (Partial Eta Squared = 0.392), the power within the 8–60 Hz frequency band (Partial Eta Squared = 0.384), and the power within the 100–400 Hz frequency band (Partial Eta Squared = 0.331) in response to increasing fatigue levels ($p < 0.05$). Following post hoc analysis, we identified the following trends:

(a) RMS: Compared to the onset of the motion, there was a significant increase at time points T4 through T10 ($p < 0.05$). Nevertheless, a significant decrease was observed at T10 when compared to T8.
(b) MF: A significant decline was observed at all time points from T5 through T10 when compared to the beginning of the action ($p < 0.05$). However, there were no significant changes noted when comparing T9 and T10 to T8.
(c) Low frequency band (8–60 Hz): Compared to the onset of the motion, power significantly increased at time points T4 through T10 ($p < 0.05$). However, significant power decrease was observed at T10 when compared to T8 ($p < 0.05$).
(d) Frequency band above 100 Hz: Compared to the onset of the motion, power significantly increased at time points T5 through T10 ($p < 0.05$), while it significantly decreased at T10 when compared to T8 ($p < 0.05$).

3.3. Co-Activation and EMG-EMG Coherence

Throughout the entire fatigue process, no significant changes were observed in the co-activation ratio (Partial Eta Squared = 0.031) as the level of fatigue deepened ($p > 0.05$), as shown in Figure 4.

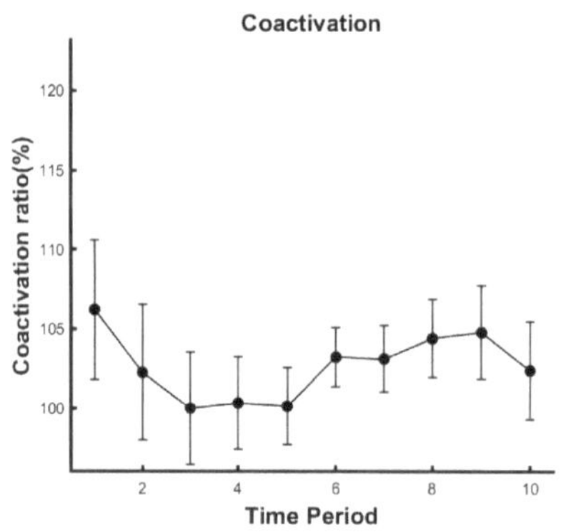

Figure 4. The co-activation changes between GL and TA muscles. Data are presented as mean ± standard error (SE).

During the fatigue process, significant changes were observed in the beta band coherence (Partial Eta Squared = 0.192) and gamma band coherence (Partial Eta Squared = 0.277) of GL and TA with the intensification of fatigue ($p < 0.05$). However, the alpha band coherence (Partial Eta Squared = 0.047) did not exhibit significant changes with the increase in fatigue ($p > 0.05$), as depicted in Figure 5.

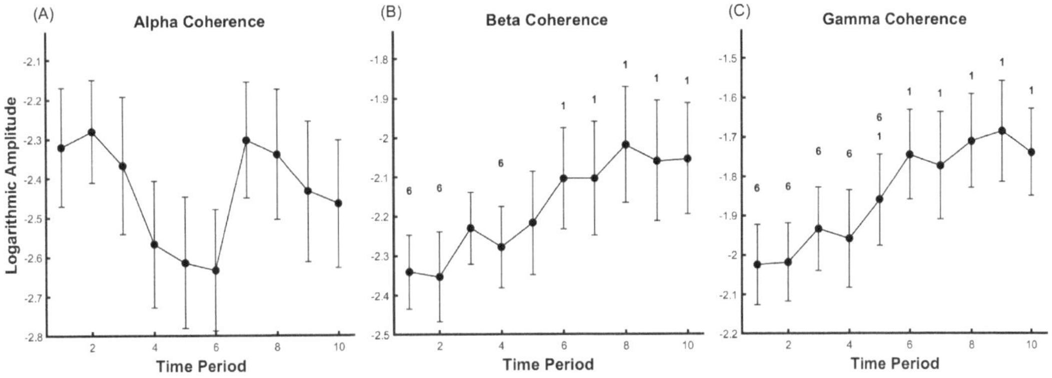

Figure 5. Representing the coherence of alpha (**A**), beta (**B**), and gamma (**C**) in the EMG of GL and TA. Data are presented as mean ± standard error (SE). '1' indicates significance compared to time point 1 ($p < 0.05$), '6' indicates significance compared to time point 6 ($p < 0.05$).

Upon post hoc analysis, we made the following observations.

In the beta frequency band: Compared to the onset of the motion, there was a significant increase from T6 to T10 ($p < 0.05$). Significant differences were observed when comparing T1, T2, and T4 to T6.

In the gamma frequency band: There was a significant increase from T5 to T10 when compared to the beginning of the action ($p < 0.05$). Significant differences were noted when comparing T1 to T5 with T6.

4. Discussion

Our study reveals a consistent decline in the MF of the GL muscle during fatigue tasks, aligning with previous findings in fatigue research [11,13,32,33]. We also observed a gradual increase in the RMS of the EMG signals from GL, accompanied by a sustained drop in MF. According to the Joint Analysis of EMG Spectrum and Amplitude (JASA) method [34], these results suggest the successful induction of fatigue in the GL.

The primary findings of this study underscore distinct variations in the RMS and power of the low-frequency band of the GL and TA at the T10 stage during fatigue contractions. Notably, significant increases in the beta and gamma intermuscular coherence of TA and GL were detected at the midpoint of the fatigue task (T6). However, no significant alterations were observed in the alpha intermuscular coherence and co-activation ratio throughout the task.

4.1. EMG Time Domain Changes in the Agonist GL and Antagonist TA Muscles

According to Henneman's size principle [35], MUs are recruited in an orderly manner, from those exerting the weakest force to those generating the most substantial force. When MUs are enlisted, there is a sudden increase in the discharge rate, resulting in an enhancement of EMG power. Our results indicate a significant increase in the RMS of the GL from T5-T8, consistent with previous studies [13]. This may suggest the onset of fatigue accumulation and the recruitment of new MUs. Subsequently, as observed in Figure 3A, the EMG activity of GL remained unchanged during T9-T10, potentially due to signal loss caused by amplitude cancellation of muscle Motor Unit Action Potentials (MUAPs) prior to task failure [36].

In the current study, the RMS of the TA peaked at T8 and then significantly fell, while the RMS of the prime mover remained unchanged, as seen in Figure 2A,C. This parallels previous findings on antagonist muscles in the calf during isometric contraction fatigue experiments [13]. Some researchers [17,37] propose this as a spinal-level control mechanism that balances the activity between the agonist and antagonist muscles. In the absence of regulation by this mechanism, the torque generated by the antagonist muscle would increase with excitability and counteract the torque of the agonist muscle [14], potentially leading to task failure.

Therefore, according to our results, the activation level of TA continues to increase in the early-to-mid stages of the fatigue task, maintaining a similar trend as GL. However, prior to the end of the fatigue task (T9-T10), the RMS of TA decreases while that of GL remains constant. This could be due to a decrease in GL's force-generating capability, with the cortex maintaining a constant co-activation level by inhibiting TA's activity, thereby preserving movement stability.

4.2. EMG Frequency Domain Changes in the Agonist GL and Antagonist TA Muscles

The average and median frequencies of electromyography (EMG) are primarily influenced by the conduction velocity of Motor Unit Action Potentials (MUAPs) along muscle fibers and the degree of MUAPs clustering [15,38]. As fatigue increases, the conduction velocity of MUAPs along muscle fibers decreases, or when the MUAPs cluster more tightly, the median frequency of the EMG will drop. Our results reveal a continuous decline in the Median Frequency (MF) of the Gastrocnemius Lateralis (GL) throughout the fatigue process, consistent with previous findings on dynamic fatigue [32,39]. Based on the theory proposed by von Tscharner et al. [38], this might indicate an increase in the degree of clustering of GL's MUAPs and a decrease in their conduction velocity with the onset of fatigue.

Our study shows a continuous decrease in the MF of the Tibialis Anterior (TA), with a significant drop occurring from T5 to T10. This aligns with the trend observed in the MF changes of antagonist muscles during elbow isometric fatigue tasks by Wang et al. [40]. Coupled with their findings on changes in the conduction velocity of antagonist muscles, they attributed this to peripheral fatigue caused by the accumulation of peripheral metabolites due to prolonged co-activation. Contrarily, a study [41] found no fatigue in the antagonist muscles after completing an isokinetic knee extension fatigue task, as there was no decline in torque and EMG amplitude. However, our results indicate a decrease in the MF of the antagonist muscles and a drop in RMS from T8 to T10, suggesting that TA, as an antagonist muscle, experienced peripheral fatigue. The effects of dynamic and static fatigue, as well as fatigue processes at different force levels, on antagonist muscles remain unclear and warrant further investigation.

Many studies [21,42–44] have pointed out that the power in the beta and gamma frequency bands of EMG is related to the synchronization activity between Motor Units (MUs). Simulated EMG study by von Tscharner et al. [38] also indicated that the higher the power in the part below 60 Hz, the stronger the synchronous activity of MUs. Our results show that during T1-T8, the power in the low-frequency range of both GL and TA increased, which is consistent with previous research [45,46]. This may suggest that during the early and middle parts of the fatigue task (T1–T8), neural drive to MUs becomes more synchronized to adapt to the effects of fatigue [45,47]. However, just before the end of the fatigue task (T10), the power in the low-frequency range of TA showed a significant decline. Few studies have reported on the changes in the power of different frequency bands of antagonist muscles throughout the fatigue task. Combined with the decline in RMS of the antagonist muscle, we speculate that this is due to the central nervous system reducing the activity of TA by inhibiting the synchronization of its MUs. This might serve as another piece of evidence for the differentiated control of antagonist muscles at the spinal level [14].

The power of EMG > 100 Hz might be related to the number of MUAPs. Our results show that the power in the high-frequency part of GL and TA EMG continuously increased during T1–T8, and began to decrease during T9-T10. Some research [48] suggests that the power of high-frequency components of EMG could be due to the rapid generation of MUAPs by fast motor units. The EMG model by von Tscharner et al. [38] also pointed out that the power in the high-frequency part of EMG is proportional to the number of MUAPs. The fatigue model by Potvin et al. [8] also predicts that the discharge rate of MUs continues to increase during fatigue, and suddenly decreases just before the task ends, which is consistent with our observations. This may indicate that the MU firing rate of TA and GL begins to decrease a short time before exhaustion (T9-T10), until task failure. Interestingly, the RMS of EMG did not show this change. A possible explanation is that when the discharge rate of MUs decreases and the number of MUAPs decreases close to task failure, the power of high-frequency EMG [38], which is proportional to the number of MUAPs, decreases. When the density of randomly distributed MUAPs is high, the cancellation effect will reduce the effective amplitude of EMG. When the density of MUAPs decreases, the cancellation effect also weakens, which might be reflected as unchanged EMG amplitude. Therefore, the power in the high-frequency range of EMG might indirectly reflect the discharge rate of MUs.

4.3. Coactivation of the Agonist GL and Antagonist TA Muscles

Coactivation, the phenomenon of antagonist and agonist muscle groups being activated simultaneously, is intended to maintain a specific level of neuromuscular function [49]. The phenomenon of coactivation is often explained by De Luca's "common drive" theory [2]. This hypothesis suggests that when two muscles are involved in a specific task, the central nervous system can control the motor neuron pool of each muscle through a single input. This means that when one muscle contracts, the antagonist muscle will also contract simultaneously to coordinate and balance movement. On the contrary, some researchers believe that the central nervous system controls the agonist and antagonist muscles through different mechanisms: one that simultaneously activates the agonist and antagonist muscles, and another that activates the agonist while inhibiting the antagonist [13,14].

Our study results show that during fatigue, the coactivation level of TA and GL remains unchanged throughout the fatigue process. This is consistent with previous fatigue studies [11,13,50]. However, our results during T10 show that although the level of coactivation remains unchanged, the RMS of TA significantly decreases, and the power in the low-frequency range of TA significantly decreases as well, while there are no significant changes in the RMS and low-frequency power of GL. The "common drive" hypothesis has difficulty explaining this phenomenon; our results seem to support the view that the central nervous system controls the activities of agonist and antagonist muscles through multiple mechanisms.

4.4. EMG-EMG Coherence of the Agonist GL and Antagonist TA Muscles

The coherence in the alpha band originates from the subcortex and may reflect the ascending or feedback interactions in neuromuscular control [5]. A study [11] found that fatigue in healthy adults resulted in a significant increase in coherence in the alpha band during isometric fatigue tasks at the elbow. However, in contrast, our results show that there were no significant changes in the coherence of the alpha band during the fatigue process. This might suggest that the ascending or feedback interaction is stable in dynamic fatigue. Another possibility is that the activity of Renshaw cells in the spinal cord could influence the coherence in the alpha band, leading to inconsistent observations of coherence in the alpha band in many studies [5].

Evidence suggests that the coherence in the beta and gamma bands is mainly controlled by the motor cortex and can reflect the coupled action of the cortex and muscles [51,52]. Many previous studies have explored the corticomuscular and intermuscular coherence in the beta band [44,53–55]. Baker [55] observed in monkeys that the coherence between the cortex and muscles in the beta band significantly increased when completing a posture maintenance task. The same results were found in humans, where the intermuscular coherence in the beta band among FDI [44], index finger flexor [10], knee extensor [52], and elbow antagonist [11] increased during muscle fatigue caused by sustained isometric maximal or dynamic movements. The coherence in the gamma band is often observed in dynamic tasks. Previous studies have observed significant EEG-EMG coherence in the gamma band during maximum voluntary contraction [51], and the coherence in the gamma band is related to the level of force [56].

Our results show that the EMG-EMG coherence in the beta and gamma bands gradually increases, with significant changes appearing in the middle of the fatigue task (T5-T6) and reaching a peak at T8-T9, but without significant changes. Similar results have been found in previous studies [7]. High coherence between EMG-EMG might indicate that the two muscles share neural drive [54], and fatigue has been observed to enhance the EMG-EMG coherence between agonist and antagonist muscles. Based on this, we speculate that in the middle of our fatigue task (T5–T6), fatigue leads to a decline in muscle function, and the central nervous system needs to enhance the driving effect on the agonist and antagonist muscles to compensate for the impact of fatigue.

Contrary to our results, Wang et al. [19] found that during a 30% MVC isometric fatigue task, there was no significant change in the intermuscular coherence in the beta and gamma bands between the agonist and antagonist muscles of the elbow before and after fatigue. This may be related to the choice of muscles; the study chose proximal elbow muscles, while other research suggests that distal muscles have stronger cortical neuron connections. This could lead to different levels of coupling between distal and proximal muscles and the central nervous system [10]. Another possible reason is the difference in coherence analysis methods; the study used the Fourier method to study coherence, which might affect the reliability of the results [23].

Interestingly, the EMG-EMG coherence in the beta band of the GL and TA showed significant changes at T6, while the EMG-EMG coherence in the gamma band showed significant changes at T5. These changes do not coincide with the time when the RMS of the antagonist or agonist muscles reached their maximum. EMG-EMG coherence is often considered a tool to quantify neuromuscular coupling [1]. The continuous increase in EMG-EMG coherence from T1 to T6, along with the increase in RMS of GL and TA, might suggest that as fatigue occurs, neuromuscular coupling continues to strengthen. This effect is manifested at the muscle activity level as an increase in EMG amplitude. However, after T6, the increase in EMG-EMG coherence slowed down significantly, and no significant changes appeared. But the RMS of GL and TA continues to increase until T8. The reason for this change is not clear. A study on synergistic muscles found that personalized control of muscles leads to a decrease in beta coherence between muscles [57]. This might indicate that at T6, the central nervous system began to differentiate control of the antagonist muscles, and this control appeared before the strongest activity of the antagonist muscles. However, we did not collect direct indicators of central fatigue, so this connection is speculative. Future research can provide more evidence and explanations for this issue.

Indeed, since coherence can reflect the communication between the cortex and muscles, it is often used to monitor the recovery status of neuromuscular function [1,58]. Moderate fatigue can enhance athletic performance [19], but excessive fatigue can reduce performance and increase the risk of injury [59]. Therefore, detecting when fatigue occurs and the extent of fatigue is of great significance. Our study identified a potentially significant transition point at 60% of the fatigue task, where neuromuscular control strategies appear to shift. However, current research on this specific time point is limited, making it challenging to provide concrete training recommendations. We suggest that future research focus on this 60% time point, studying the effects of different training interventions. This could lead to more effective, individualized training protocols that optimize performance while minimizing injury risk. In the meantime, trainers could consider monitoring EMG-EMG coherence and EMG changes of antagonist muscles during training sessions to gain insights into fatigue development and neuromuscular control strategies.

4.5. Limitation

There are several limitations in this study. Firstly, our EMG collector has a built-in 20–450 Hz LTI filter, and the frequency bands (alpha, beta) we are interested in overlap with this frequency segment. According to the research of Chen et al. [23], theoretically, preprocessing with an LTI filter will not improve or affect coherence estimation. The factors that affect coherence are mainly the calculation methods. Using the Welch method to estimate spectral density for coherence analysis may lead to increased variance or spectral leakage in results, thereby affecting the outcome. In our study, we adopted wavelet coherence analysis. Secondly, this study did not collect indicators that can directly reflect central fatigue, so we cannot directly discuss the state of central fatigue. Future research can measure corticospinal excitability to discuss the effect of central fatigue on neuromuscular control. Thirdly, the sample in this study is limited to healthy male college students aged 18–25 without investigating their physical activity levels. Future research could include participants of different ages, genders, and fitness levels (ranging from sedentary to highly active) to understand how population characteristics affect neuromuscular control during fatigue.

5. Conclusions

Our research results reveal changes in neuromuscular control during the dynamic fatigue process. In the first 60% of the fatigue task, the beta and gamma band EMG-EMG coherence of the calf's agonist and antagonist muscles continuously increase, and this increase presents a synchronous trend with the changes in muscle EMG amplitude. This finding may suggest that during the first 60% of the task, the central nervous system maintains the stability of the exercise level by continuously enhancing the driving effect on the muscles. Interestingly, after the task progresses to 60%, the ENG-EMG coherence remains stable, while the EMG amplitude of the GL and TA continues to increase. This might be because the central nervous system maintains the level of co-activation by differentiating the control of the antagonist muscles through mechanisms other than co-driving, and this differentiated control appears before muscle activity. This implies that the central nervous system adopts more complex and refined strategies to regulate muscle activity under fatigue. Based on our research results, we believe that monitoring EMG-EMG coherence and the EMG changes of antagonist muscles may be of significant importance in the process of athletic training. This monitoring method may provide a basis for the evaluation and monitoring of training effects, thereby helping trainers better understand the development of muscle fatigue and adopt corresponding training adjustment measures.

Author Contributions: Conceptualization, X.N., L.I. and Y.L.; methodology, X.N., L.I. and Y.L.; validation, L.I., Y.L. and X.N.; formal analysis, X.N., M.X. and L.I.; investigation, X.N. and L.I.; data curation, X.N. and L.I.; writing—original draft preparation, X.N.; writing—review and editing, X.N., L.I., M.X. and Y.L.; visualization, X.N. and M.X.; supervision, X.N., L.I. and Y.L.; project administration, X.N. All authors have read and agreed to the published version of the manuscript.

Funding: This research received no external funding.

Institutional Review Board Statement: The study was conducted in accordance with the Declaration of Helsinki, and approved by the Ethics Committee of Beijing Sport University (protocol code 2023104H, 2023-04-21).

Informed Consent Statement: Informed consent was obtained from all subjects involved in the study.

Data Availability Statement: The data presented in this study are available on request from the corresponding author. The data are not publicly available due to privacy and ethical restrictions, as they contain information that could compromise the privacy of research participants.

Acknowledgments: The authors would like to express our sincere gratitude to the study participants for their valuable contributions to this research. The authors are also thankful to the individuals who assisted with the experiments for their invaluable support.

Conflicts of Interest: The authors declare no conflicts of interest.

References

1. Liu, J.; Sheng, Y.; Liu, H. Corticomuscular Coherence and Its Applications: A Review. *Front. Hum. Neurosci.* **2019**, *13*, 100. [CrossRef] [PubMed]
2. De Luca, C.J.; Mambrito, B. Voluntary Control of Motor Units in Human Antagonist Muscles: Coactivation and Reciprocal Activation. *J. Neurophysiol.* **1987**, *58*, 525–542. [CrossRef] [PubMed]
3. Yang, Q.; Fang, Y.; Sun, C.-K.; Siemionow, V.; Ranganathan, V.K.; Khoshknabi, D.; Davis, M.P.; Walsh, D.; Sahgal, V.; Yue, G.H. Weakening of Functional Corticomuscular Coupling during Muscle Fatigue. *Brain Res.* **2009**, *1250*, 101–112. [CrossRef] [PubMed]
4. Xie, P.; Cheng, S.; Zhang, Y.; Liu, Z.; Liu, H.; Chen, X.; Li, X. Direct Interaction on Specific Frequency Bands in Functional Corticomuscular Coupling. *IEEE Trans. Biomed. Eng.* **2020**, *67*, 762–772. [CrossRef] [PubMed]
5. Mehrkanoon, S.; Breakspear, M.; Boonstra, T.W. The Reorganization of Corticomuscular Coherence during a Transition between Sensorimotor States. *NeuroImage* **2014**, *100*, 692–702. [CrossRef]
6. Qiu, Q.; Cao, L.; Hao, D.; Yang, L.; Hillstrom, R.; Zheng, D. Muscle Extremely Low Frequency Magnetic Stimulation Eliminates the Effect of Fatigue on EEG-EMG Coherence during the Lateral Raise Task: A Pilot Quantitative Investigation. *BioMed Res. Int.* **2018**, *2018*, 1–8. [CrossRef]
7. Yang, Q.; Siemionow, V.; Yao, W.; Sahgal, V.; Yue, G.H. Single-Trial EEG-EMG Coherence Analysis Reveals Muscle Fatigue-Related Progressive Alterations in Corticomuscular Coupling. *IEEE Trans. Neural Syst. Rehabil. Eng.* **2010**, *18*, 97–106. [CrossRef]
8. Potvin, J.R.; Fuglevand, A.J. A Motor Unit-Based Model of Muscle Fatigue. *PLoS Comput. Biol.* **2017**, *13*, e1005581. [CrossRef]
9. Al-Mulla, M.R.; Sepulveda, F.; Colley, M.; Al-Mulla, F. Statistical Class Separation Using sEMG Features Towards Automated Muscle Fatigue Detection and Prediction. In Proceedings of the 2009 2nd International Congress on Image and Signal Processing, Tianjin, China, 17–19 October 2009; IEEE: New York, NY, USA, 2009; pp. 1–5. [CrossRef]
10. Kattla, S.; Lowery, M.M. Fatigue Related Changes in Electromyographic Coherence between Synergistic Hand Muscles. *Exp. Brain Res.* **2010**, *202*, 89–99. [CrossRef]
11. Wang, L.; Lu, A.; Zhang, S.; Niu, W.; Zheng, F.; Gong, M. Fatigue-Related Electromyographic Coherence and Phase Synchronization Analysis between Antagonistic Elbow Muscles. *Exp. Brain Res.* **2015**, *233*, 971–982. [CrossRef]
12. Latash, M.L. Muscle Coactivation: Definitions, Mechanisms, and Functions. *J. Neurophysiol.* **2018**, *120*, 88–104. [CrossRef] [PubMed]
13. Lévénez, M.; Kotzamanidis, C.; Carpentier, A.; Duchateau, J. Spinal Reflexes and Coactivation of Ankle Muscles during a Submaximal Fatiguing Contraction. *J. Appl. Physiol.* **2005**, *99*, 1182–1188. [CrossRef] [PubMed]
14. Lévénez, M.; Garland, S.J.; Klass, M.; Duchateau, J. Cortical and Spinal Modulation of Antagonist Coactivation During a Submaximal Fatiguing Contraction in Humans. *J. Neurophysiol.* **2008**, *99*, 554–563. [CrossRef]
15. Al-Mulla, M.R.; Sepulveda, F.; Colley, M. A Review of Non-Invasive Techniques to Detect and Predict Localised Muscle Fatigue. *Sensors* **2011**, *11*, 3545–3594. [CrossRef] [PubMed]
16. Grosse, P.; Cassidy, M.J.; Brown, P. EEG–EMG, MEG–EMG and EMG–EMG Frequency Analysis: Physiological Principles and Clinical Applications. *Clin. Neurophysiol.* **2002**, *113*, 1523–1531. [CrossRef] [PubMed]
17. Hansen, S.; Hansen, N.L.; Christensen, L.O.D.; Petersen, N.T.; Nielsen, J.B. Coupling of Antagonistic Ankle Muscles during Co-Contraction in Humans. *Exp. Brain Res.* **2002**, *146*, 282–292. [CrossRef]
18. Chen, Y.-T.; Li, S.; Magat, E.; Zhou, P.; Li, S. Motor Overflow and Spasticity in Chronic Stroke Share a Common Pathophysiological Process: Analysis of Within-Limb and Between-Limb EMG-EMG Coherence. *Front. Neurol.* **2018**, *9*, 795. [CrossRef]
19. Wang, L.-J.; Yu, X.-M.; Shao, Q.-N.; Wang, C.; Yang, H.; Huang, S.-J.; Niu, W.-X. Muscle Fatigue Enhance Beta Band EMG-EMG Coupling of Antagonistic Muscles in Patients With Post-Stroke Spasticity. *Front. Bioeng. Biotechnol.* **2020**, *8*, 1007. [CrossRef]
20. Sato, S.; Choi, J.T. Increased Intramuscular Coherence Is Associated with Temporal Gait Symmetry during Split-Belt Locomotor Adaptation. *J. Neurophysiol.* **2019**, *122*. [CrossRef]
21. Brown, P. Cortical Drives to Human Muscle: The Piper and Related Rhythms. *Prog. Neurobiol.* **2000**, *60*, 97–108. [CrossRef]
22. Gwin, J.T.; Ferris, D.P. Beta- and Gamma-Range Human Lower Limb Corticomuscular Coherence. *Front. Hum. Neurosci.* **2012**, *6*, 258. [CrossRef] [PubMed]
23. Chen, M.; Lu, Z.; Zhou, P. A Dilemma for Coherence Calculation: Should Preprocessing Filters Be Applied? *Front. Neurosci.* **2022**, *16*, 838627. [CrossRef] [PubMed]
24. Karlsson, S.; Yu, J.; Akay, M. Time-Frequency Analysis of Myoelectric Signals during Dynamic Contractions: A Comparative Study. *IEEE Trans. Biomed. Eng.* **2000**, *47*, 228–238. [CrossRef] [PubMed]
25. Hari, L.M.; Venugopal, G.; Ramakrishnan, S. Dynamic Contraction and Fatigue Analysis in Biceps Brachii Muscles Using Synchrosqueezed Wavelet Transform and Singular Value Features. *Proc. Inst. Mech. Eng. Part H* **2022**, *236*, 208–217. [CrossRef]
26. Webster, K.A.; Pietrosimone, B.G.; Gribble, P.A. Muscle Activation During Landing Before and After Fatigue in Individuals With or Without Chronic Ankle Instability. *J. Athl. Train.* **2016**, *51*, 629–636. [CrossRef]
27. Trappe, S.; Trappe, T.; Lee, G.; Costill, D. Calf Muscle Strength in Humans. *Int. J. Sports Med.* **2001**, *22*, 186–191. [CrossRef]
28. Borg, G.A. Psychophysical Bases of Perceived Exertion. *Med. Sci. Sports Exerc.* **1982**, *14*, 377–381. [CrossRef]
29. Österberg, U.; Svantesson, U.; Takahashi, H.; Grimby, G. Torque, Work and EMG Development in a Heel-Rise Test. *Clin. Biomech.* **1998**, *13*, 344–350. [CrossRef]
30. Merletti, R.; Cerone, G.L. Tutorial. Surface EMG Detection, Conditioning and Pre-Processing: Best Practices. *J. Electromyogr. Kinesiol.* **2020**, *54*, 102440. [CrossRef]

31. Maraun, D.; Kurths, J.; Holschneider, M. Nonstationary Gaussian Processes in Wavelet Domain: Synthesis, Estimation, and Significance Testing. *Phys. Rev. E* **2007**, *75*, 016707. [CrossRef]
32. Gonzalez-Izal, M.; Malanda, A.; Navarro-Amezqueta, I.; Gorostiaga, E.; Mallor, F.; Ibanez, J.; Izquierdo, M. EMG Spectral Indices and Muscle Power Fatigue during Dynamic Contractions. *J. Electromyogr. Kinesiol.* **2010**, *20*, 233–240. [CrossRef] [PubMed]
33. Hunter, S.K.; Lepers, R.; MacGillis, C.J.; Enoka, R.M. Activation among the Elbow Flexor Muscles Differs When Maintaining Arm Position during a Fatiguing Contraction. *J. Appl. Physiol.* **2003**, *94*, 2439–2447. [CrossRef] [PubMed]
34. Hägg, G.M.; Luttmann, A.; Jäger, M. Methodologies for Evaluating Electromyographic Field Data in Ergonomics. *J. Electromyogr. Kinesiol.* **2000**, *10*, 301–312. [CrossRef] [PubMed]
35. Henneman, E. Relation between Size of Neurons and Their Susceptibility to Discharge. *Science* **1957**, *126*, 1345–1347. [CrossRef]
36. Keenan, K.G.; Farina, D.; Maluf, K.S.; Merletti, R.; Enoka, R.M. Influence of Amplitude Cancellation on the Simulated Surface Electromyogram. *J. Appl. Physiol.* **2005**, *98*, 120–131. [CrossRef]
37. Benwell, N.M.; Sacco, P.; Hammond, G.R.; Byrnes, M.L.; Mastaglia, F.L.; Thickbroom, G.W. Short-Interval Cortical Inhibition and Corticomotor Excitability with Fatiguing Hand Exercise: A Central Adaptation to Fatigue? *Exp. Brain Res.* **2006**, *170*, 191–198. [CrossRef]
38. von Tscharner, V. A Model Computation of How Synchronization and Clustering of Motor Unit Action Potentials Alter the Power Spectra of Electromyograms. *Biomed. Signal Process. Control.* **2019**, *47*, 344–349. [CrossRef]
39. Walker, S.; Davis, L.; Avela, J.; Häkkinen, K. Neuromuscular Fatigue during Dynamic Maximal Strength and Hypertrophic Resistance Loadings. *J. Electromyogr. Kinesiol.* **2012**, *22*, 356–362. [CrossRef]
40. Wang, L.; Song, X.; Yang, H.; Wang, C.; Shao, Q.; Tao, H.; Qiao, M.; Niu, W.; Liu, X. Are the Antagonist Muscle Fatigued during a Prolonged Isometric Fatiguing Elbow Flexion at Very Low Forces for Young Adults? *Front. Physiol.* **2022**, *13*, 956639. [CrossRef]
41. Kotzamanidis, C. Are the Antagonist Muscles Fatigued during a Fatigue Task of Agonist Muscles? *IES* **2004**, *12*, 167–171. [CrossRef]
42. Dietz, V. Spinal Cord Pattern Generators for Locomotion. *Clin. Neurophysiol.* **2003**, *114*, 1379–1389. [CrossRef] [PubMed]
43. Norton, J.A.; Gorassini, M.A. Changes in Cortically Related Intermuscular Coherence Accompanying Improvements in Locomotor Skills in Incomplete Spinal Cord Injury. *J. Neurophysiol.* **2006**, *95*, 2580–2589. [CrossRef] [PubMed]
44. Conway, B.A.; Halliday, D.M.; Farmer, S.F.; Shahani, U.; Maas, P.; Weir, A.I.; Rosenberg, J.R. Synchronization between Motor Cortex and Spinal Motoneuronal Pool during the Performance of a Maintained Motor Task in Man. *J. Physiol.* **1995**, *489*, 917–924. [CrossRef]
45. Liu, X.; Zhou, M.; Geng, Y.; Meng, L.; Wan, H.; Ren, H.; Zhang, X.; Dai, C.; Chen, W.; Ye, X. Changes in Synchronization of the Motor Unit in Muscle Fatigue Condition during the Dynamic and Isometric Contraction in the Biceps Brachii Muscle. *Neurosci. Lett.* **2021**, *761*, 136101. [CrossRef] [PubMed]
46. Chowdhury, S.K.; Nimbarte, A.D.; Jaridi, M.; Creese, R.C. Discrete Wavelet Transform Analysis of Surface Electromyography for the Fatigue Assessment of Neck and Shoulder Muscles. *J. Electromyogr. Kinesiol.* **2013**, *23*, 995–1003. [CrossRef] [PubMed]
47. Kline, J.C.; De Luca, C.J. Synchronization of Motor Unit Firings: An Epiphenomenon of Firing Rate Characteristics Not Common Inputs. *J. Neurophysiol.* **2016**, *115*, 178–192. [CrossRef]
48. von Tscharner, V.; Ullrich, M.; Mohr, M.; Comaduran Marquez, D.; Nigg, B.M. Beta, Gamma Band, and High-Frequency Coherence of EMGs of Vasti Muscles Caused by Clustering of Motor Units. *Exp. Brain Res.* **2018**, *236*, 3065–3075. [CrossRef]
49. Hogan, N. Adaptive Control of Mechanical Impedance by Coactivation of Antagonist Muscles. *IEEE Trans. Autom. Control.* **1984**, *29*, 681–690. [CrossRef]
50. Benitez, B.; Dinyer-McNeely, T.K.; McCallum, L.; Kwak, M.; Succi, P.J.; Bergstrom, H.C. Load-Specific Performance Fatigability, Coactivation, and Neuromuscular Responses to Fatiguing Forearm Flexion Muscle Actions in Women. *J. Strength Cond. Res.* **2023**, *37*, 769–779. [CrossRef]
51. Brown, P.; Farmer, S.; Halliday, D.; Marsden, J.; Rosenberg, J. Coherent Cortical and Muscle Discharge in Cortical Myoclonus. *Brain* **1999**, *122*, 461–472. [CrossRef]
52. Chang, Y.-J.; Chou, C.-C.; Chan, H.-L.; Hsu, M.-J.; Yeh, M.-Y.; Fang, C.-Y.; Chuang, Y.-F.; Wei, S.-H.; Lien, H.-Y. Increases of Quadriceps Inter-Muscular Cross-Correlation and Coherence during Exhausting Stepping Exercise. *Sensors* **2012**, *12*, 16353–16367. [CrossRef] [PubMed]
53. Aumann, T.D.; Prut, Y. Do Sensorimotor β-Oscillations Maintain Muscle Synergy Representations in Primary Motor Cortex? *Trends Neurosci.* **2015**, *38*, 77–85. [CrossRef]
54. Laine, C.M.; Valero-Cuevas, F.J. Intermuscular Coherence Reflects Functional Coordination. *J. Neurophysiol.* **2017**, *118*, 1775–1783. [CrossRef]
55. Baker, S.N.; Olivier, E.; Lemon, R.N. Coherent Oscillations in Monkey Motor Cortex and Hand Muscle EMG Show Task-Dependent Modulation. *J. Physiol.* **1997**, *501*, 225–241. [CrossRef]
56. Chakarov, V.; Naranjo, J.R.; Schulte-Mönting, J.; Omlor, W.; Huethe, F.; Kristeva, R. Beta-Range EEG-EMG Coherence with Isometric Compensation for Increasing Modulated Low-Level Forces. *J. Neurophysiol.* **2009**, *102*, 1115–1120. [CrossRef] [PubMed]
57. Reyes, A.; Laine, C.M.; Kutch, J.J.; Valero-Cuevas, F.J. Beta Band Corticomuscular Drive Reflects Muscle Coordination Strategies. *Front. Comput. Neurosci.* **2017**, *11*, 17. [CrossRef] [PubMed]

58. Velázquez-Pérez, L.; Tünnerhoff, J.; Rodríguez-Labrada, R.; Torres-Vega, R.; Ruiz-Gonzalez, Y.; Belardinelli, P.; Medrano-Montero, J.; Canales-Ochoa, N.; González-Zaldivar, Y.; Vazquez-Mojena, Y.; et al. Early Corticospinal Tract Damage in Prodromal SCA2 Revealed by EEG-EMG and EMG-EMG Coherence. *Clin. Neurophysiol.* **2017**, *128*, 2493–2502. [CrossRef]
59. Sun, J.; Liu, G.; Sun, Y.; Lin, K.; Zhou, Z.; Cai, J. Application of Surface Electromyography in Exercise Fatigue: A Review. *Front. Syst. Neurosci.* **2022**, *16*, 893275. [CrossRef]

life

Article

Impact of Stiffness of Quadriceps on the Pedaling Rate of Maximal Cycling

Loi Ieong [1], Xindi Ni [1], Huisong Xie [2] and Ye Liu [1,*]

1 School of Sport Science, Beijing Sport University, Beijing 100084, China; loi_326@hotmail.com (L.I.)
2 School of Track and Field, Beijing Sport University, Beijing 100084, China
* Correspondence: liuye_tg@163.com

Abstract: Propulsive power is one of the factors that determine the performance of sprint cycling. Pedaling rate is related to power output, and stiffness is associated with improving performance in athletic tasks. Purpose: to investigate the relationship between musculoarticular stiffness and pedaling rate. Methods: twenty-two healthy, untrained male volunteers (19 ± 2 years, 175 ± 6 cm, 74 ± 16 kg) were divided into two groups after their musculoarticular (MA) stiffness was tested, and these groups were the stiffness group (SG) and compliant group (CG). A 6-s maximal cycling test was conducted in four cycling modes, which were levels 5 and 10 air-resistance, and levels 3 and 7 magnetic-resistance. Peak and average cadence, peak power output (PO_{peak}), crank force (CF_{peak}), peak rate of crank force development (RCFD), and the angle of peak crank force were collected. The significance of differences between the two groups for these variables was assessed using an independent samples t-test. Pearson product–moment correlations were calculated to analyze the relationship between MA stiffness and each performance variable. Results: the SG had significantly higher peak cadence and average cadence at level 3 magnetic-resistance, peak crank force, and peak power output at level 10 air-resistance, peak rate of crank force development at levels 5 air-resistance, 10 air-resistance, and 3 magnetic-resistance ($p < 0.05$). MA stiffness was significantly correlated with average cadence at levels 5 and 10 air-resistance, peak crank force in all 4 modes, and RCFD and peak power output at level 10 air-resistance. There were no significant relationships between MA stiffness and the angle of peak crank force in each cycling mode. Conclusion: results indicate that participants with relatively higher MA stiffness seemed to have a higher pedaling rate during a 6-s sprint cycling in these conditions. They also performed a superior crank force and rate of crank force development, producing greater power output when sprint cycling. Optimizing cycling resistance or gear ratio to enhance both RCFD and musculotendinous stiffness may be crucial for improving sprint cycling performance.

Keywords: muscle stiffness; pedaling rate; rate of force development; cycling; power output

Citation: Ieong, L.; Ni, X.; Xie, H.; Liu, Y. Impact of Stiffness of Quadriceps on the Pedaling Rate of Maximal Cycling. *Life* **2024**, *14*, 956. https://doi.org/10.3390/life14080956

Academic Editor: Ines Drenjančević

Received: 25 June 2024
Revised: 22 July 2024
Accepted: 29 July 2024
Published: 30 July 2024

1. Introduction

Maximal cycling power output is a significant determinant of propulsive power in sprint cycling, which is influenced by factors such as pedaling rate, crank torque, muscle size, muscle fiber-type distribution, cycling position, and fatigue [1]. The performance of sprint cycling is governed by the interplay between propulsive power and resistance forces. The pedaling rate is a dominant influencing factor in performance, which is determined by muscle shortening velocity and excitation–relaxation kinetics [1–3].

The relationship between muscle shortening velocity and force production, first described by Hill, demonstrates an inverse correlation. As shortening velocity increases, force production decreases, resulting in a quadratic power–velocity relationship. During maximal sprint cycling, power output initially increases with increased pedaling rate, reaching a maximum, and then it decreases with further increases in the pedaling rate [2,4]. The pedaling rate determines the time available for muscle excitation–relaxation cycles. In maximal sprint cycling, cyclists can perform pedaling cadences of up to 155 rpm [5], where the extension–flexion phases occur within 194 to 423 milliseconds [6]. However, previous research has reported that the time to peak tension for the quadriceps femoris [7] and triceps surae [8] ranges from 121 to 400 milliseconds, with half relaxation taking up to 76 milliseconds. Martin [6] indicated that this time might not be sufficient to achieve maximal force production, and the muscle force depression during the cycle would be more significant if the muscle did not reach peak tension.

Cycling efficiency seems to be related to pedaling rate. Lucia [9] found that a relatively higher preferred cadence resulted in better efficiency in professional cyclists, whereas recreational cyclists might tend to prefer lower cadences [10]. Considering discrepancies in preferred cadence, Takaishi [11] found that the cycling skills of professional cyclists might facilitate better utilization of knee flexors at higher cadences, contributing to decreased peak pedal force and lower muscle activity for the knee extensors. Additionally, higher pedaling rates may influence fiber-type recruitment patterns, favoring type I fibers and minimizing the involvement of type II fibers [12]. In summary, the optimal pedal speed is determined by muscle shortening velocity, and may exhibit different preferred pedaling rates under the same conditions. Cycling skills might alter pedal force and muscle activity levels during cycling, leading to peak power output occurring at different pedaling rates.

Recent evidence suggests that muscle stiffness is related to sports performance, especially in the utilization of the stretch-shortening cycle (SSC). Stiffness is a term used to describe the force required to achieve a certain deformation of a structure [13]. Musculoarticular (MA) stiffness comprehensively considers the stiffness of the muscle–tendon unit, surrounding articular surfaces, ligaments, and skin [14]. MA stiffness was assessed using a free oscillation technique, which has been described as a valid and reliable method for quantifying stiffness [15–17]. It has been reported that stiffness is related to the efficiency of storage and a release of elastic energy during the stretch-shortening cycle movements [18,19]. Increased stiffness is associated with improved performance in athletic tasks such as jumping, hopping, sprinting, and throwing. It could facilitate improved high ground-reaction forces on impact, increased ground-contact frequency, and shorter ground-contact times [19]. Previous studies have found that relatively stiffer cyclists exhibit higher crank force and rate of crank torque development (RCTD) during 6-s sprint cycling [20]. The rate of force development has recently become a popular indicator of explosive strength in athletes, which is the increase in force or torque per unit time during an explosive muscle contraction. Thus, higher stiffness might potentially enable a higher pedal frequency through improving the efficient use of elastic muscle energy.

The present study aims to investigate the relationship between musculoarticular stiffness and pedaling rate. It has previously been observed that muscle stiffness can be improved through proper resistance training [21–24]. Therefore, our findings should provide important insights into sprint cycling performance and training, which will be of interest to cyclists and coaches.

2. Materials and Methods

2.1. Subjects

Twenty-two healthy, untrained male volunteers with no prior specialized sports training background (age: 19 ± 2 years, height: 175 ± 6 cm, weight: 74 ± 16 kg) were recruited. All subjects reported engaging in physical activities 1–3 times per week. Inclusion criteria required that participants had not suffered any injury for at least 6 months prior to data collection. While participants were familiar with cycling, they were not competitive cyclists or engaged in regular cycling training. Exclusion criteria included any history of chronic diseases, current use of medications that could affect physical performance, and engagement in professional activities with a significant physical component. Participants were instructed to refrain from intense exercise for three days before the experiment. Each participant provided written informed consent, and ethical approval was obtained from Beijing Sport University.

2.2. Experimental Protocol

The testing involved two sessions: a 6-s sprint cycling exercise and a musculoarticular (MA) stiffness test. During the 6-s sprint cycling exercise, participants performed on a Wattbike ergometer, and the following parameters were assessed: peak cadence, average cadence, peak power output, peak crank force, and the angle at which peak crank force occurred in four cycling modes. In the MA stiffness test session, participants' preferred legs were assessed for maximal isometric torque and MA stiffness of the quadriceps.

2.3. Warm-Up

Participants were instructed to perform 10 repetitions of knee-extension exercises at each of the 5 kg, 10 kg, and 15 kg loads before the maximal isometric torque and musculoarticular (MA) stiffness tests. For the 6-s sprint-cycling exercise, participants completed a comprehensive warm-up protocol. The warm-up began with a 6-min cycling session, consisting of 3 min at level 3 air-resistance followed by 3 min at level 2 magnetic-resistance. After the initial cycling, participants performed a series of dynamic stretching exercises targeting the major muscle groups involved in cycling. Following the dynamic stretches, participants executed a submaximal 6-s sprint to familiarize themselves with the test conditions and to further prepare their bodies for the maximal effort required in the formal test.

2.4. Cycling Performance

A 6-s cycling performance test was conducted on a Wattbike ergometer (Wattbike, Atom: Wattbike Trainer, Firmware: Model B Monitor, England). Saddle height and handlebar position were customized for each participant based on their individual anthropometric measurements to ensure a safe, comfortable, and standardized cycling position. The saddle height was set so that when the participant stood next to the bicycle, the top of the saddle was parallel to their anterior superior iliac spine (ASIS). The handlebar position was adjusted such that when the participant was sitting on the bicycle in a riding posture, a vertical line from their elbow would extend to their knee joint. The vertical distance between the handlebar and the saddle was maintained at no more than 10 cm. Participants used the same crank length. In the current experiment, the use of this same crank length on the Wattbike ergometer allowed cycling velocity to be reflected by pedaling cadence, with higher cadence indicating higher velocity.

Participants started in a static position, with the seat in the cycling position and the preferred leg at a 90-degree downstroke. Upon a start signal, participants were instructed to ride as hard as possible for 6 s while receiving strong verbal encouragement. Four cycling modes were tested: levels 5 and 10 air-resistance (level 5: pedaling rate 130 RPM = 400 W, level 10: pedaling rate 130 RPM = 595 W), and levels 3 and 7 magnetic-resistance (level 3: pedaling rate 130 RPM = 225 W, level 7: pedaling rate 130 RPM = 810 W). The air-resistance adjustment lever regulated the airflow into the flywheel, with more airflow resulting in faster pedaling and higher resistance. The magnetic-resistance adjustment lever simulated the force of gravity for a "climbing feel", applying a fixed resistance based on the airflow into the flywheel.

Each participant completed eight maximal trials (two trials for each mode), with at least 3 min of rest between trials and 5 min between modes to avoid fatigue. The collected data included peak cadence, average cadence, peak power output (POpeak), and peak crank force (CFpeak), which were utilized for analysis. Additionally, the peak rate of crank force development (RCFD) was calculated as a relevant index to performance for analysis. RCFD was computed as the rate of change in the crank force (CF) values (from minimum to maximum, $RFD = \Delta F/\Delta T$) for the second pedal-revolution. The experiment assessed CFpeak and RCFD on the participants' preferred leg. The angle at which peak crank force occurred was also collected.

2.5. Isometric Contraction Strength

Knee extensor-strength of the participant's preferred leg was assessed on a leg-extension dynamometer (DAVID F200, Helsinki, Finland). The participant sat in the seat with the seatback and ankle cushion adjusted for maximum comfort when exerting maximum force on the device. After familiarizing themselves with the device, participants were instructed to produce maximum isometric force with the quadriceps at a knee angle of 100 degrees as quickly as possible for approximately 5 s. Participants fastened the seat belt and held onto the handles to prevent hip extension. During the testing process, strong verbal encouragement was provided, and participants were instructed to keep their buttocks in contact with the seat and their back against the seatback. Each participant completed at least three maximal trials, with the best maximal isometric torque (MIT) used for analysis.

2.6. MA Stiffness

A free oscillation technique [16,25–27] was used to assess the unilateral musculoarticular (MA) stiffness of the quadriceps. The free oscillation technique models the human body as a damped, single-degree-of-freedom "spring-mass" system. It considers the viscoelastic properties of muscle–tendon components while eliminating the influence of unrelated tissues. The technique is based on the following second-order differential equation:

$$\mathrm{m}\frac{d^2\mathrm{x}}{\mathrm{d}t^2} + c\frac{dx}{dt} + kx = mg \tag{1}$$

where m is mass, c is damping coefficient, k is stiffness, and g is gravitational acceleration. This method has been widely used and validated over the past 40 years, proving to be an effective way to measure muscle–tendon unit (MTU) stiffness in vivo without requiring sophisticated equipment.

The test was also performed on the leg-extension dynamometer, with the same position as the isometric test (i.e., 100 degrees at the knee angle). The dynamometer was elevated to satisfy the experimental requirements. Participants wore a customized weight-bearing boot on the distal portion of the lower leg, and carried a load approximately 50% of their maximum isometric force. A uniaxial accelerometer (YIYANG, YSV2303S, Beijing, China) recorded the oscillations, and it was attached to the distal end of the weight-bearing boot (heel of the boot). Data were sampled at 5000 Hz and recorded on a computer using data-acquisition software (YIYANG, YSV 8004 24-bits, Beijing, China). Each participant completed three trials, which were averaged for analysis. One minute of rest was provided between each trial. Data were processed using YIYANG analysis software (HDSample v7.0, Beijing, China), utilizing a band-pass filter (0–6 Hz).

2.7. Statistical Analysis

Participants were divided into two groups, the stiffness group (SG) and the compliant group (CG), according to their quadriceps' stiffness ranking. The median value of the 22 subjects' stiffness results was calculated, and the participants were divided into groups based on this value. The highest 11 stiffness values were assigned to the SG, and the lowest 11 stiffness values were assigned to the CG. An independent-samples t-test was conducted to compare the two groups and ensure that they had different stiffness characteristics.

After establishing the group division, the differences in all cycling performance variables and isometric contraction variables between the two groups were assessed using independent-samples t-tests. The effect size (ES) was calculated to quantify the magnitude of differences between the groups, using Cohen's d.

Pearson product–moment correlation was used to analyze the relationship between musculoarticular (MA) stiffness and each performance value. Furthermore, the correlation between the rate of crank force development (RCFD) and maximal power output for each cycling level was also examined.

For all statistical analyses, the alpha level was set at $p < 0.05$. Effect sizes were interpreted as very small (Cohen's d < 0.2), small (0.2 ≤ Cohen's d < 0.5), medium (0.5 < Cohen's d < 0.8), or large (Cohen's d ≥ 0.8).

3. Results

There was a significant difference in MA stiffness between the SG and the CG, with that of the SG being 36% higher than that of the CG ($p < 0.01$). In addition, there were no differences in age, height, or weight between the two groups (Table 1). Furthermore, the MIT of the SG was, significantly, 30.4% higher than that of the CG.

Table 1. Summary of the parameters (±SD) for anthropometric stiffness and isometric torque.

Basic Variables	All Subjects (n = 22)	CG (n = 11)	SG (n = 11)	p	ES (Cohen d)
Age (year)	19 ± 2	20 ± 3	19 ± 1	0.44	0.34
Height (cm)	175 ± 6	175 ± 7	175 ± 4	0.97	0.02
Weight (kg)	74 ± 16	68 ± 12	79 ± 17	0.09	0.75
MA stiffness	331.52 ± 91.24	258.60 ± 32.05	404.43 ± 68.98 *	<0.01	2.7
MIT	212.82 ± 53.88	174.64 ± 27.18	251 ± 46.36 *	<0.01	2.01

* Significant difference from the CG.

The SG reported that peak cadence, average cadence, peak crank force (CFpeak), RCFD, and peak power output (POpeak) were higher than their CG equivalents. The SG had significantly higher peak cadence (5.25%) ($p < 0.05$, Cohen d > 0.8) and average cadence (4.73%) ($p < 0.05$, Cohen d >0.8) at level 3 magnetic-resistance, peak crank force (13.7%) ($p < 0.05$, Cohen d > 0.8), and peak power output (14.42%) ($p < 0.05$, Cohen d > 0.8) at level 10 air-resistance, peak rate of crank force development (21%, 19%, 17.5%) at level 5 air-resistance ($p < 0.05$, Cohen d > 0.8), 10 air-resistance ($p < 0.05$, Cohen d > 0.8), and level 3 magnetic-resistance ($p < 0.05$, Cohen d > 0.8) (Figure 1).

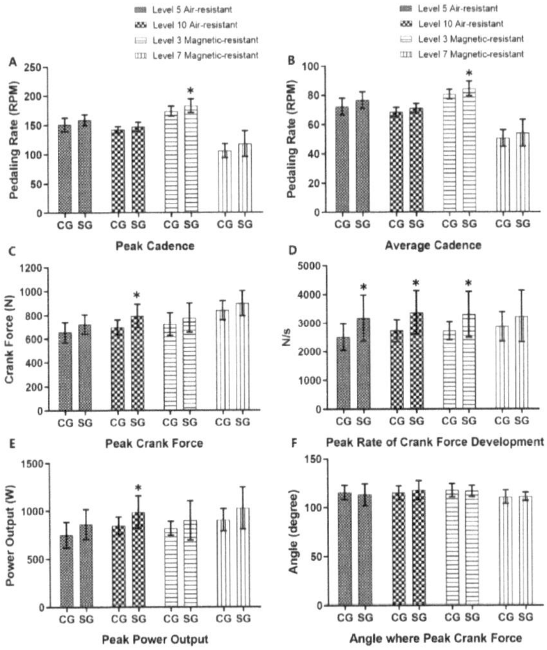

Figure 1. Comparison of cycling performance parameters between the compliant group (CG) and stiffness group (SG). Statistical analysis was performed using independent samples t-tests. (**A**) Peak

cadence: SG 5.25% higher than the CG at level 3 magnetic-resistance ($p < 0.05$). (**B**) Average cadence: SG 4.73% higher than the CG at level 3 magnetic-resistance ($p < 0.05$). (**C**) Peak crank force: SG 13.7% higher than the CG at level 10 air-resistance ($p < 0.05$). (**D**) Rate of crank force development (RCFD): SG 21%, 19%, and 17.5% higher than the CG at level 5 air-resistance, level 10 air-resistance, and level 3 magnetic-resistance, respectively (all $p < 0.05$). (**E**) Peak power output: SG 14.42% higher than CG at level 10 air-resistance ($p < 0.05$). (**F**) Angle at peak crank force. * Indicates significant difference between SG and CG.

MA stiffness was a significant and largely positive relationship with MIT (r = 0.88, $p < 0.01$) in isometric contraction-strength. The significant relationship between MA stiffness and average cadence during cycling at levels 5 and 10 air-resistance is displayed in Figure 2A,B. MA stiffness was significantly correlated with peak crank force during cycling at 4 cycling levels (Figure 2C–F). In addition, MA stiffness was also significantly correlated with RCFD and peak power output during cycling at level 10 air-resistance (Figure 2G). Furthermore, there were no significant relationships between MA stiffness and the angle of peak crank force during cycling. The relationship between RCFD and peak power output during cycling was found to be a largely positive, significant relationship at each cycling level, namely, r = 0.85 at level 5 air-resistance, r = 0.81 at level 10 air-resistance, r = 0.81 at level 3 magnetic-resistance, and r = 0.83 at level 7 magnetic-resistance.

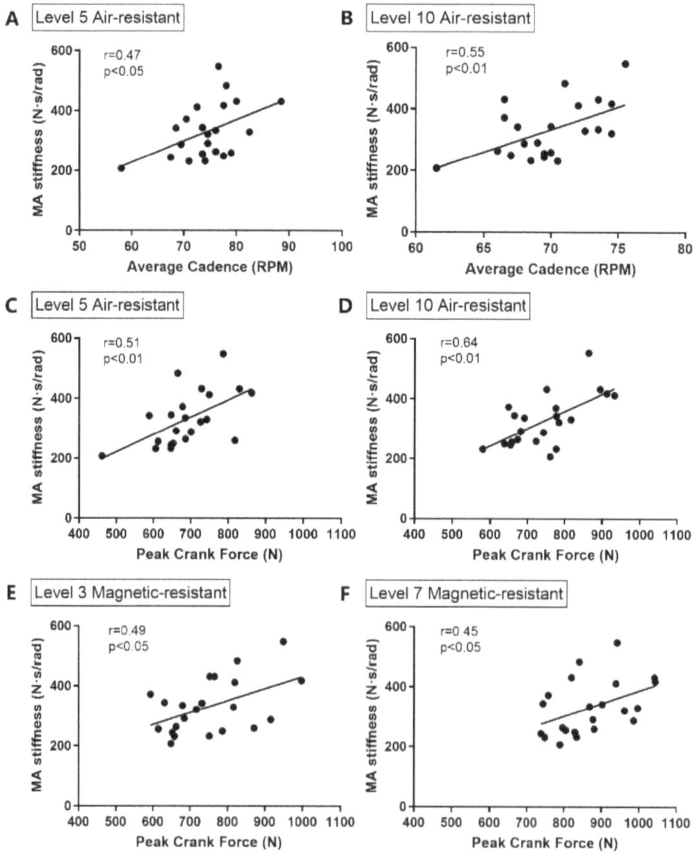

Figure 2. *Cont.*

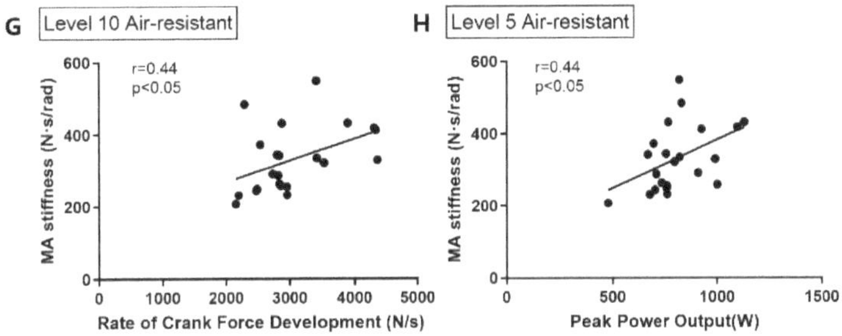

Figure 2. Correlations between MA stiffness and cycling performance parameters. Statistical analysis was performed using Pearson's correlation coefficient. (**A**,**B**) Average cadence at levels 5 and 10 air-resistance. (**C**–**F**) Peak crank force at four cycling levels. (**G**) Rate of crank force development (RCFD) at level 10 air-resistance. (**H**) peak power output at level 5 air-resistance. Significant positive correlations were observed for all displayed relationships ($p < 0.05$).

4. Discussion

Cycling is a unique racing event that requires cyclists to exert force on the cranks of their bicycles. The performance of sprint cycling is influenced by factors such as power output, air-resistance, rolling-resistance, and bearing drag. Power output (P) is the amount of work (W) transferred per unit of time (t), represented by the equation $P = W/t = F \cdot v$, where F is force and v is velocity.

Participants with relatively higher musculoarticular (MA) stiffness exhibited significantly higher peak and average cadence under the level 3 magnetic-resistance condition during the 6-s cycling sprint (Figure 1), and a trend of increased cadence was observed in the other three resistance conditions. Furthermore, there were significant, medium, positive correlations between MA stiffness and average cadence in the levels 5 ($r = 0.47$) and 10 ($r = 0.55$) air-resistance conditions. This evidence suggests that relatively stiffer musculotendinous units may be associated with higher pedaling cadence, with significant differences observed under certain conditions. This study is the first to reveal the relationship between MA stiffness and pedaling cadence.

Muscular stiffness enhances the transmission of elastic energy from tendons during muscle contraction [28]. Our study supports this view. Participants in the stiffness group (SG) exhibited higher rates of crank-force development (RCFD) during the 6-s sprint cycling, with significant differences observed in the levels 5 and 10 air-resistance, and level 3 magnetic-resistance, conditions (Figure 1). The rate of force development (RFD) is an important index characterizing explosive strength, which is the capacity to produce maximal voluntary activation in the early phase of explosive contractions [22]. RCFD is a specific application in cycling, computed as the rate of change in crank force (CF) values. An efficient pedaling stroke depends on the stretch-shortening cycle (SSC). During the upstroke, the quadriceps contracts eccentrically for knee flexion, followed by a concentric contraction for knee extension during the downstroke. Cyclists with relatively higher musculoarticular (MA) stiffness may enhance concentric performance by better utilizing stored elastic energy from the eccentric phase [18–20] during the SSC movement, thereby increasing muscle contraction velocity [29]. Furthermore, it has been reported that the series elastic element could increase the maximum fiber shortening velocity and reduce tendon slack length [29]. Previous studies have noted a correlation between tendon slack length and electromechanical delay (EMD) [30], indicating that a lower EMD in higher MA stiffness could reduce the time required to stretch the series elastic component, improving force transmission from the contractile element to the bone, and thus increasing RFD [31]. Consequently, higher MA stiffness may contribute to superior RCFD performance, enhancing pedaling rate during 6-s sprint cycling.

The results of this study also revealed a significant moderate correlation between RCFD and MA stiffness ($r = 0.44$, $p < 0.05$). These findings support the review by Maffiuletti [22], which suggested that an increase in RFD with training-induced adaptations may be related to changes in musculotendinous stiffness, likely affecting the capacity for rapid force increases at the onset of contraction [31]. The initial acceleration response is critical for sprint cyclists, and it may be improved by enhancing RCFD. Accordingly, selecting an appropriate cycling resistance or gear ratio to improve both RCFD and musculotendinous stiffness might be crucial. Future research is needed to determine whether increasing RCFD influences MA stiffness, and whether different loads used to achieve higher speed-training influences MA stiffness.

Furthermore, there was no difference in the angle at which peak crank force occurred between the compliant group (CG) and the stiffness group (SG), and no correlation between the angle at which peak crank force occurred and MA stiffness. This result is inconsistent with Watsford [20], who observed a negative correlation between MA stiffness and the angle at which peak crank force occurred, with a lower angle of peak crank force occurring in higher MA stiffness. The possible reasons are as follows. Firstly, there are differences in the experimental design. In this experiment, the pedaling cadence during cycling was not fixed, while Watsford adopted a fixed frequency of 80 rpm. Whether the pedaling cadence is fixed or not may affect the force-generation pattern and energy output of the muscles, and subsequently may influence the angle at which the peak crank force occurs. Secondly, the selection of participants varies. The experiment selected trained male cyclists, while this experiment chose male college students without systematic training experience. These untrained male college students might have had difficulty maintaining a stable force application and rhythm during cycling, which could have made the angle at which the peak crank force occurs more random, thereby resulting in a difference from Watsford's experimental results.

Relatively stiffer musculotendinous units have a higher capacity for force or torque production [32–34]. In accordance with previous studies [16,17,20], our results showed that the SG exhibited significantly higher maximal isometric torque (MIT), and MIT was significantly, largely correlated with MA stiffness. Wilson [35] assumed that higher MA stiffness displays a better optimal resting length of the muscle. The popular Hill three-component model [36] depicts muscle as consisting of a contractile component (CC), serial elastic components (SEC), and parallel elastic components (PEC). When the CC is inactive or has low activity, the tension of the musculotendinous unit is borne by the PEC, consisting of muscle fasciae, sarcolemma, and interactions between filaments and residual cross-bridge attachments [37]. Based on the sarcomere length–tension relation of muscle, maximal contraction strength is achieved at the optimal resting length. A greater optimal resting length may promote a relatively higher force output.

Musculoarticular (MA) stiffness also appears to be related to power output during 6-s sprint cycling. The results of this study show that the stiffness group (SG) exhibited higher peak power output (POpeak) than the compliant group (CG), with a significantly larger difference observed in the level 10 air-resistance condition. Cadence and power output represent riding velocity, and the combination of forces and velocity during the 6-s sprint cycling, respectively. Cycling performance is determined by the sources of resistance and the rider's power output, which includes factors such as riding position, body mass, rolling resistance, air-resistance, and the individual's aerobic and anaerobic power and capacity for skeletal muscle contraction. Our research suggests a relative relationship between MA stiffness and cadence, and power output in sprint cycling. The cycling cadence and power output represent the overall performance, incorporating physical, physiological, and cycling technique elements of sprint cyclists. Sprint cyclists with higher MA stiffness might have a greater optimal resting length and a shorter electromechanical delay (EMD) to produce higher force output and rate of crank force development (RCFD), allowing them to store more energy for pedal strokes with higher peak crank force (CFpeak), resulting in higher cadence and power output. Our findings also revealed that RCFD was significantly, largely positively, correlated with POpeak in each cycling mode, and the average cadence and POpeak were significantly, moderately positively, correlated with MA stiffness at levels 5 and 10 air-resistance, and level 5 air-resistance, respectively. This suggests that modifying musculotendinous stiffness may be considered to improve sprint cycling performance. Foure [38] and Kubo [39,40] indicated that the stiffness of the muscle–tendon complex increased after long-term plyometric, isometric, and weight training. However, high levels of stiffness may be associated with increased peak forces, impact forces, and reduced joint motion, increasing the risk of bony injuries [19,41].

The resistant mode was considered in the selection of cycling resistance for our experiment. In the level 3 magnetic-resistance condition, both the SG and CG exhibited the highest peak and average pedaling cadences, and significant differences were reported between the two groups. The difference between the two resistance modes is that the cycling resistance varies with cycling frequency for air-resistance but not for magnetic-resistance. The change in magnetic-resistance is smaller than that in air-resistance. This might better reflect the pedaling cadence performance of sprint cyclists under conditions of lower initial resistance and smaller frequency-dependent changes. However, when cycling at the highest resistance (level 7 magnetic-resistance), there were no significant differences in any parameter between the two groups. This lack of difference at high resistance may be attributed to the substantial resistance requiring subjects to recruit more muscle fibers to generate sufficient force to drive the crank. Consequently, the explosive power advantage of the high-stiffness group may have been diminished. Furthermore, significant differences between the two stiffness groups were observed in maximum peak force and maximum power at air-resistance level 10. The difference in maximum force-loading rate was also more pronounced at air-resistance level 10, compared to air-resistance level 5 and magnetic-resistance level 3. These findings may be explained by the variable nature of air-resistance, where pedaling force and power change with pedaling speed. In contrast, magnetic-resistance provides a fixed resistance, requiring subjects to exert a consistently high force to drive the crank. Therefore, the air-resistance mode may better reflect the explosive power capacity of cyclists. Our results suggest that the air-resistance mode may be more suitable for developing cyclists' explosive power ability, while the magnetic-resistance mode may be more appropriate for improving speed endurance capabilities. Researchers should consider these findings when selecting the appropriate resistance mode for their experimental needs.

Several limitations of this study should be acknowledged. Firstly, from a physiological perspective, individuals with higher body weight typically have greater muscle mass, which may influence joint stiffness. Future research could increase sample size and employ statistical methods to mitigate the impact of body weight on results, thereby more accurately assessing the relationships between target variables. Secondly, our study primarily focused on the 'anatomical' or inherent stiffness of the muscle–tendon unit, without directly measuring or controlling for the effects of voluntary modulation of joint stiffness through neural control and muscle co-activation. Recent literature [42,43] has indicated that joint stiffness can be voluntarily modulated through the co-activation of antagonist muscles. This neural control aspect could potentially impact our results, as athletes might compensate for lower inherent stiffness through more efficient neural control and muscle co-ordination.

5. Conclusions

This study suggests that participants with relatively higher musculoarticular (MA) stiffness demonstrated higher pedaling rates during 6-s sprint cycling conditions. These participants also exhibited superior crank force and rate of crank force development, resulting in greater power output during sprint cycling. However, there was no correlation observed between the angle at which peak crank force occurred and MA stiffness. These findings suggest that selecting an appropriate cycling resistance or gear ratio to improve both rate of crank force development (RCFD) and musculotendinous stiffness might be crucial for enhancing sprint cycling performance.

Author Contributions: Conceptualization, L.I., H.X. and Y.L.; methodology, L.I. and Y.L.; validation, L.I., Y.L., H.X. and X.N.; formal analysis, L.I. and X.N.; investigation, X.N. and L.I.; data curation, X.N. and L.I.; writing—original draft preparation, L.I.; writing—review and editing, X.N., H.X. and Y.L.; visualization, L.I.; supervision, L.I. and Y.L.; project administration, L.I. All authors have read and agreed to the published version of the manuscript.

Funding: This research received no external funding.

Institutional Review Board Statement: The study was conducted in accordance with the Declaration of Helsinki, and approved by the Ethics Committee of Beijing Sport University (protocol code 2020153H, 24 November 2020).

Informed Consent Statement: Informed consent was obtained from all subjects involved in the study.

Data Availability Statement: The data presented in this study are available on request from the corresponding author. The data are not publicly available due to privacy and ethical restrictions, as they contain information that could compromise the privacy of research participants.

Acknowledgments: The authors would like to express their sincere gratitude to the laboratory staff for their support with experimental facilities and equipment. We are deeply appreciative of the assistance and support provided by the coaches and staff of the National Cycling Team. Our heartfelt thanks also go to all the participants who volunteered their time and effort for this study. Their contribution was invaluable to the success of this research.

Conflicts of Interest: The authors declare no conflicts of interest.

References

1. Martin, J.C.; Davidson, C.J.; Pardyjak, E.R. Understanding sprint-cycling performance: The integration of muscle power, resistance, and modeling. *Int. J. Sports Physiol. Perform.* **2007**, *2*, 5–21. [CrossRef] [PubMed]
2. Martin, J.C.; Spirduso, W.W. Determinants of maximal cycling power: Crank length, pedaling rate and pedal speed. *Eur. J. Appl. Physiol.* **2001**, *84*, 413–418. [CrossRef] [PubMed]
3. Connolly, S.; Peeling, P.; Binnie, M.J.; Goods, P.S.; Latella, C.; Taylor, J.L.; Blazevich, A.J.; Timmerman, W.P.; Abbiss, C.R. Sprint cycling rate of torque development associates with strength measurement in trained cyclists. *Eur. J. Appl. Physiol.* **2023**, *123*, 1215–1227. [CrossRef] [PubMed]
4. Gardner, A.S.; Martin, J.C.; Martin, D.T.; Barras, M.; Jenkins, D.G. Maximal torque- and power-pedaling rate relationships for elite sprint cyclists in laboratory and field tests. *Eur. J. Appl. Physiol.* **2007**, *101*, 287–292. [CrossRef] [PubMed]
5. Dorel, S.; Hautier, C.A.; Rambaud, O.; Rouffet, D.; Van Praagh, E.; Lacour, J.R.; Bourdin, M. Torque and power-velocity relationships in cycling: Relevance to track sprint performance in world-class cyclists. *Int. J. Sports Med.* **2005**, *26*, 739–746. [CrossRef]
6. Martin, J.C. Muscle power: The interaction of cycle frequency and shortening velocity. *Exerc. Sport Sci. Rev.* **2007**, *35*, 74–81. [CrossRef] [PubMed]
7. Aagaard, P.; Simonsen, E.B.; Andersen, J.L.; Magnusson, P.; Dyhre-Poulsen, P. Increased rate of force development and neural drive of human skeletal muscle following resistance training. *J. Appl. Physiol.* **2002**, *93*, 1318–1326. [CrossRef] [PubMed]
8. Davies, C.T.; Young, K. Effect of temperature on the contractile properties and muscle power of triceps surae in humans. *J. Appl. Physiol. Respir. Environ. Exerc. Physiol.* **1983**, *55*, 191–195. [CrossRef] [PubMed]
9. Lucia, A.; San Juan, A.F.; Montilla, M.; CaNete, S.; Santalla, A.; Earnest, C.; Perez, M. In professional road cyclists, low pedaling cadences are less efficient. *Med. Sci. Sports Exerc.* **2004**, *36*, 1048–1054. [CrossRef]
10. Carnevale, T.J.; Gaesser, G.A. Effects of pedaling speed on the power-duration relationship for high-intensity exercise. *Med. Sci. Sports Exerc.* **1991**, *23*, 242–246. [CrossRef]
11. Takaishi, T.; Yasuda, Y.; Moritani, T. Neuromuscular fatigue during prolonged pedalling exercise at different pedalling rates. *Eur. J. Appl. Physiol. Occup. Physiol.* **1994**, *69*, 154–158. [CrossRef] [PubMed]
12. Faria, E.W.; Parker, D.L.; Faria, I.E. The science of cycling: Factors affecting performance—Part 2. *Sports Med.* **2005**, *35*, 313–337. [CrossRef] [PubMed]
13. Baumgart, E. Stiffness—An unknown world of mechanical science? *Injury* **2000**, *31* (Suppl. S2), S-B14-23. [CrossRef] [PubMed]
14. Ochala, J.; Lambertz, D.; Van Hoecke, J.; Pousson, M. Changes in muscle and joint elasticity following long-term strength training in old age. *Eur. J. Appl. Physiol.* **2007**, *100*, 491–498. [CrossRef] [PubMed]
15. Ditroilo, M.; Watsford, M.; Murphy, A.; De Vito, G. Assessing musculo-articular stiffness using free oscillations: Theory, measurement and analysis. *Sports Med.* **2011**, *41*, 1019–1032. [CrossRef]
16. Murphy, A.J.; Watsford, M.L.; Coutts, A.J.; Pine, M.J. Reliability of a test of musculotendinous stiffness for the triceps-surae. *Phys. Ther. Sport* **2003**, *4*, 175–181. [CrossRef]
17. Walshe, A.D.; Wilson, G.J.; Murphy, A.J. The validity and reliability of a test of low of lower body musculotendinous stiffness. *Eur. J. Appl. Physiol.* **1995**, *73*, 332–339. [CrossRef]

18. Pruyn, E.C.; Watsford, M.; Murphy, A. The relationship between lower-body stiffness and dynamic performance. *Appl. Physiol. Nutr. Metab.* **2014**, *39*, 1144–1150. [CrossRef]
19. Brazier, J.; Maloney, S.; Bishop, C.; Read, P.J.; Turner, A.N. Lower extremity stiffness: Considerations for testing, performance enhancement, and injury risk. *J. Strength Cond. Res.* **2019**, *33*, 1156–1166. [CrossRef]
20. Watsford, M.; Ditroilo, M.; Fernandez-Pena, E.; D'Amen, G.; Lucertini, F. Muscle stiffness and rate of torque development during sprint cycling. *Med. Sci. Sports Exerc.* **2010**, *42*, 1324–1332. [CrossRef]
21. Waugh, C.M.; Korff, T.; Fath, F.; Blazevich, A.J. Effects of resistance training on tendon mechanical properties and rapid force production in prepubertal children. *J. Appl. Physiol.* **2014**, *117*, 257–266. [CrossRef] [PubMed]
22. Maffiuletti, N.A.; Aagaard, P.; Blazevich, A.J.; Folland, J.; Tillin, N.; Duchateau, J. Rate of force development: Physiological and methodological considerations. *Eur. J. Appl. Physiol.* **2016**, *116*, 1091–1116. [CrossRef] [PubMed]
23. Wannop, J.W.; Worobets, J.T.; Madden, R.; Stefanyshyn, D.J. Influence of Compression and Stiffness Apparel on Vertical Jump Performance. *J. Strength Cond. Res.* **2016**, *30*, 1093–1101. [CrossRef] [PubMed]
24. Kawama, R.; Takahashi, K.; Ikifune, H.; Tozawa, H.; Obata, T.; Ito, R.; Hojo, T.; Wakahara, T. Can Eccentric-only Resistance Training Decrease Passive Muscle Stiffness while Increasing Size and Strength of Hamstrings? *Med. Sci. Sports Exerc.* **2023**. [CrossRef] [PubMed]
25. Granata, K.P.; Padua, D.A.; Wilson, S.E. Gender differences in active musculoskeletal stiffness. Part II. Quantification of leg stiffness during functional hopping tasks. *J. Electromyogr. Kinesiol.* **2002**, *12*, 127–135. [CrossRef] [PubMed]
26. McLachlan, K.A.; Murphy, A.J.; Watsford, M.L.; Rees, S. The interday reliability of leg and ankle musculotendinous stiffness measures. *J. Appl. Biomech.* **2006**, *22*, 296–304. [CrossRef] [PubMed]
27. McNair, P.J.; Wood, G.A.; Marshall, R.N. Stiffness of the hamstring muscles and its relationship to function in anterior cruciate ligament deficient individuals. *Clin. Biomech.* **1992**, *7*, 131–137. [CrossRef] [PubMed]
28. Wilson, J.M.; Flanagan, E.P. The role of elastic energy in activities with high force and power requirements: A brief review. *J. Strength Cond. Res.* **2008**, *22*, 1705–1715. [CrossRef] [PubMed]
29. Van Soest, A.J.; Huijing, P.A.; Solomonow, M. The effect of tendon on muscle force in dynamic isometric contractions: A simulation study. *J. Biomech.* **1995**, *28*, 801–807. [CrossRef]
30. Muraoka, T.; Muramatsu, T.; Fukunaga, T.; Kanehisa, H. Influence of tendon slack on electromechanical delay in the human medial gastrocnemius in vivo. *J. Appl. Physiol.* **2004**, *96*, 540–544. [CrossRef]
31. Rodriguez-Rosell, D.; Pareja-Blanco, F.; Aagaard, P.; Gonzalez-Badillo, J.J. Physiological and methodological aspects of rate of force development assessment in human skeletal muscle. *Clin. Physiol. Funct. Imaging* **2018**, *38*, 743–762. [CrossRef] [PubMed]
32. Wilson, G.J.; Wood, G.A.; Elliott, B.C. Optimal stiffness of series elastic component in a stretch-shorten cycle activity. *J. Appl. Physiol.* **1991**, *70*, 825–833. [CrossRef] [PubMed]
33. Moran, J.; Liew, B.; Ramirez-Campillo, R.; Granacher, U.; Negra, Y.; Chaabene, H. The effects of plyometric jump training on lower-limb stiffness in healthy individuals: A meta-analytical comparison. *J. Sport Health Sci.* **2023**, *12*, 236–245. [CrossRef] [PubMed]
34. Ando, R.; Suzuki, Y. Positive relationship between passive muscle stiffness and rapid force production. *Hum. Mov. Sci.* **2019**, *66*, 285–291. [CrossRef] [PubMed]
35. Wilson, G.J.; Murphy, A.J.; Pryor, J.F. Musculotendinous stiffness: Its relationship to eccentric, isometric, and concentric performance. *J. Appl. Physiol.* **1994**, *76*, 2714–2719. [CrossRef] [PubMed]
36. Hill, A.V. The series elastic component of muscle. *Proc. R. Soc. Lond. B Biol. Sci.* **1950**, *137*, 273–280. [CrossRef] [PubMed]
37. Shorten, M.R. Muscle Elasticity and Human Performance. In *Current Research in Sports Biomechanics: Selected Topics*; Karger Publishers: Berlin, Germany, 1987; Volume 25, pp. 1–18. [CrossRef]
38. Foure, A.; Nordez, A.; Cornu, C. Plyometric training effects on Achilles tendon stiffness and dissipative properties. *J. Appl. Physiol.* **2010**, *109*, 849–854. [CrossRef] [PubMed]
39. Kubo, K.; Yata, H.; Kanehisa, H.; Fukunaga, T. Effects of isometric squat training on the tendon stiffness and jump performance. *Eur. J. Appl. Physiol.* **2006**, *96*, 305–314. [CrossRef]
40. Kubo, K.; Morimoto, M.; Komuro, T.; Yata, H.; Tsunoda, N.; Kanehisa, H.; Fukunaga, T. Effects of plyometric and weight training on muscle-tendon complex and jump performance. *Med. Sci. Sports Exerc.* **2007**, *39*, 1801–1810. [CrossRef]
41. Butler, R.J.; Crowell, H.P.; Davis, I.M. Lower extremity stiffness: Implications for performance and injury. *Clin. Biomech.* **2003**, *18*, 511–517. [CrossRef] [PubMed]
42. Hug, F.; Avrillon, S.; Ibáñez, J.; Farina, D. Common synaptic input, synergies and size principle: Control of spinal motor neurons for movement generation. *J. Physiol.* **2023**, *601*, 11–20. [CrossRef] [PubMed]
43. Borzelli, D.; Vieira, T.M.; Botter, A.; Gazzoni, M.; Lacquaniti, F.; d'Avella, A. Synaptic inputs to motor neurons underlying muscle coactivation for functionally different tasks have different spectral characteristics. *J. Neurophysiol.* **2024**, *131*, 1126–1142. [CrossRef] [PubMed]

life

MDPI

Article

The Impact of Birth Season and Sex on Motor Skills in 2-Year-Old Children: A Study in Jinhua, Eastern China

Yuanye Zhu [1], Shuying Wang [2], Yongdong Qian [1], Jiahui Hu [1], Huiling Zhou [1], Mallikarjuna Korivi [1], Weibing Ye [1,*] and Rong Zhu [3,*]

[1] Institute of Human Movement and Sports Engineering, College of Physical Education and Health Sciences, Zhejiang Normal University, Jinhua 321004, China; zhuyuanye@zjnu.edu.cn (Y.Z.); tyxyqyd@zjnu.cn (Y.Q.); hujiahui@zjnu.edu.cn (J.H.); huiling@zjnu.edu.cn (H.Z.); mallik.k5@gmail.com (M.K.)

[2] College of Physical Education, North University of China, Taiyuan 030051, China; 2111014205@st.nuc.edu.cn

[3] Department of Sports Science, Wenzhou Medical University, Wenzhou 325035, China

* Correspondence: ywbls@zjnu.cn (W.Y.); zhurong@wmu.edu.cn (R.Z.); Tel.: +86-13757995718 (W.Y.)

Abstract: Background: This study investigates the effects of birth season and sex on the development of gross and fine motor skills in 2-year-old children in Jinhua, Eastern China. Methods: Conducted in Jinhua, a city in central Zhejiang Province, Eastern China, this research involved 225 children, assessing their gross and fine motor skills using the Peabody Developmental Motor Scales, Second Edition. Scores were adjusted for age in months to avoid the relative age effect. Statistical analyses included MANOVA to evaluate the impacts of season and sex. Results: Sex had no significant impact on overall motor development scores ($p > 0.05$). However, the season of birth significantly affected fine motor quotient (FMQ) and total motor quotient (TMQ) ($p < 0.05$). Boys' motor skills were generally unaffected by season, whereas girls born in winter exhibited superior fine motor skills compared to those born in summer. Conclusions: Seasonal environmental factors significantly influence early motor development, particularly fine motor skills in girls. These findings highlight the importance of considering seasonal variations in early childhood interventions aimed at enhancing exercise physiology and sports performance.

Keywords: children; seasonal effect; motor development; PDMS-2; gross motor; fine motor

Citation: Zhu, Y.; Wang, S.; Qian, Y.; Hu, J.; Zhou, H.; Korivi, M.; Ye, W.; Zhu, R. The Impact of Birth Season and Sex on Motor Skills in 2-Year-Old Children: A Study in Jinhua, Eastern China. *Life* **2024**, *14*, 836. https://doi.org/10.3390/life14070836

Academic Editor: Laikang Yu

Received: 29 May 2024
Revised: 26 June 2024
Accepted: 27 June 2024
Published: 30 June 2024

1. Introduction

The 270-day gestation period and the subsequent two years (730 days) of life together form an essential and sensitive period for human development [1,2]. Evidence suggests that high-quality early childhood development can significantly enhance academic outcomes [3], improve quality of life [4], bolster executive functioning [5], and even contribute to reductions in societal crime rates [6] and increased societal earnings [7]. Conversely, children with developmental challenges often exhibit academic underperformance and are more prone to conditions like depression [8], anxiety [8], and obesity [9]. Numerous factors influence early childhood development, including the home environment [10], caregiver response style [11], neighborhood built environment [12], and feeding type [13].

Given the sensitivity to environmental factors during early life, seasonal factors such as temperature variations, the availability of fresh vegetables, the prevalence of diseases, and the duration of sunlight exposure can all impact infants' growth [14]. Prior research has linked birth season to various developmental outcomes, such as emotional and behavioral regulation [15] and childhood stunting [16]. For example, in rural northwest China, infants born in winter have been found to have higher hemoglobin (Hb) concentrations compared to those born in summer. Similarly, cognitive development scores and psychomotor development scores were significantly higher among winter-born infants [17].

Motor skills, which refer to the nervous system's ability to control and coordinate movements, can be broadly categorized into gross and fine motor abilities [18]. The development of these skills, which underscores a child's capacity to engage with their surroundings, is paramount during early childhood [19]. While numerous studies have delved into the relationship between birth season and motor development, the findings have been inconclusive. Some research suggests that birth season can influence young children's motor skills [14,20,21], while a study by Gil [22] found no such association. A 3-year follow-up study of 23 healthy infants who received 400 IU/d of supplemental vitamin D found that no significant effects with respect to the season at birth were observed in the gross motor quotient (GMQ) [23]. Yasumitsu-Lovell, K. et al. [14] found that summer-born Japanese infants had the worst outcomes at both 6 months and 12 months of age. Tsuchiya, K.J. et al. [24] found that among Japanese infants aged 6 months, those born in spring showed better neuromotor development compared to those born in autumn. However, by 14 months, these differences disappeared. Notably, the majority of these studies predominantly focus on gross motor skills, often overlooking the potential relationship between birth season and children's fine motor development.

Scientific evidence indicates that sex can influence motor competence. For instance, boys often demonstrate greater proficiency in gross motor skills, whereas girls tend to excel in fine motor skills, such as drawing and writing [25]. Additionally, the trimester of birth may affect motor competence due to the relative age effect, where older children within the same age group exhibit more advanced motor skills than their younger peers [26,27]. Considering the presence of the relative age effect, it is important to evaluate motor skills using month-standardized scores. Furthermore, within the same class, younger children may experience the "Matthew effect," where children who initially lag in skills due to being younger miss out on opportunities to practice and improve, thus falling further behind over time. To address these factors, we chose to test 2-year-old children who have just entered kindergarten. This allows us to assess their motor skills before the potential impact of formal preschool settings.

Standardized tests to measure motor competence include a variety of tools such as the Alberta Infant Motor Scale, the Gesell Developmental Schedules, the Bruininks-Oseretsky Test of Motor Proficiency (BOT-2), the Movement Assessment Battery for Children (MABC-2), and the Peabody Developmental Motor Scales, Second Edition (PDMS-2) [28]. The Alberta Infant Motor Scale is often used for assessing motor skills in children with cerebral palsy, making it unsuitable for our study's population. The Gesell Developmental Schedules is now rarely used. While the BOT-2 and MABC-2 are widely recognized and assess both gross and fine motor skills, they have limitations in comprehensively assessing motor skills in 2-year-old children. Additionally, the Bayley Scales of Infant Development-III has not yet been applied in China. The PDMS-2 was chosen for this study because it comprehensively assesses both gross and fine motor skills and has been validated for use in 2-year-old children [29].

In the present study, our objective was to assess the development of both gross and fine motor skills in 2-year-old children. We sought to discern the potential variations in these skills based on the season of birth and further aimed to elucidate any sex-specific influences on motor skills relative to the birth season. To guide our research, we posed the following questions: Are there differences between 2-year-old boys and girls in motor skills depending on the season of birth? What impact does the season of birth have on the various motor skill subtests at this age? By addressing these questions, we aim to contribute to a deeper understanding of the factors influencing early motor skill development.

2. Materials and Methods

2.1. Recruitment Strategy and Participants

This study, adopting a cross-sectional design, was conducted in Jinhua, a city in Eastern China's Zhejiang Province with an approximate population of 1 million. We initially recruited 228 healthy children from two childcare institutions and a kindergarten affiliated with Zhejiang Normal University using a random cluster sampling technique.

The study spanned from 18 October 2022 to 20 December 2022. Over this period, 3 children were excluded due to age constraints, culminating in a final sample size of 225 children.

Child participants' fundamental attributes, such as age, birth date, and sex, were documented at their respective institutions. The cohort comprised 114 boys and 111 girls. The season of birth for each child was deduced from their birth dates. The details of this study were provided with the Chinese version of the written informed consent form. The Chinese version of the informed consent form was signed by the parents or guardians of the children. We also verbally explained the details of this study to the directors of the kindergarten, teachers, parents or grandparents, and students before their voluntary participation. The research design and all associated evaluation procedures received approval from the Institutional Ethical Committee of Zhejiang Normal University under approval number ZSRT2022097.

2.2. Assessment of Children's Basic Characteristics

Children's demographic details, specifically sex and birth date, were sourced from their teachers and subsequently recorded for analysis. The season of birth was determined based on each child's birth date. According to astronomical division, spring (21 March–20 June) begins with the vernal equinox and ends with the summer solstice. Summer (21 June–22 September) begins with the summer solstice and ends with the autumn equinox. Autumn (23 September–20 December) begins with the autumn equinox and ends with the winter solstice. Winter (21 December–20 March) begins at the winter solstice and ends at the spring equinox.

2.3. Evaluation of Children's Motor Skills

The motor proficiency of the 2-year-old participants was gauged using the Peabody Developmental Motor Scales, Second Edition [30]. The PDMS-2 encompasses two primary skill categories: gross and fine motor skills. Gross motor skills are further divided into 8 reflection skill (Re:) tasks (0–11 months), 30 stationary skill (St) tasks, 89 locomotion skill (Lo) tasks, and 24 object manipulation skill (Ob) tasks (12–72 months). Fine motor skills comprise 24 grasping skill (Gr) tasks and 26 visual–motor integration skill (Vi) tasks. Each task's evaluation score spans three levels: 2 points for correct execution, 1 point for partial completion, and 0 points for non-compliance with developmental criteria [30]. The cumulative points constitute the raw scores for each skill, which are then converted to standard scores, gross motor quotient (GMQ), fine motor quotient (FMQ), and total motor quotient (TMQ) using tables in the PDMS-2 manual. Importantly, these standard scores are adjusted for the child's age in months, avoiding the relative age effect. The evaluations took place in vacant halls within the kindergartens and professional training centers during regular hours. The rooms used for evaluation were well lit and quiet, minimizing distractions. Children were dressed in their regular attire, which typically included comfortable, movement-friendly clothing.

The PDMS-2, employed in this research, stands as a credible and reliable tool for gauging motor skills in Chinese children. It boasts commendable test–retest reliability, content and structural validity, criterion validity, and internal consistency, as corroborated by multiple studies [29,31,32]. Fourteen professionals, encompassing four sports science researchers and ten postgraduate testers, facilitated the assessments. Each PDMS-2 assessment session lasted approximately 30 min per child. Dual independent testers concurrently observed each child's performance, adhering strictly to PDMS-2 guidelines [30]. All requisite equipment was readied ahead of the assessment day. On the day of evaluation, participants received verbal briefings, supplemented with precise demonstrations on skill scoring. The tests were administered at similar times during the day across all participating locations to control for potential diurnal variations in children's performance.

2.4. Statistical Analysis

The obtained data (Supplementary Materials) were analyzed using Microsoft® Excel® 2019. Descriptive statistics for participant characteristics were presented as n (%). For the analysis, children were categorized based on sex (two groups) and season of birth (four groups). To appropriately assess the influence of both sex and season of birth on motor development, two separate two-factor multivariate analyses of variance (MANOVA) were conducted. The first MANOVA assessed the effects on GMQ, FMQ, and TMQ, while the second MANOVA focused on the scores of St, Lo, Ob, Gr, Vi, and other subskills. This separation accounts for the differences in the magnitude of these measures. The Bonferroni correction was applied to control for multiple comparisons, and eta squared (η^2) was calculated to determine the effect size and statistical power of the results. The level of significance was set at $p < 0.05$. This approach ensures that the potential interactions between sex and season of birth are appropriately considered in the analysis, providing a more accurate assessment of their effects on motor development.

3. Results

3.1. Characteristics of Children

In this study, a total of 225 children aged 2 years participated, of which 114 (50.67%) were boys and 111 (49.33%) were girls. The number of children born in each season was 19 boys and 11 girls born in spring, 42 boys and 32 girls born in summer, 32 boys and 34 girls born in autumn, and 21 boys and 34 girls born in winter (Table 1).

Table 1. Basic characteristics of the 2-year-old children.

Variables	Boys (n = 114)	Girls (n = 111)	Total (n = 225)
Spring	19 (16.7%)	11 (9.9%)	30 (13.3%)
Summer	42 (36.8%)	32 (28.8%)	74 (32.9%)
Autumn	32 (28.1%)	34 (30.6%)	66 (29.3%)
Winter	21 (18.4%)	34 (30.6%)	55 (24.4%)

Values expressed in numbers (percentage).

3.2. Effects of Sex and Season of Birth on GMQ, FMQ, and TMQ

We first conducted a MANOVA on the GMQ, FMQ, and TMQ because these comprehensive measures capture both gross and fine motor development, providing a holistic view of a child's motor skills. The analysis revealed that the intercept was highly significant ($p < 0.001$, $\eta^2 = 0.999$), indicating that almost all variations in motor development scores are explained by the model. Sex had no significant impact on motor development scores ($p = 0.106$, $\eta^2 = 0.028$), indicating minimal influence. However, the season of birth had a significant effect on motor development scores ($p = 0.017$, $\eta^2 = 0.030$), with a modest but meaningful impact. The interaction between sex and season of birth was not significant ($p = 0.735$, $\eta^2 = 0.009$), suggesting the minimal combined influence of these factors on motor development. These results highlight that while the season of birth significantly affects motor development, sex alone and its interaction with the season do not.

We examined the individual effects on each motor skill measure (Table 2). There were no significant effects of sex on the GMQ, FMQ, and TMQ, indicating similar motor development scores between boys and girls. Significant effects of the season of birth were found on the FMQ and TMQ ($p < 0.05$), suggesting that motor development scores vary with the season of birth. No significant interaction effects were observed between sex and the season of birth, indicating consistent seasonal influences across sexes. These findings highlight that while the season of birth significantly impacts fine and total motor skills, sex and its interaction with the season do not.

Table 2. Seasonal differences in the scores of motor skills for boys and girls.

Variables	GMQ	FMQ	TMQ	St	Lo	Ob	Gr	Vi
Boys (M ± SE)								
Spring	113.63 ± 3.05	106.63 ± 3.20	111.47 ± 3.03	12.47 ± 0.62	11.11 ± 0.51	12.79 ± 0.62	10.89 ± 0.67	11.32 ± 0.55
Summer	113.40 ± 2.05	108.36 ± 2.15	112.14 ± 2.04	13.07 ± 0.42	10.81 ± 0.35	12.38 ± 0.42	11.62 ± 0.45	11.00 ± 0.37
Autumn	119.81 ± 2.35	110.78 ± 2.47	117.28 ± 2.33	13.84 ± 0.48	11.78 ± 0.40	13.59 ± 0.48	11.88 ± 0.52	11.72 ± 0.43
Winter	112.05 ± 2.90	112.00 ± 3.05	112.86 ± 2.88	12.86 ± 0.59	11.05 ± 0.49	11.71 ± 0.59	12.29 ± 0.64	11.71 ± 0.53
Total	114.72 ± 1.31	109.44 ± 1.38	113.44 ± 1.30	13.06 ± 0.27	11.19 ± 0.22	12.62 ± 0.27	11.67 ± 0.29	11.44 ± 0.24
Girls (M ± SE)								
Spring	111.91 ± 4.00	110.64 ± 4.21	112.27 ± 3.98	13.00 ± 0.81	11.73 ± 0.68	10.82 ± 0.82	11.91 ± 0.67	11.64 ± 0.73
Summer	113.16 ± 2.35	108.06 ± 2.47	111.88 ± 2.33	13.06 ± 0.48	11.59 ± 0.40	11.50 ± 0.48	11.31 ± 0.51	11.38 ± 0.43
Autumn	117.32 ± 2.28	114.56 ± 2.39	117.50 ± 2.26	13.94 ± 0.46	12.09 ± 0.39	12.09 ± 0.47	12.59 ± 0.50	12.26 ± 0.41
Winter	117.94 ± 2.28	120.29 ± 2.39	120.41 ± 2.26	14.21 ± 0.46	11.94 ± 0.39	12.21 ± 0.47	13.85 ± 0.50	12.91 ± 0.41
Total	115.08 ± 1.41	113.39 ± 1.48	115.51 ± 1.40	13.55 ± 0.29	11.84 ± 0.24 #	11.65 ± 0.29 #	12.42 ± 0.31	12.05 ± 0.26
Effect								
Sex								
F	0.035	3.799	1.177	1.583	4.018	6.032	3.085	3.027
p	0.852	0.053	0.279	0.21	0.046 *	0.015 *	0.080	0.083
η^2	0.000	0.017	0.005	0.007	0.018	0.027	0.014	0.014
Season								
F	2.236	3.808	2.717	1.710	1.262	1.78	3.584	2.586
p	0.085	0.011 *	0.046 *	0.166	0.288	0.152	0.015 *	0.054
η^2	0.030	0.050	0.036	0.023	0.017	0.024	0.047	0.035
Sex × Season								
F	1.075	0.963	1.062	0.752	0.202	1.772	1.102	0.35
p	0.361	0.411	0.366	0.522	0.895	0.153	0.349	0.789
η^2	0.015	0.013	0.014	0.01	0.003	0.024	0.015	0.005

St, stationary skills; Lo, locomotion skills; Ob, object manipulation skills; GMQ, gross motor quotient; Gr, grasping skills; Vi, visual–motor integration skills; FMQ, fine motor quotient; and TMQ, total motor quotient. The scores were significant at * $p < 0.05$; #, different from boys ($p < 0.05$).

3.3. *Effects of Sex and Season of Birth on Gross and Fine Motor Subskills*

Further analyses of individual motor subskills (stationary, locomotion, object manipulation, grasping, and visual–motor integration) revealed that sex significantly impacts motor skills ($p < 0.001$, $\eta^2 = 0.131$). The season of birth showed a modest impact on these subskills, with some tests indicating significance ($p = 0.088$, $\eta^2 = 0.034$). However, no significant interaction effects were observed between sex and season of birth ($p > 0.7$, $\eta^2 = 0.017$), indicating that the influence of the season of birth on motor skills is consistent across sexes. These results suggest that while sex significantly affects motor skills, the impact of seasonal factors is relatively modest.

The subskills reveal that sex significantly affects locomotion and object manipulation skills (Table 2), while the season of birth significantly impacts grasping skills and may have a modest effect on visual–motor integration skills. No significant interaction effects were observed, indicating that the influence of the season of birth on motor skills is consistent across the sexes.

3.4. *The Motor Skills Scores of Girls Born in Winter Compared to Those Born in the Summer*

Figure 1 shows the influence of the season of birth on the development of motor skills in boys and girls. Initially, we noticed that the motor skills scores (St, Lo, Ob, Gr, and Vi) and motor developmental quotient (GMQ, FMQ, and TMQ) of boys were not significant influenced by the season of birth. We also found that there was no significant difference in the season of birth difference on the St, Lo, Ob, and Vi standard scores and the GMQ and TMQ of girls. However, it is worth noting that girls born in winter showed superior Gr and FMQ ($p < 0.01$) than girls born in summer (Figure 1C,D).

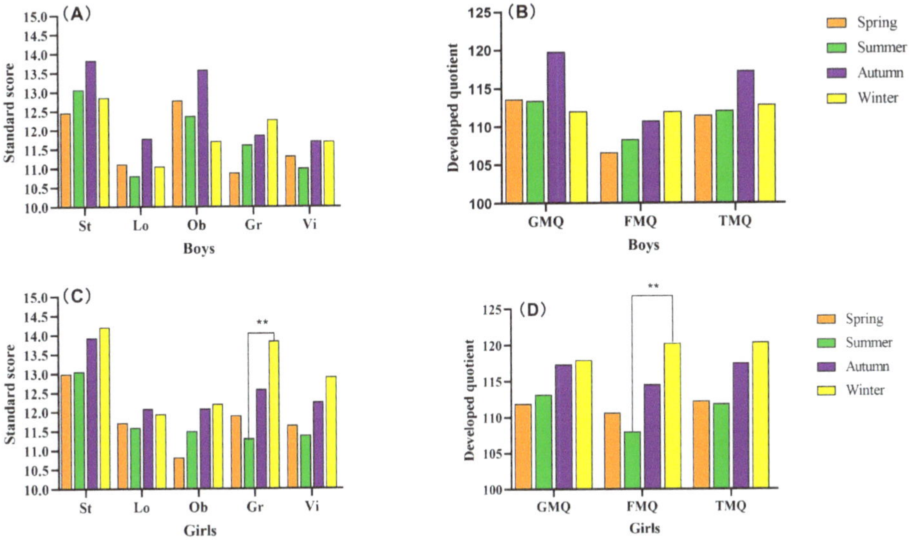

Figure 1. Changes in motor skills scores with different seasons of birth. (**A**), The standard score for boys; (**B**), The developed quotient for boys; (**C**), The standard score for girls; (**D**), The developed quotient for girls; St, stationary skills; Lo, locomotion skills; Ob, object manipulation skills; Gr, grasping skills; Vi, visual–motor integration skills; GMQ, gross motor quotient; FMQ, fine motor quotient; TMQ, total motor quotient. The scores were significant at ** $p < 0.01$ among spring, summer, autumn, and winter.

4. Discussion

To our knowledge, this study represents a pioneering effort to discern the potential influence of birth season on the motor skills of 2-year-old children. We delved into the motor skill nuances among 2-year-old children from Zhejiang Province, China, juxtaposing the variations against their sex and birth season. Our results underscore that a child's birth season does indeed sway their motor skill development, and intriguingly, the extent of this effect varies between boys and girls. Specifically, the birth season markedly impacted the fine motor skills of girls, whereas the effect of birth season on motor skills of boys was not significant. Notably, girls born in winter exhibited superior grasping skills compared to their summer-born counterparts.

The sex-based differentiation in the motor skills of 2-year-old children offers a compelling avenue for exploration. Our results show no significant statistical differences in the Gr, Vi, and FMQ scores among two-year-olds. Similarly to our findings, a study on Saudi Arabian children found no significant differences in fine motor skills between boys and girls [33]. However, studies from Greece [34], Iran [35], and India [36] confirmed that girls tend to outperform boys in fine motor skills. The discrepancies in findings across different studies may be attributed to the varying countries where these studies were conducted [37]. Additionally, these differences might also be due to the distinct characteristics exhibited by different age groups [38].

Our analysis also highlighted that boys scored higher in Ob skills than girls. This aligns with findings from China [39], USA [40], and Spain [41], which highlighted boys' superior ball skills [42]. Differences in proficiency in object control skills between boys and girls seem to be influenced by evolutionary factors. Early humans lived by throwing stones and swinging clubs. Women invested more resources into reproduction, and men were more likely to be hunters and warriors. These kinds of patterns are inherited through natural selection [43]. Gender differences among children in learning motor skills depend on their family, elder siblings, peers, and teachers with respect to socialization and imitation, and consequently, they participate in activities that fit these gender norms [44]. The better ball skills of boys in our study may also be explained by boys spending more time practicing object-control-related activities, like ball games, thereby developing their ball skills. Additionally, sociocultural factors may also influence the development of children's motor skills [25].

Our results highlighted that the motor skills of children born in winter significantly surpassed those of summer-born children. This observation aligns with findings from Israel [45], China [17], and the USA [46], all of which underscored the motor developmental edge in winter-born children. Specifically, our data revealed that winter-born girls manifested superior fine motor skills (grasping) compared to summer-born girls, while no such seasonal disparity was evident in gross motor skills. For boys, the birth season did not seem to influence either gross or fine motor skills.

One study suggests that differences in motor skills across seasons are due to the relative age effect (RAE) [26]. However, our research differs from this study in several ways. Firstly, the other study involved older children, such as 4-year-olds, who may have already spent a year or more in kindergarten, thereby being more influenced by group activities and formal education settings. Secondly, the seasonal divisions in their study differ from ours, as their quarters were defined as follows: quarter 1 (q1: born from January to March); quarter 2 (q2: born from April to June); quarter 3 (q3: born from July to September); and quarter 4 (q4: born from October to December). Additionally, their measurement methods may use standard scores based on annual or semi-annual divisions, whereas we calculate standard scores on a monthly basis. This monthly calculation provides a more precise assessment of each child's development relative to their exact age.

We postulate that the ambient temperature during winter might be a pivotal factor driving superior fine motor development in children. Winter's chill necessitates bundling up children in layered clothing. Research has indicated that such bulky attire can restrict movement, potentially impeding the exploration and execution of motor skills, predominantly gross motor skills [47]. This scenario might inadvertently provide children with augmented opportunities to hone their fine motor skills. The Peabody Developmental Motor Scales posits that rapid developmental milestones like hand grasping and head movement in infants aged 0-2 months fall under the fine motor domain [30]. Thus, winter's cold might exert a minimal impact on gross motor skills but accentuate fine motor skill development. Additionally, winter's shorter days curtail outdoor activities, leading to increased parent–child interactions, which might further amplify this effect. Winter-born infants experiencing relatively fewer outdoor activities may engage more in hand-related activities at home or indoors. Additionally, in summer, environmental factors like temperature might significantly increase the likelihood of premature birth [48], and premature infants tend to lag in motor development compared to full-term infants, which might explain why winter-born infants exhibit better motor skills than those born in summer [49]. The sex-based variations in motor skills across birth seasons might be tethered to parental play promotions. Research indicates that parental play promotions play a pivotal role in optimizing development at six to nine months of age. Parents of boys tend to promote gross motor skills, while those of girls lean towards fine motor skills [25,50].

While our study sheds light on intriguing aspects of motor skills development in relation to birth season, it is not without limitations. Firstly, the sample size, encompassing 225 children, is relatively modest. Subsequent research with a more expansive sample is recommended to validate and expand upon our findings. Secondly, the participants in our study hail exclusively from Zhejiang Province, China, characterized by a subtropical monsoon climate. This geographic and climatic specificity may limit the generalizability of our findings to children from different climatic regions. Lastly, our investigation was confined to discerning the impact of birth season on the motor skills of 2-year-old children. The influence of birth season on motor skills across different age groups, especially those younger than a year, remains an open avenue for exploration.

5. Conclusions

Our research highlights a strong association between the season of birth and motor skills development in 2-year-old children. Notably, the fine motor skills, especially grasping, of girls born in winter surpassed those of summer-born girls. This suggests that targeted interventions and motivational strategies might be beneficial for 2-year-old girls born in summer to ensure the balanced development of both gross and fine motor skills during their formative years.

Supplementary Materials: The following supporting information can be downloaded at https://www.mdpi.com/article/10.3390/life14070836/s1, Data.xls.

Author Contributions: Y.Z., J.H., W.Y., Y.Q. and H.Z.: data collection. Y.Z. and S.W.: data analysis and original draft preparation. M.K., W.Y., Y.Z. and R.Z.: review and editing. Y.Q., W.Y. and R.Z.: funding acquisition. All authors have read and agreed to the published version of the manuscript.

Funding: This research endeavor was financially supported by the project titled "Early Development of 0–3 years old infants in China: Characteristics and evaluation of physical and motor development", bearing grant number KYH06Y21383.

Institutional Review Board Statement: The research protocols involving human participants were meticulously reviewed and subsequently approved by the Faculty of Medicine Ethics Committee at Zhejiang Normal University, under approval code ZSRT2022097.

Informed Consent Statement: Informed consent was obtained from all subjects involved in the study.

Data Availability Statement: The primary data and contributions presented in this study are encapsulated within the article. For additional details or inquiries, interested parties are encouraged to reach out to the lead authors.

Acknowledgments: The authors extend their heartfelt gratitude to the affiliated kindergarten of Zhejiang Normal University for their unwavering support throughout the research process.

Conflicts of Interest: The authors unequivocally assert that this research was undertaken while devoid of any commercial or financial affiliations that might be perceived as potential conflicts of interest.

References

1. Karakochuk, C.D.; Whitfield, K.C.; Green, T.J.; Kraemer, K. *The Biology of the First 1000 Days*; Taylor & Francis: Boca Raton, FL, USA, 2018. [CrossRef]
2. Alves, J.; Alves, L.V. Early-life nutrition and adult-life outcomes. *J. Pediatr.* **2024**, *100* (Suppl. 1), S4–S9. [CrossRef] [PubMed]
3. Ulferts, H.; Wolf, K.M.; Anders, Y. Impact of Process Quality in Early Childhood Education and Care on Academic Outcomes: Longitudinal Meta-Analysis. *Child Dev.* **2019**, *90*, 1474–1489. [CrossRef] [PubMed]
4. Zwicker, J.G.; Harris, S.R.; Klassen, A.F. Quality of life domains affected in children with developmental coordination disorder: A systematic review. *Child Care Health Dev.* **2013**, *39*, 562–580. [CrossRef] [PubMed]
5. Thompson, A.; Steinbeis, N. Sensitive periods in executive function development. *Curr. Opin. Behav. Sci.* **2020**, *36*, 98–105. [CrossRef] [PubMed]
6. Garcia, J.L.; Heckman, J.J.; Ziff, A.L. Early childhood education and crime. *Infant Ment. Health J.* **2019**, *40*, 141–151. [CrossRef] [PubMed]
7. Clark, H.; Coll-Seck, A.M.; Banerjee, A.; Peterson, S.; Dalglish, S.L.; Ameratunga, S.; Balabanova, D.; Bhan, M.K.; Bhutta, Z.A.; Borrazzo, J.; et al. A future for the world's children? A WHO-UNICEF-Lancet Commission. *Lancet* **2020**, *395*, 605–658. [CrossRef] [PubMed]
8. Wang, D.; Jiang, Q.; Yang, Z.; Choi, J.K. The longitudinal influences of adverse childhood experiences and positive childhood experiences at family, school, and neighborhood on adolescent depression and anxiety. *J. Affect. Disord.* **2021**, *292*, 542–551. [CrossRef] [PubMed]
9. De Wolf, A.; Nauwynck, E.; Vanbesien, J.; Staels, W.; De Schepper, J.; Gies, I. Optimizing Childhood Obesity Management: The Role of Edmonton Obesity Staging System in Personalized Care Pathways. *Life* **2024**, *14*, 319. [CrossRef] [PubMed]
10. Rahman, F.; Tuli, S.N.; Mondal, P.; Sultana, S.; Hossain, A.; Kundu, S.; Clara, A.A.; Hossain, A. Home environment factors associated with early childhood development in rural areas of Bangladesh: Evidence from a national survey. *Front. Public Health* **2023**, *11*, 1209068. [CrossRef] [PubMed]

11. Connery, A.K.; Raghunathan, R.S.; Colbert, A.M.; Erdodi, L.; Warschausky, S.; Huth-Bocks, A.; Gerry, T.H.; Raghunathan, T.; Berglund, P.; Staples, A.D.; et al. The influence of sociodemographic factors and response style on caregiver report of infant developmental status. *Front. Pediatr.* **2022**, *10*, 1080163. [CrossRef] [PubMed]
12. Alderton, A.; Villanueva, K.; Connor, M.O.; Boulangé, C.; Badland, H. Reducing inequities in early childhood mental health: How might the neighborhood built environment help close the gap? A systematic search and critical review. *Int. J. Environ. Res. Public Health* **2019**, *16*, 1516. [CrossRef] [PubMed]
13. Escolano-Perez, E.; Sanchez-Lopez, C.R.; Herrero-Nivela, M.L. Early Environmental and Biological Influences on Preschool Motor Skills: Implications for Early Childhood Care and Education. *Front. Psychol.* **2021**, *12*, 725832. [CrossRef] [PubMed]
14. Yasumitsu-Lovell, K.; Thompson, L.; Fernell, E.; Eitoku, M.; Suganuma, N.; Gillberg, C.; Group, J.E.A.C. Birth month and infant gross motor development: Results from the Japan Environment and Children's Study (JECS). *PLoS ONE* **2021**, *16*, e0251581. [CrossRef] [PubMed]
15. Asano, R.; Tsuchiya, K.J.; Harada, T.; Kugizaki, Y.; Nakahara, R.; Nakayasu, C.; Okumura, A.; Suzuki, Y.; Takagai, S.; Mori, N.; et al. Season of Birth Predicts Emotional and Behavioral Regulation in 18-Month-Old Infants: Hamamatsu Birth Cohort for Mothers and Children (HBC Study). *Front. Public Health* **2016**, *4*, 152. [CrossRef] [PubMed]
16. Linger, E.M.; Diress, G.; Linger, E.B.; Wagaye, B.; Almaw, H. Does birth season correlate with childhood stunting? An input for astrological nutrition. *BMC Pediatr.* **2022**, *22*, 306. [CrossRef]
17. Bai, Y.; Shang, G.; Wang, L.; Sun, Y.; Osborn, A.; Rozelle, S. The relationship between birth season and early childhood development: Evidence from northwest rural China. *PLoS ONE* **2018**, *13*, e0205281. [CrossRef] [PubMed]
18. Sutapa, P.; Pratama, K.W.; Rosly, M.M.; Ali, S.; Karakauki, M. Improving Motor Skills in Early Childhood through Goal-Oriented Play Activity. *Children* **2021**, *8*, 994. [CrossRef] [PubMed]
19. Leo, I.; Leone, S.; Dicataldo, R.; Vivenzio, C.; Cavallin, N.; Taglioni, C.; Roch, M. A Non-Randomized Pilot Study on the Benefits of Baby Swimming on Motor Development. *Int. J. Environ. Res. Public Health* **2022**, *19*, 9262. [CrossRef]
20. Hermes, L.; Martins, F.K.; Righi, N.C.; Böck, T.H.O.; Trevisan, C.M.; Weinmann, A.R.M.; Valentini, N.C. Seasonality and risk factors associated with motor development of full-term infants. *Cad. Bras. Ter. Ocup.* **2021**, *29*, e2811. [CrossRef]
21. Bodensteiner, E.G. *The Correlation between the Season of Birth and Early Child Development in Healthy Children: Evidence from the Midwestern Region of the United States: A Pilot Study*; Physical Therapy Scholarly Projects 762; University of North Dakota: Grand Forks, ND, USA, 2023.
22. Gil, J. Factors Association with Motor Milestone Development in Infants and Young Children. Master's Thesis, Emory University, Atlanta, GA, USA, 2014. Available online: https://etd.library.emory.edu/concern/etds/2r36tx747?locale=es (accessed on 15 November 2023).
23. Weiler, H.A.; Hazell, T.J.; Majnemer, A.; Vanstone, C.A.; Gallo, S.; Rodd, C.J. Vitamin D supplementation and gross motor development: A 3-year follow-up of a randomized trial. *Early Hum. Dev.* **2022**, *171*, 105615. [CrossRef]
24. Tsuchiya, K.J.; Tsutsumi, H.; Matsumoto, K.; Takei, N.; Narumiya, M.; Honda, M.; Thanseem, I.; Anitha, A.; Suzuki, K.; Matsuzaki, H.; et al. Seasonal variations of neuromotor development by 14 months of age: Hamamatsu Birth Cohort for mothers and children (HBC Study). *PLoS ONE* **2012**, *7*, e52057. [CrossRef] [PubMed]
25. Dinkel, D.; Snyder, K. Exploring gender differences in infant motor development related to parent's promotion of play. *Infant Behav. Dev.* **2020**, *59*, 101440. [CrossRef] [PubMed]
26. Navarro-Patón, R.; Arufe-Giráldez, V.; Sanmiguel-Rodríguez, A.; Mecías-Calvo, M. Differences on motor competence in 4-year-old boys and girls regarding the quarter of birth: Is there a relative age effect? *Children* **2021**, *8*, 141. [CrossRef] [PubMed]
27. Li, K.; Chen, S.; Ma, J.; Martins, C.; Duncan, M.; Sheng, X.; Liu, S.; Cai, Y. The relative age effect on fundamental movement skills in Chinese children aged 3–5 years. *BMC Pediatr.* **2023**, *23*, 150. [CrossRef] [PubMed]
28. Lee, E.J.; Zwicker, J.G. Early identification of children with/at risk of developmental coordination disorder: A scoping review. *Dev. Med. Child Neurol.* **2021**, *63*, 649–658. [CrossRef] [PubMed]
29. Zhu, Y.; Hu, J.; Ye, W.; Korivi, M.; Qian, Y. Assessment of the measurement properties of the Peabody Developmental Motor Scales-2 by applying the COSMIN methodology. *Ital. J. Pediatr.* **2024**, *50*, 87. [CrossRef] [PubMed]
30. Folio, M.; Fewell, R. *Peabody Developmental Motor Scales*, 2nd ed.; Pearson: London, UK, 2000.
31. Alvarez-Gonzalo, V.; Pandiella-Dominique, A.; Kurlander-Arigon, G.; Simo-Segovia, R.; Caballero, F.F.; Miret, M. Validation of the PDMS-2 scale in the Spanish population. Evaluation of physiotherapy intervention and parental involvement in the treatment of children with neurodevelopmental disorders. *Rev. Neurol.* **2021**, *73*, 81–88. [CrossRef] [PubMed]
32. Zanella, L.W.; Valentini, N.C.; Copetti, F.; Nobre, G.C. Peabody Developmental Motor Scales—Second Edition (PDMS-2): Reliability, content and construct validity evidence for Brazilian children. *Res. Dev. Disabil.* **2021**, *111*, 103871. [CrossRef] [PubMed]
33. Alsakhawi, R.; Aleidi, J.; Almutairi, R.; Alrifaei, S.; Alnutifi, A.; Almutiri, R. *The Difference in Fine Motor Performance between Saudi Boys and Girls in Preschool Age: A Pilot Study*; Research Square Platform LLC: Durham, NC, USA, 2023.
34. Vlachos, F.; Papadimitriou, A.; Bonoti, F. An investigation of age and gender differences in preschool children's specific motor skills. *Eur. Psychomot. J.* **2014**, *6*, 12–21.
35. Pahlevanian, A.A.; Ahmadizadeh, Z. Relationship between gender and motor skills in preschoolers. *Middle East J. Rehabil. Health* **2014**, *1*, e20843. [CrossRef]

36. Manna, S.; Pal, A.; Dhara, P.C. Fine motor skills and its growth pattern in variation to age and gender on Bengali (Indian) primary school children: A cross sectional study. *Age* **2018**, *5*, 10–11.
37. Kendić, K.; Šarić, E.; Džananović, D. Gender Differences In Fundamental Motor Skills of Preschool Children from Bosnia and Herzegovina and Germany. *Res. Educ. Rehabil.* **2023**, *6*, 70–75. [CrossRef]
38. Kokstejn, J.; Musalek, M.; Tufano, J.J. Are sex differences in fundamental motor skills uniform throughout the entire preschool period? *PLoS ONE* **2017**, *12*, e0176556. [CrossRef] [PubMed]
39. Hu, J.; Zhang, S.; Ye, W.; Zhu, Y.; Zhou, H.; Lu, L.; Chen, Q.; Korivi, M. Influence of different caregiving styles on fundamental movement skills among children. *Front. Public Health* **2023**, *11*, 1232551. [CrossRef] [PubMed]
40. Robinson, L.E. The relationship between perceived physical competence and fundamental motor skills in preschool children. *Child Care Health Dev.* **2011**, *37*, 589–596. [CrossRef] [PubMed]
41. Honrubia-Montesinos, C.; Gil-Madrona, P.; Losada-Puente, L. Motor Development among Spanish Preschool Children. *Children* **2021**, *8*, 41. [CrossRef] [PubMed]
42. Bolger, L.E.; Bolger, L.A.; O'Neill, C.; Coughlan, E.; O'Brien, W.; Lacey, S.; Burns, C.; Bardid, F. Global levels of fundamental motor skills in children: A systematic review. *J. Sports Sci.* **2021**, *39*, 717–753. [CrossRef]
43. Young, R.W. The ontogeny of throwing and striking. *Human_Ontogenetics* **2009**, *3*, 19–31. [CrossRef]
44. Zheng, Y.; Ye, W.; Korivi, M.; Liu, Y.; Hong, F. Gender Differences in Fundamental Motor Skills Proficiency in Children Aged 3–6 Years: A Systematic Review and Meta-Analysis. *Int. J. Environ. Res. Public Health* **2022**, *19*, 8318. [CrossRef] [PubMed]
45. Atun-Einy, O.; Cohen, D.; Samuel, M.; Scher, A. Season of birth, crawling onset, and motor development in 7-month-old infants. *J. Reprod. Infant Psychol.* **2013**, *31*, 342–351. [CrossRef]
46. McGrath, J.J.; Saha, S.; Lieberman, D.E.; Buka, S. Season of birth is associated with anthropometric and neurocognitive outcomes during infancy and childhood in a general population birth cohort. *Schizophr. Res.* **2006**, *81*, 91–100. [CrossRef] [PubMed]
47. Benson, J.B. Season of birth and onset of locomotion: Theoretical and methodological implications. *Infant Behav. Dev.* **1993**, *16*, 69–81. [CrossRef]
48. Barreca, A.; Schaller, J. The impact of high ambient temperatures on delivery timing and gestational lengths. *Nat. Clim. Chang.* **2020**, *10*, 77–82. [CrossRef]
49. de Kieviet, J.F.; Piek, J.P.; Aarnoudse-Moens, C.S.; Oosterlaan, J. Motor development in very preterm and very low-birth-weight children from birth to adolescence: A meta-analysis. *JAMA* **2009**, *302*, 2235–2242. [CrossRef] [PubMed]
50. Matarma, T.; Lagstrom, H.; Loyttyniemi, E.; Koski, P. Motor Skills of 5-Year-Old Children: Gender Differences and Activity and Family Correlates. *Percept. Mot. Ski.* **2020**, *127*, 367–385. [CrossRef] [PubMed]

life

MDPI

Article

Associations Between Sprint Mechanical Properties and Change of Direction Ability and Asymmetries in COD Speed Performance in Basketball and Volleyball Players

Zhihao Zhang [1], Mingchen Jiang [2,3], Yifan Jing [4,5], Mingjia Li [6], Yanchun Li [4,7,8,*] and Xiaolin Yang [4]

1. Sport Science School, Beijing Sport University, Beijing 100084, China; zhangzhihao@bsu.edu.cn
2. China Volleyball College, Beijing Sport University, Beijing 100084, China; jiangmingchen@cup.edu.cn
3. College of Physical Education and Arts Humanities, China University of Petroleum, Beijing 100100, China
4. China Institute of Sport and Health Science, Beijing Sport University, Beijing 100084, China; 1004320180587@bsu.edu.cn (Y.J.); yangxiaolin@bsu.edu.cn (X.Y.)
5. Caofeidian College of Technology, Tangshan 063205, China
6. China Basketball College, Beijing Sport University, Beijing 100084, China; limingjia@bsu.edu.cn
7. Beijing Key Laboratory of Sports Performance and Skill Assessment, Beijing 100084, China
8. Key Laboratory for Performance Training & Recovery of General Administration of Sport, Beijing 100084, China

* Correspondence: liyanchun@bsu.edu.cn; Tel.: +86-010-62989306

Abstract: This study aimed to assess the associations between sprint force–velocity profile variables with change of direction (COD) performance and to investigate the impact of these variables on asymmetries in COD speed performance. Ninety-nine participants (volleyball players: n = 44, basketball players: n = 55) performed 40 m sprints for Fv relationship calculation, two COD tests (Modified Agility T-test and 505 test). A partial least squares (PLS) regression analysis was conducted to determine the relationships between the variables. The V_0 was the most influential variable; it was negatively associated with COD performance variables (β = −0.260, −0.263 and −0.244 for MAT, 505-D and 505-ND, respectively), and F_0 (β = 0.169, 0.163) was associated with the COD performance variables (COD deficit D and COD deficit ND, respectively), slightly larger than the effects of Fv_{slope} (β = −0.162, −0.146), D_{RF} (β = −0.159, −0.142) and P_{max} (β = −0.162, −0.146). For COD deficit imbalance, the D_{RF} (β = −0.070) was the most influential variable followed by Fv_{slope} (β = −0.068), F_0 (β = 0.046) and gender (β = 0.031). V_0 and RF_{max} were the critical variables for improving COD performance that includes linear sprints, while D_{RF}, Fv_{slope}, F_0 and P_{max} collectively influence 180° COD performance. Meanwhile, D_{RF} and Fv_{slope} were important factors for asymmetries in COD speed performance. It is recommended to use the Fv profile to diagnose different COD movement patterns and then develop training plans accordingly for team sports played on smaller courts, such as basketball and volleyball.

Keywords: force–velocity profile; agility performance; asymmetries in COD speed performance

Citation: Zhang, Z.; Jiang, M.; Jing, Y.; Li, M.; Li, Y.; Yang, X. Associations Between Sprint Mechanical Properties and Change of Direction Ability and Asymmetries in COD Speed Performance in Basketball and Volleyball Players. *Life* **2024**, *14*, 1434. https://doi.org/10.3390/life14111434

Academic Editor: Nicola Smania

Received: 20 September 2024
Revised: 28 October 2024
Accepted: 2 November 2024
Published: 6 November 2024

1. Introduction

Team sports players must frequently perform change of direction (COD) movements such as running back and forth repeatedly. Compared to linear sprinting, COD movements involve additional acceleration and deceleration to overcome inertia and quickly generate propulsive forces in new directions. The ability to change direction is particularly critical in basketball and volleyball [1–3]. Characteristic gameplay in basketball involves short and rapid offenses, with an average maximum movement distance of 9.48 m [4]. In volleyball, 83.7% of game durations are less than 10 s. Regarding covered distance, 45.7% of movements range between 5 and 10 m, while 85.3% are less than 15 m [5]. In both aforementioned sports, due to the small area of the competition court and the frequent changes in competition intensity, athletes are required to repeatedly engage in high-intensity exercise accompanied by frequent COD movements [4,5].

Due to the similarity between the movement patterns of the Modified Agility Test (MAT) and the sprinting and lateral movement demands in volleyball and basketball, as well as its ability to better replicate the typical multi-directional, high-intensity movement distances in team sports, a series of studies have used the MAT to assess the COD performance of basketball and volleyball players [6–9]. The 505 COD test also widely used to measure COD performance for each leg in volleyball and basketball [10,11]. Time is recorded at the 10-m mark of the sprint and ends with the 5-m sprint following the turn. This timing omits the acceleration process from a stationary start, reflecting the phase where athletes move at higher speeds during a game [10]. However, the relevance of the total time of the 505 COD test in evaluating COD ability has recently been questioned, as literature shows that only 31% of the 505 COD test time is devoted to changing direction [12,13]. Therefore, to minimize the influence of physical fitness (body weight, size and body's center of gravity), linear speed and acceleration on 505 COD test results, a COD deficit, which is an evaluation method for COD ability—where a participant's 10 m time is subtracted from the 505 COD time—is considered an effective method to independently assess COD ability while avoiding these confounding factors [2,5,14,15]. The MAT and 505 COD test provide a comprehensive assessment of multidirectional (90° and 180°) movement, such as a 180° backdoor cut to sprint away from defensive pressure, with performance influenced by the linear and lateral speeds [10]. Conversely, COD deficit offers an isolated evaluation of COD performance [16].

In recent years, studies of lower limb asymmetry in sports performance have garnered considerable attention. The COD deficit is an effective indicator for assessing lower limb asymmetry [14,17]. It appears that team sports players who demonstrate greater limb symmetry (assessed via unilateral vertical jump or distance reached during a dynamic balance test) are faster than their asymmetrical counterparts during COD sprint tests [18]. Spiteri et al. reported that professional female basketball players demonstrate better lower limb balance than collegiate female basketball players [19]. Additionally, the injured side of recreational team sport players was found to be 2% slower than the healthy side in the 505 COD test [18]. Among basketball players, combined unilateral training effectively reduced lower limb asymmetry and enhanced COD capability [20]. It has also been observed that during double-leg landing tasks, volleyball athletes may be predisposed to unilaterally higher ground reaction or muscle forces, ultimately increasing the risk of injury during landing [21]. These findings underscore the importance of reducing lower limb imbalances to execute more proficient on-court movements.

Samozino et al. have recently developed a method to assess the entire force–velocity (Fv) spectrum during sprint acceleration (sprint Fv profile), profiling the mechanical capabilities of the neuromuscular system [22,23]. Since the sprint Fv relationship is linearity, the maximal capacities of muscles to produce force (F_0), velocity (V_0), power (P_{max}), Fv_{slope} (the ratio between F_0 and V_0), maximal ratio of force (RF_{max}), maximum speed during a sprint (V_{max}) and the index of force application technique (D_{RF}) can be determined within a linear regression model [23,24]. The sprint Fv profile has already been established as a reliable theoretical basis for devising personalized training guidance for athletes [22,25–28]. An athlete engaged in sports with lower speed demands may benefit from more maximal speed sprint training, while those exhibiting lower horizontal force outputs may require additional horizontal force training [29]. Additionally, studies have indicated that sprint Fv profile variables contribute independently to explaining COD performance in basketball and volleyball [30,31]. Thus far, only three studies have explored the relationship between COD ability and the Fv profile variables in volleyball and basketball. One study found that the F_0, Fv_{slope} and RF_{max} had low to moderate correlations (0.32–0.54) with the COD deficit in volleyball players [30]. Another study revealed a moderate to strong negative correlation ($r = -0.569$ to -0.794) between the 505 COD test times of the dominant and non-dominant legs and the variables F_0, V_0, P_{max} and F_0 within the Fv profile in 15 basketball players [31].

Furthermore, research on basketball players showed that F_0, RF_{max} and P_{max} are the most determinant sprint Fv profile variables for greater COD performance and minimizing the COD deficit [32]. Studies on the association between COD performance and sprint Fv profile variables in basketball players have produced varying conclusions. In contrast, the limited number of studies on volleyball players has left the relationship between COD performance and sprint Fv profile variables unclear and inconclusive. Meanwhile, currently no research has examined the connection between asymmetries in COD speed performance and the sprint Fv profile in basketball and volleyball players. However, improving COD ability and reducing lower limb imbalance during the COD performance are crucial for volleyball and basketball players. The purpose of this study was to assess the associations between sprint Fv profile variables with COD performance and to investigate the impact of these variables on the asymmetries in COD speed performance.

2. Materials and Methods

2.1. Participants

Ninety-nine team sports collegiate players were selected, including forty-four volleyball players (age: 20.55 ± 1.88 years; height: 176.56 ± 5.21 cm; body mass: 82.55 ± 9.20 kg; BMI: 23.24 ± 2.55; training years: 7.64 ± 1.56) and fifty-five basketball players (age: 20.32 ± 2.41 years; height: 179.95 ± 8.41 cm; body mass: 75.89 ± 10.15 kg; BMI: 23.38 ± 2.10; training years: 8.48 ± 2.83). All the participants participated in an average of 10 h per week of combined team practice and technical skills, plus one competitive match per week. Additionally, they participated as key players in provincial-level or higher competitions in China, achieving top-three finishes. Exclusion criteria were musculoskeletal injuries within the last 6 months and traumatic surgeries within the last 12 months. None of the participants had any injuries or limitations that could affect their testing performance. Written informed consent was obtained from all subjects, and the study was approved by the ethics committee of Beijing Sport University (approval number 2023211H).

2.2. Study Design

A descriptive cross-sectional design was used to determine the relationships between the sprint Fv profile (F_0, V_0, P_{max}, Fv_{slope}, RF_{max}, D_{RF} and V_{max}) and COD performance. Before the testing, all participants were familiarized with the experimental procedures and completed two tests separated by at least 24 h and no more than 7 days. During the first test, participants performed the sprint Fv profile test. During test two, participants performed the COD performance test (the order of the COD performance tests was randomized, and sufficient rest was ensured between each test). All tests were performed indoors at a similar time of day to avoid effect of the circadian rhythm and under controlled conditions (i.e., temperature: min 20 °C, max 33 °C; atmospheric pressure: 1016 hPa). Participants were required not to engage in strenuous exercise within the 24-h period preceding the testing (i.e., no professional practice; only dynamic mobility was allowed). All participants completed three 40 m sprints, three MATs and six attempts of 505 COD tests with both the dominant and non-dominant legs. Before testing, all participants performed a standardized warm-up, starting with a 5-min jog, followed by 5 min of low-intensity sprints and ending with 5 min of dynamic stretching.

2.3. Sprint Force–Velocity Profile Test

A standardized warm-up was performed before the test. Six pairs of photocells (Smart Speed; Fusion Sport, Brisbane, Australia) were positioned at the starting line and at distances of 10, 20, 25, 30 and 40 m at approximately 1.2 m high, to measure the intervals of 0–10 m, 0–20 m, 0–25 m, 0–30 m and 0–40 m [24,33]. The sprint test was conducted on an indoor running track. Participants started from a standing position 0.5 m behind the starting gate and then sprinted at full effort through the finish line, performing the test three times with a 5 min rest interval between each, and the time was recorded to the nearest 0.001 s. The sprint Fv profile characteristics (F_0, V_0, P_{max}, Fv_{slope}, RF_{max}, D_{RF}, V_{max})

were computed with the mean of three sprint split times according to Samozino's specific spreadsheet [23,24]. The mean 0–10 m split time was used to calculated the COD deficit and asymmetry index, the process of which will be detailed later.

2.4. COD Performance Tests

2.4.1. Modified Agility T-Test (MAT)

A pair of photocells (Smart Speed; Fusion Sport, Brisbane, Australia) was positioned at the start line. Participants started from 0.5 m behind the starting gate and performed the MAT, which included liner sprinting and multidirectional running (Figure 1) [7]. Initially, participants completed a forward 5 m linear sprint to touch the landmark by hand, followed by a lateral, leftward shuffle for 2.5 m to touch the landmark, then a lateral, rightward shuffle for 5 m to touch the landmark, a lateral, leftward shuffle for 2.5 m to touch the middle landmark and, finally, a linear backpedal for 5 m. Each test was performed three times, and a 2 min rest interval was allowed between tests; the time was recorded to the nearest 0.001 s. The fastest times to complete the test were used for the subsequent analysis.

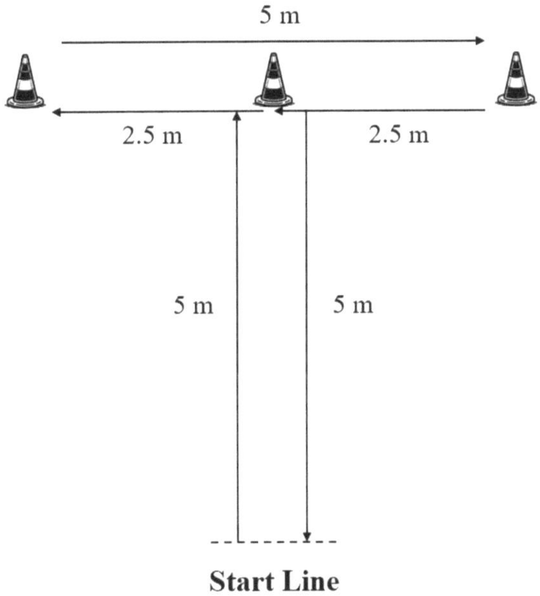

Figure 1. The structure of the Modified Agility T-test, m = meters.

2.4.2. 505 COD Test

The participants were required to sprint with full effort to a line placed 15 m from the start line and performed a 180° turn, and then sprint 5 m through the finish line (Figure 2). Each participant completed six trials with 2 min of recovery between trials (three turning off the right leg and three off the left leg); the order of trials was randomized amongst the participants [16]. A pair of photocells (Smart Speed; Fusion Sport, Brisbane, Australia) was positioned at the 10 m mark, and time was recorded to the nearest 0.001 s. Participants started from 0.5 m behind the starting line, sprinted at full speed to the turning line, then performed a 180° turn with either the left or right leg randomly, and sprinted at full speed to the 5 m finish line to complete the trial. If a participant changed direction or turned off the incorrect foot before reaching the turning line, the result of that attempt would be disregarded. Participants were required to fully rest before beginning the test again [14]. The fastest times to complete the distance with each leg were calculated for the COD deficit and asymmetry index.

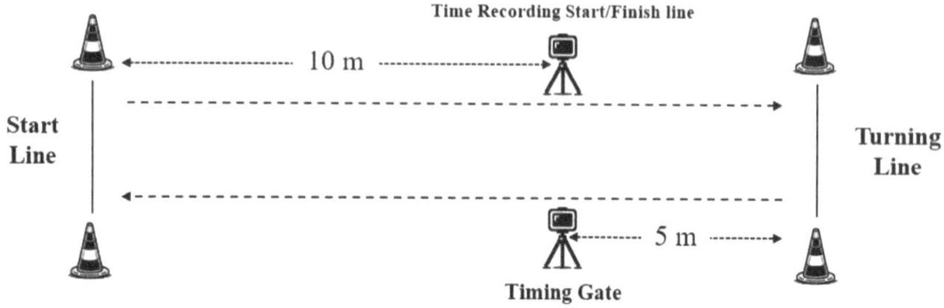

Figure 2. The structure of the 505 Change of Direction test, m = meters.

2.5. Data Processing

The COD deficit was calculated by subtracting the 10 m sprint from the 505 COD test time [14].The dominant (D) COD speed performance was defined as the leg side with the fastest completion time, while the non-dominant (ND) COD speed performance was the leg side with the slower completion time [14]. The COD deficit for dominant and non-dominant limbs was calculated using the following formula: 505 COD test time—10 m sprint time; the 10 m sprint time was taken from the 40 m split (the 0–10 m split time) [15,17,34]. Meanwhile, the asymmetry index was determined via the COD deficit of each leg. The COD deficit imbalance were asymmetry indices to evaluate lower limb balance during COD in participants. Calculated using the COD deficits of the D and ND limbs, the asymmetry index calculation formula is as follows: $(D - ND)/D \times 100$ [14,34,35].

2.6. Statistical Analyses

Participants' baseline data were reported as mean $\pm$ standard deviation (SD), and other continuous variables were presented as medians with interquartile range (IQR) depending on the distribution of the data. The Shapiro–Wilk test was used to assess the normality of the variables. Collinearity was evaluated for each variable using the variance inflation factor (VIF), and variables with VIF ≥ 10 were considered collinear [36].

Due to the non-normality and multicollinearity of the datasets, a partial least squares (PLS) regression analysis was conducted. This analysis included explanatory variables from the mechanical characteristics of the Fv sprint profile (F_0, V_0, P_{max}, RF_{max}, D_{RF}, Fv_{slope} and V_{max}) and gender. It also considered COD performance indicators such as COD deficit imbalance and the completion times of the MAT, 505-D, 505-ND, COD deficit D and COD deficit ND tests as different response variables. This approach aimed to explore the deeper relationship between the Fv sprint profile and COD performance. Composite variables constructed by PLS were used to build a linear regression model to identify which of these composite variables, encompassing different potential variables, significantly predict COD performance, with a p-value < 0.050 considered significant for composite variables [37].

Different composite variables significantly predicted various response variables for COD performance. The PLS regression coefficients (β) for the original explanatory variables within these composite variables were then estimated, along with standard error (SE) and bias-corrected and accelerated (bca) 95% confidence intervals (CIs) after bootstrapping. The SE and 95%CIs for coefficients in the PLS regression were estimated using the bootstrap method, an agnostic estimation method that does not rely on model form assumptions [38–40]. The number of bootstrap iterations was set to 2000 [41,42]. Explanatory variables were considered significant if zero was not included in the 95% bcaCIs [43,44]. The larger the absolute value of the coefficient of an explanatory variable, the greater the influence on the response variable relative to other Fv profile explanatory variables in the composite variable [45]. The optimal number of components in the PLS regression was determined by comparing the cross-validation root mean square error of prediction (RMSEP) across different numbers of components [45]. Statistical analyses were performed using the R (version 4.3.2). The following packages were utilized:pls, ggplot2, showtext, gridExtra.

3. Results

Table 1 shows the variables within the sprint Fv and COD performance tests.

Table 1. Descriptive statistics for COD performance variables and Fv sprint profile variables (n = 99).

Variables	Medians (IQR)
Sprint Fv profile	
F_0 ($N{\cdot}kg^{-1}$)	12.96 (8.33, 28.12)
V_0 ($m{\cdot}s^{-1}$)	7.97 (5.47, 9.33)
P_{max} ($W{\cdot}kg^{-1}$)	26.43 (12.17, 56.10)
RF_{max}	0.50 (0.37, 0.55)
D_{RF}	−0.15 (−0.35, −0.10)
Fv_{slope} ($N{\cdot}s{\cdot}m^{-1}kg^{-1}$)	−1.64 (−3.61, −1.09)
V_{max} (m/s^{-1})	7.84 (5.41, 9.02)
COD Performance tests	
Time to 10 m (s)	1.82 (1.50, 2.41)
MAT (s)	5.73 (4.84, 7.67)
505-D (s)	2.36 (1.99, 2.94)
505-ND (s)	2.43 (2.11, 3.02)
COD deficit D (s)	0.53 (0.23, 1.07)
COD deficit ND (s)	0.61 (0.24, 1.08)
505 COD imbalance (%)	−2.58 (−7.4, −0.04)
COD deficit imbalance (%)	−11.22 (−34.89, −0.12)

IQR: interquartile range; F_0: theoretical maximal force production; V_0: theoretical maximal running velocity; P_{max}: theoretical maximal mechanical power in the horizontal direction; Fv_{slope}: force–velocity slope; RF_{max}: maximum ratio value of horizontal component to resultant force; D_{RF}: index of force application technique; V_{max}: maximal velocity; COD: change of direction; D: dominant; ND: non-dominant; m: meters.

The RMSEP was plotted against the number of components used in the PLS regression analysis in Figure 3, which suggested that two components should be included in the PLS regression model (Figure 3); meanwhile, two components were chosen because adding a third component did not lead to an increased adjusted R^2. Component 1 comprised a linear combination of gender, V_0, P_{max}, RF_{max}, D_{RF}, Fv_{slope} and V_{max}, while component 2 was formed from a linear combination of F_0, P_{max}, RF_{max}, D_{RF} and Fv_{slope}.

Table 2 presents results from a linear model predicting COD performance using two PLS components. The results revealed that the effect of component 1 was significantly negatively associated with the performance of MAT, 505-D and 505-ND ($p < 0.05$). The effect of component 2 was significantly positively associated with COD deficit D and COD deficit ND ($p < 0.05$). Additionally, the effects of components 1 and 2 were both significantly correlated with COD deficit imbalance ($p < 0.05$).

Figures 4 and 5 present the relationship between the sprint Fv profile variables and COD performance in PLS regression. Figure 4 displays the correlations between the response variables significantly related to component 1 and the explanatory variables that composed component 1 in the PLS model. Gender (95% bcaCl = [0.08 to 0.13], [0.08 to 0.13], [0.07 to 0.12]) was significantly correlated with COD performance variables (MAT, 505-D and 505-ND). In contrast, V_0 (95% bcaCl = [−0.30 to −0.21], [−0.30 to −0.21], [−0.29 to −0.19]), RF_{max} (95% bcaCl = [−0.28 to −0.20], [−0.29 to −0.20], [−0.27 to −0.17]), V_{max} (95% bcaCl = [−0.23 to −0.16], [−0.23 to −0.16], [−0.22 to −0.15]) and D_{RF} (95% bcaCl = [−0.07 to −0.01], [−0.08 to −0.02], [−0.08 to −0.02]) showed significant negative correlations with COD performance variables (MAT, 505-D and 505-ND). The V_0 was the most influential variable negatively associated with COD performance variables (β = −0.260, −0.263 and −0.244 for MAT, 505-D and 505-ND, respectively), followed by RF_{max}, V_{max}, gender and D_{RF}.

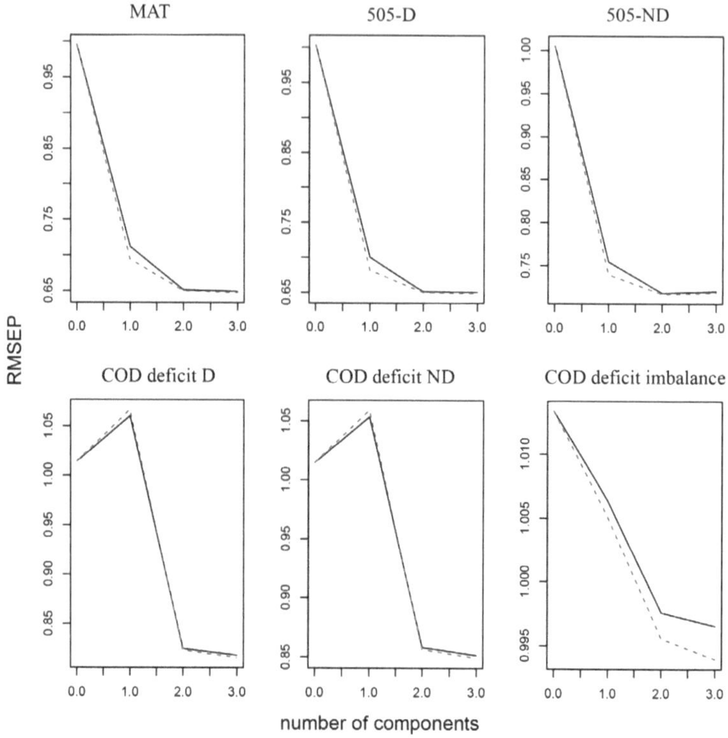

Figure 3. Root mean square error of prediction (RMSEP) for different numbers of components of partial least squares (PLS) regression.

Table 2. Coefficients from linear regression model of partial least squares (PLS) composite variables predicting COD performance.

		β (SE)	T	*p*-Value	adj-R^2
MAT	Intercept	−0.034 (0.06)	−0.535	0.594	0.59
	Comp1	−0.451 (0.04)	−11.94	<0.001 *	
	Comp2	−0.028 (0.03)	−0.842	0.402	
505-D	Intercept	−0.019 (0.06)	−0.294	0.770	0.59
	Comp1	−0.456 (0.04)	−12.053	<0.001 *	
	Comp2	−0.010 (0.03)	−0.303	0.763	
505-ND	Intercept	−0.017 (0.07)	−0.239	0.812	0.50
	Comp1	−0.423 (0.04)	−10.083	<0.001 *	
	Comp2	−0.011 (0.04)	−0.305	0.761	
COD deficit D	Intercept	−0.005 (0.08)	−0.057	0.954	0.36
	Comp1	0.004 (0.05)	0.089	0.929	
	Comp2	0.322 (0.04)	7.571	<0.001 *	
COD deficit ND	Intercept	−0.002 (0.08)	−0.025	0.980	0.31
	Comp1	0.050 (0.05)	1.01	0.315	
	Comp2	0.297 (0.04)	6.727	<0.001 *	
505 COD imbalance	Intercept	−0.008 (0.10)	−0.081	0.935	0.05
	Comp1	−0.151 (0.06)	−2.582	0.011 *	
	Comp2	−0.005 (0.05)	−0.098	0.922	
COD deficit imbalance	Intercept	−0.009 (0.10)	−0.088	0.930	0.07
	Comp1	−0.122 (0.06)	−2.115	0.037 *	
	Comp2	0.109 (0.05)	2.103	0.038 *	

Comp: component; β: standardized regression coefficient; SE: standard error; adj-R^2: adjusted coefficient of determination; COD: change of direction; D: dominant; ND: non-dominant. * Significant correlation (0 was not included in the 95% bcaCI).

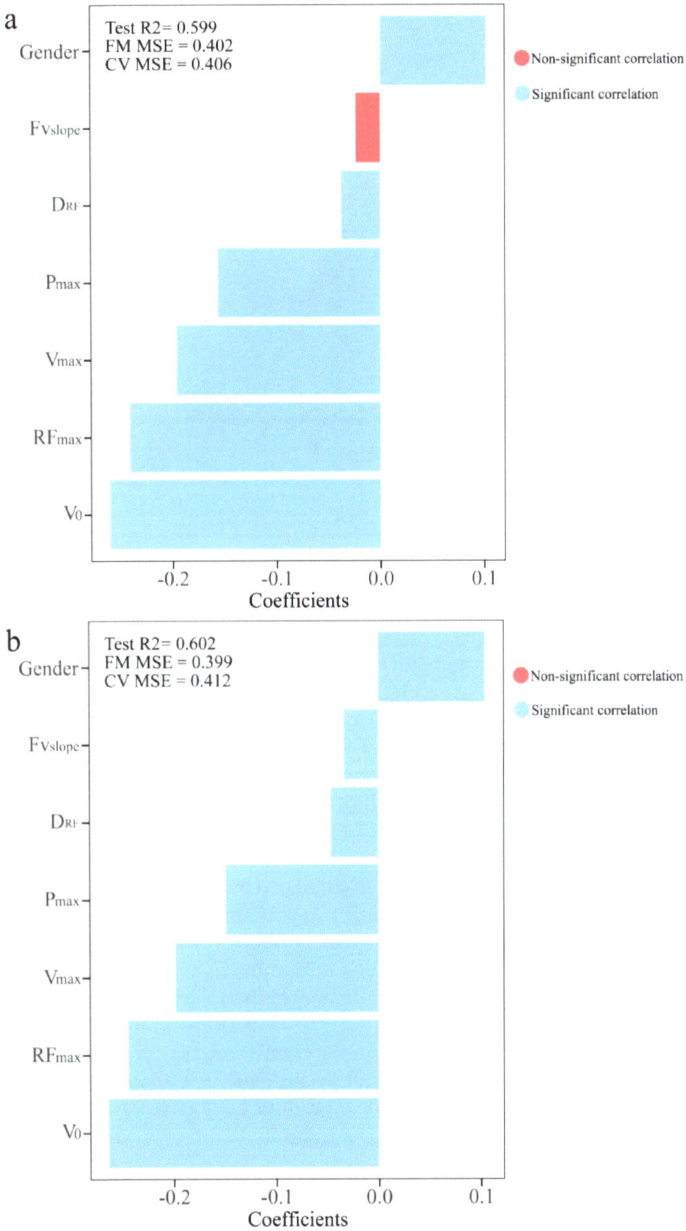

Figure 4. *Cont.*

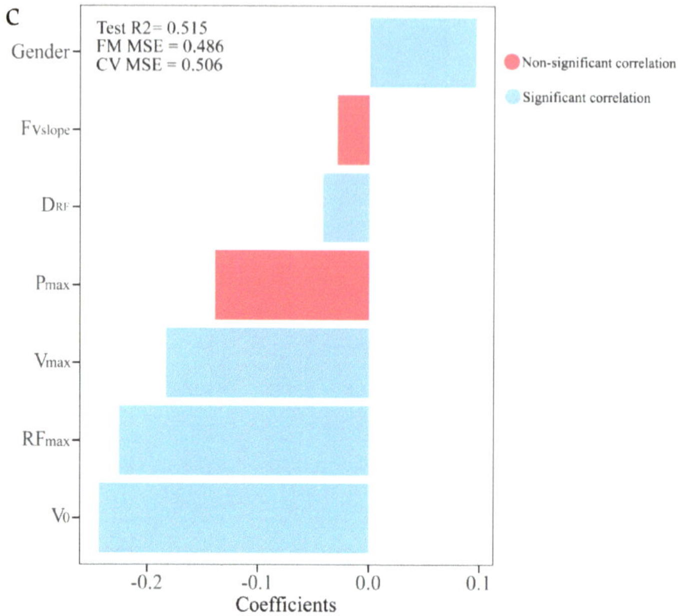

Figure 4. PLS regression coefficients between sprint Fv profile variables and COD performance. (**a**) MAT time, (**b**) 505−D time, (**c**) 505−ND time. R^2: coefficient of determination; FM MSE: folded mean square error; CV MSE: cross-validated mean square error.

Figure 5a,b show the correlations between the response variables significantly related to component 2 and the explanatory variables that comprised component 2 in the PLS model. Except for RF_{max}, all the variables had significant associations with COD performance. F_0 (95% bcaCl = [0.13 to 0.22], [0.12 to 0.22]) and P_{max} (95% bcaCl = [0.11 to 0.21], [0.11 to 0.22]) were significantly positively associated with COD performance variables (COD deficit D and COD deficit ND), while D_{RF} (95% bcaCl = [−0.21 to −0.12], [−0.19 to −0.10]) and Fv_{slope} (95% bcaCl = [−0.21 to −0.13], [−0.20 to −0.11]) were negatively significant associated with COD performance variables (COD deficit D and COD deficit ND). The effect of F_0 (β = 0.169, 0.163) on the COD performance variables (COD deficit D and COD deficit ND) was slightly larger than the effects of Fv_{slope} (β = −0.162, −0.146), D_{RF} (β = −0.159, −0.142) and P_{max} (β = −0.162, −0.146).

Figure 5c displays the correlations between the response variables significantly related to two components and the explanatory variables that comprised these two components in the PLS model. Gender (95% bcaCl = [0.00 to 0.60]) and F_0 (95% bcaCl = [0.00 to 0.10]) were significantly positively associated with COD deficit imbalance, while D_{RF} (95% bcaCl = [−0.12 to −0.03]) and Fv_{slope} (95% bcaCl = [−0.11 to −0.02]) were significantly negatively associated with COD deficit imbalance. The D_{RF} (β = −0.070) was the most influential variable followed by Fv_{slope} (β = −0.068), F_0 (β = 0.046) and gender (β = 0.031).

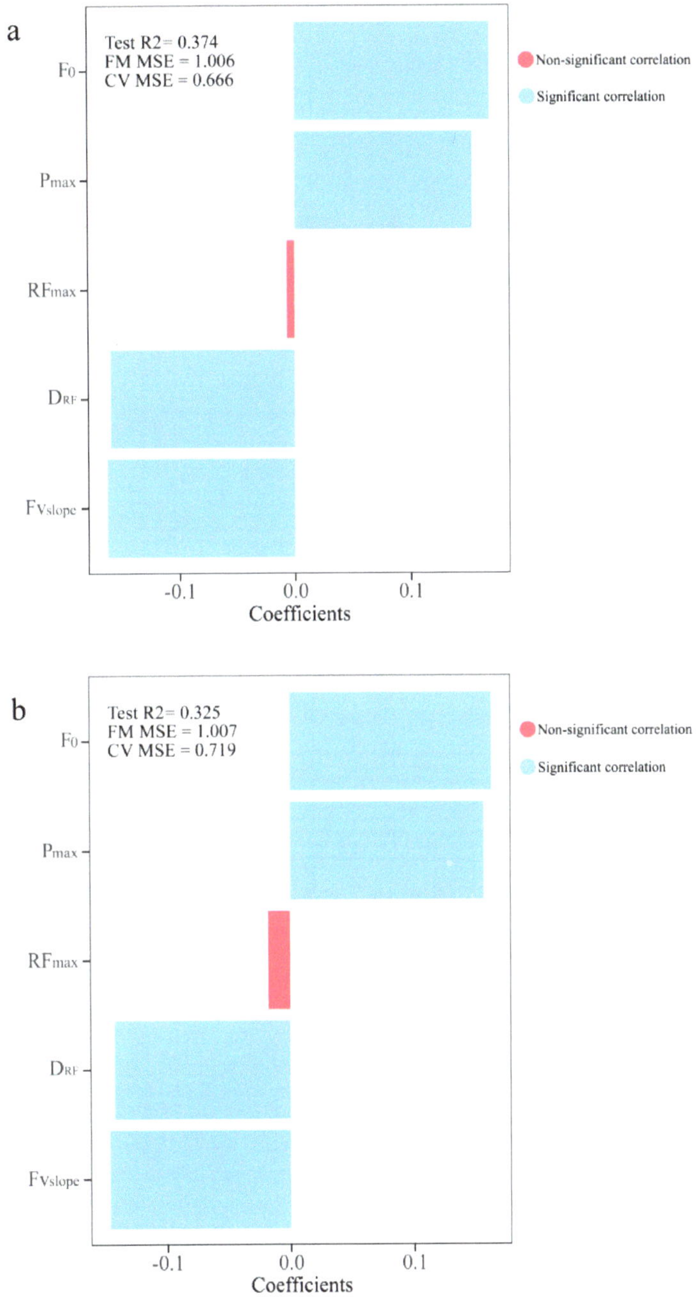

Figure 5. *Cont.*

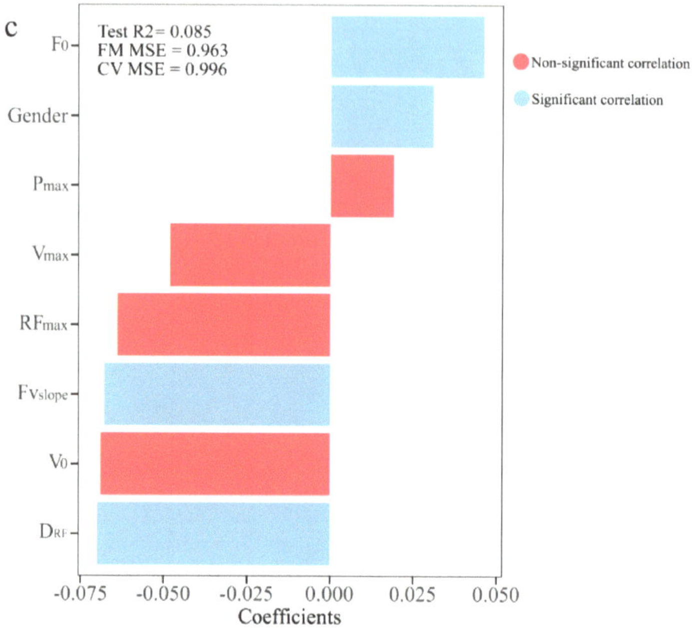

Figure 5. PLS regression coefficients between sprint Fv profile variables and COD performance. (**a**) COD deficit D time; (**b**) COD deficit ND time; (**c**) COD deficit imbalance time. R^2: coefficient of determination; FM MSE: folded mean square error; CV MSE: cross-validated mean square error.

4. Discussion

This aim of this study was to evaluate associations between sprint force–velocity profile variables and COD performance and to investigate the impact of these variables on the asymmetries in COD speed performance for basketball and volleyball players using PLS regression analysis. The main findings indicated that, within the Fv profile variables, V_0 and RF_{max} were significant predictors affecting performance in COD movement patterns that included linear sprints, while F_0, P_{max}, D_{RF} and Fv_{slope} were significant predictors impacting COD deficit. D_{RF} and Fv_{slope} were significant predictors for assessing asymmetry in COD speed performance using the COD deficit time. Sprint Fv profile variables appeared to be helpful in explaining COD capability.

The MAT and 505 COD test involved athletes performing a linear sprint before initiating a COD movement at angles of 90° and 180°, followed by a subsequent linear sprint [7,14]. The COD deficit was calculated by subtracting the time for a 10 m sprint from the total 505 COD test time, which independently evaluates the performance of the COD movement [16]. This might also have been the reason why the predictive variables for COD deficit and 505 COD test time differ in this study. Studies have shown that the speeds of 10 m and 30 m linear sprints significantly impacted the total time of the 505 COD test. Meanwhile, in the MAT, both COD speed and linear sprinting jointly determine the final MAT time [16,46]. Linear sprinting involves the movement of the body's center of mass (CoM); the forward acceleration of the CoM from one step to another is directly related to the net force developed by the athlete onto the ground in the horizontal, anteroposterior direction. Research has shown that greater agility and T-test athletes demonstrated significantly greater propulsive impulse compared with slower athletes. Faster athletes during the modified 505 test produced greater horizontal propulsive force in shorter ground contact times, and under constant conditions, the greater the net horizontal force relative to body weight, the higher the forward acceleration of the body, and the anteroposterior force was identified as a critical factor in enhancing sprint acceleration and related performance metrics [47–49]. Studies indicated that while the acceleration phase was highly related to

V_0, as well as to averaged velocity and power measured in the forward direction obtained in the Fv relationship, F_0 was not significantly correlated with performance parameters during the acceleration phase. Additionally, the Fv relationship showed that elite sprinters are able to produce higher horizontal force than sub-elite sprinters at any velocity [50]. Therefore, RF_{max} and V_0 were particularly important factors affecting the performance of COD movement patterns that included linear sprints.

In badminton, boys performed better than girls in the modified 505 test [51]. And in rugby, male athletes significantly outperformed female athletes in all COD tests (5-10-5, L-drill) [52]. Similar to these findings, we found that the times for the MAT as well as the 505 COD tests for dominant and non-dominant legs showed a significant positive correlation with gender, with males having shorter times than females; this may be related to differences between males and females in muscle cross-sectional area and the capacity to recruit muscle fibers [53,54]. Another study examining the impact of gender on the correlation between sprint Fv variables and COD capability in basketball players indicated that F_0, RF_{max} and Pmax affected the 505-D and 505-ND performance of both male and female basketball players, and V_0 was also significantly correlated with 505-D and 505-ND performance in female basketball players [32]. In a study examining the relationship between the Fv profile and linear sprint and COD performance across multiple sports (soccer, tennis and basketball), the 505-D and 505-ND times were significantly negatively correlated with sprint Fv-related variables F_0, V_0 and P_{max} [31]. These results were generally consistent with the findings of the present experiment, except for the correlation with the variable F_0, which may be due to differences in the team sports, as the sports in this study focused on team sports with intensive directional changes on smaller fields.

When independently evaluating the 180° COD ability using the COD deficit metric, it was found that F_0 and P_{max} were negatively correlated with COD capability, while D_{RF} and Fv_{slope} were positively correlated. When changing direction, athletes had to rapidly apply force during the braking phase (eccentric), plant phase (isometric) and propulsive phase (concentric) of movement [47]. Braking and propulsive forces are crucial factors affecting the performance of 180° COD movements. In the COD 505 test, faster athletes produced significantly greater braking and propulsive force compared with slower athletes [47]. Increasing force application during the braking and propulsive phases of COD movements has been shown to increase exit and starting velocities during the COD movement because muscles initially underwent eccentric contractions and were passively elongated to do negative work, which increased the storage of mechanical energy in their elastic components. Subsequently, they performed concentric contractions, enabling the muscles to provide significant braking and propulsive force during COD movements [49]. Moreover, studies have shown that compared to 90° changes of direction, 180° changes require more braking and propulsive time [47]. Consequently, in 180° COD movements, athletes need to maintain sufficient braking and propulsive force over a longer contact time (braking time and propulsive time) to ensure a smooth 180° COD. Thus, a smaller absolute value of D_{RF} can reduce COD deficit, improving performance in 180° COD movements. In basketball players, D_{RF} was significantly negatively correlated with the COD deficit in female basketball players and the dominant leg COD deficit in male basketball players, which is generally consistent with the findings of this study [32].

Research has found that faster youth netball athletes had longer COD deficit times and may not have had the capability to efficiently decelerate, change direction by 180° and reaccelerate, and that eccentric strength was essential for COD ability, especially during the braking phase [15,55]. While F_0 may optimize performance in linear sprints, it is not entirely applicable to COD movements. Additionally, a higher F_0 might lead to propulsive force exceeding braking force, thus requiring athletes to recruit more neuromuscular fibers during the braking phase to generate greater eccentric force for deceleration, which, in turn, increases COD deficit time. Furthermore, an increase in F_0 also increases P_{max}; hence, higher values of both F_0 and P_{max} were associated with a higher COD deficit. One study indicated that a higher F_0 was associated with a lower COD deficit, while another study showed that a higher F_0 was associated with a higher COD deficit on the right leg [30,32]. This discrepancy could be due to differences in the COD techniques of the participants in the experiments. Training should focus on both strength and skill enhancement to increase the utilization rate of F_0 during a 180° COD. The Fv_{slope} indicates an athlete's acceleration performance, with a theoretical optimal slope that can maximize acceleration performance and thus minimize COD time. An Fv_{slope} that is too high compared to the optimal slope can reduce the ability to maintain horizontal force, while too low a slope can decrease the average horizontal output power, both affecting the athlete's performance [33]. In this study, a smaller absolute value of Fv_{slope} was associated with a better COD deficit. This result aligns with the significant correlation found between higher F_0 and poorer COD deficit, as a higher F_0 corresponds to a larger absolute value of Fv_{slope}. One possible explanation is that, although a high F_0 increases output force, it lacks the capacity to maintain force under high-intensity output.

A study examining the association between Fv profile variables and specific performance metrics in volleyball players indicated that F_0, Fv_{slope} and RF_{max} were significantly correlated with COD deficit performance. COD right deficit was significantly correlated with D_{RF}, and P_{max} obtained during sprints was closely related to 505 left performance, but not significantly related to 505 right [30]. This was generally consistent with our results; however, in our study, the predictors for dominant and non-dominant legs under COD deficit assessment showed better uniformity. Similarly, another study found that female collegiate soccer athletes performing a COD (cutting) task with their dominant and non-dominant legs exhibited similar movement patterns [55]. Meanwhile, our findings revealed the predictors for the dominant and non-dominant sides showed differences in P_{max} during the total 505 COD time evaluation, reflecting the variations in the ability of the two legs to generate horizontal force across the entire speed range. This discrepancy may have been caused by uneven force production by the legs during linear sprints.

COD deficit asymmetries, which independently evaluate asymmetries in COD speed performance, could be interpreted as a deficiency in COD ability on one side, indicating that an athlete has a faster or slower side when performing a 180° COD [15]. This is disadvantageous for multidirectional sports, as proficiently changing direction from both limb directions equally could enhance performance in gameplay like basketball and volleyball. Our findings revealed that COD deficit asymmetries were significantly negatively correlated with D_{RF} and Fv_{slope}. However, the effect size of the PLS regression was insufficient. Currently, only one study has investigated the relationship between lower limb asymmetry in jumping and the force–velocity profile. The study indicated that there was no or a very low linear relationship between the isokinetic knee force–velocity profile and unilateral jumps in basketball players [56]. Assessments should consider specific COD in gameplay, including angles and whether they include certain short-distance linear sprints, to determine the asymmetry in COD speed performance and influencing factors [2]. Currently, no studies have reported on the relationships between asymmetries in COD speed performance and Fv profile variables. A study on lower limb injuries has reported that gender plays a significant role in knee joint mechanics during COD tasks (cutting and pivoting) and was one of the risk factors for knee injuries [57]. Consistent with the results of this study, gender was a significant predictor affecting COD deficit imbalance.

This study has limitations that must be highlighted. Firstly, it is a cross-sectional study, and these results require further comparison in future prospective studies to assess whether COD performance can be enhanced by optimizing Fv profile-related variables. Secondly, COD movements on the sports field are complex processes. Our finding has already explored the association between maximum force and maximum horizontal force produced during sprints and COD performance. Future research can combine biomechanical analysis to explore the relationships between mechanical characteristics, such as the penultimate and final foot contacts in COD performance, and the sprint Fv profile variables. Lastly, we conducted a preliminary analysis of the relationships between asymmetries in COD speed performance and sprint Fv profile variables. Future research can refine the analysis of Fv profile variables and asymmetries by incorporating biomechanical analysis.

5. Conclusions

Investigating the relationships between COD ability and sprint Fv profile variables in team sports played on smaller courts, such as basketball and volleyball, can provide new insights for personalized training based on sprint mechanical characteristics to enhance COD performance. According to our findings, RF_{max} and V_0 among the sprint Fv profile variables are significantly associated with COD performance that includes linear sprints, while D_{RF}, Fv_{slope}, F_0 and P_{max} collectively affect the 180° COD performance. Concurrently, the D_{RF} and Fv_{slope} within sprint Fv profile variables can partly explain asymmetries in COD speed performance. In the future, training interventions could be tailored by identifying the specific COD performance patterns, including the angle of change and the distance of the linear sprint, by determining the related sprint Fv profile variables, thereby providing guidance for enhancing specific COD performance.

6. Practical Application

Current research evidence suggests that developing personalized training based on force–velocity (Fv) profile variables to improve change of direction (COD) performance in basketball and volleyball players may be an effective option for coaches and researchers. This study also innovatively explores the correlation between sprint lower limb asymmetry and the sprint force–velocity profile in basketball and volleyball athletes. Among the assessed subjects, asymmetries in COD performance were indeed present. The results of this study provide preliminary theoretical evidence supporting the use of Fv profile-based personalized training by coaches to reduce asymmetries in COD performance.

Author Contributions: Z.Z. conceived and designed the study, conducted testing, collected and analyzed the data and drafted the manuscript. M.J. and Y.J. recruited participants, conducted testing and collected and analyzed the data. M.L. recruited participants, conducted testing and assisted in drafting the manuscript. X.Y. contributed to the conception and design of the study and critically reviewed the manuscript. Y.L. contributed to the writing and revision of the manuscript and approved the final version for submission. All authors have read and agreed to the published version of the manuscript.

Funding: Funding for this study was provided by the Hebei Sport University Project on Technology Innovation Platform Construction for the Integration of Sports and Health (No. 285748) and the Central University Basic Research Fund of China (No. 2024TNJN010).

Institutional Review Board Statement: The study was conducted in accordance with the Declaration of Helsinki, and was approved by the ethics committee of Beijing Sport University (approval number 2023211H, and the approval date was 10 October 2023).

Informed Consent Statement: Informed consent was obtained from all subjects involved in the study.

Data Availability Statement: The data of this study are available from the author upon request.

Acknowledgments: The author would like to thank Xiaolin Yang for assistance with experimental testing and data collecting.

Conflicts of Interest: The authors have no conflicts of interest to declare, and this study complies with the current laws of the country where it was conducted.

References

1. Scanlan, A.T.; Madueno, M.C.; Guy, J.H.; Giamarelos, K.; Spiteri, T.; Dalbo, V.J. Measuring Decrement in Change-of-Direction Speed Across Repeated Sprints in Basketball: Novel vs. Traditional Approaches. *J. Strength Cond. Res.* **2021**, *35*, 841–845. [CrossRef] [PubMed]
2. Gonzalo-Skok, O.; Dos'Santos, T.; Bishop, C. Assessing Limb Dominance and Interlimb Asymmetries Over Multiple Angles During Change of Direction Speed Tests in Basketball Players. *J. Strength Cond. Res.* **2023**, *37*, 2423–2430. [CrossRef] [PubMed]

3. Lockie, R.G.; Dawes, J.J.; Callaghan, S.J. Lower-body power, linear speed, and change-of-direction speed in division I collegiate women's volleyball players. *Biol. Sport* **2020**, *37*, 423–428. [CrossRef]
4. Scanlan, A.; Dascombe, B.; Reaburn, P. A comparison of the activity demands of elite and sub-elite Australian men's basketball competition. *J. Sports Sci.* **2011**, *29*, 1153–1160. [CrossRef]
5. Willberg, C.; Kohler, A.; Zentgraf, K. Construct Validity and Applicability of a Team-Sport-Specific Change of Direction Test. *J. Hum. Kinet.* **2023**, *85*, 115–126. [CrossRef]
6. Williams, M.N.C.; Wen, N.; Pyne, D.B.; Ferioli, D.; Conte, D.; Dalbo, V.J.; Scanlan, A.T. Anthropometric and Power-Related Attributes Differ Between Competition Levels in Age-Matched Under-19-Year-Old Male Basketball Players. *Int. J. Sports Physiol. Perform.* **2022**, *17*, 562–568. [CrossRef]
7. Scanlan, A.T.; Wen, N.; Pyne, D.B.; Stojanović, E.; Milanović, Z.; Conte, D.; Vaquera, A.; Dalbo, V.J. Power-Related Determinants of Modified Agility T-test Performance in Male Adolescent Basketball Players. *J. Strength Cond. Res.* **2021**, *35*, 2248–2254. [CrossRef]
8. Martín-López, J.; Sedliak, M.; Valadés, D.; Muñoz, A.; Buffet-García, J.; García-Oviedo, R.; Rodríguez-Aragón, M.; Pérez-López, A.; López-Samanes, Á. Impact of time-of-day and chronotype on neuromuscular performance in semi-professional female volleyball players. *Chronobiol. Int.* **2022**, *39*, 1006–1014. [CrossRef]
9. Sekulic, D.; Spasic, M.; Mirkov, D.; Cavar, M.; Sattler, T. Gender-Specific Influences of Balance, Speed, and Power on Agility Performance. *J. Strength Cond. Res.* **2013**, *27*, 802–811. [CrossRef]
10. Wen, N.; Dalbo, V.J.; Burgos, B.; Pyne, D.B.; Scanlan, A.T. Power Testing in Basketball: Current Practice and Future Recommendations. *J. Strength Cond. Res.* **2018**, *32*, 2677–2691. [CrossRef]
11. Pleša, J.; Kozinc, Ž.; Šarabon, N. Bilateral Deficit in Countermovement Jump and Its Influence on Linear Sprinting, Jumping, and Change of Direction Ability in Volleyball Players. *Front. Physiol.* **2022**, *13*, 768906. [CrossRef] [PubMed]
12. Nimphius, S.; Spiteri, T.; Geib, G.; Carlisle, D. "Change of direction deficit" measurement in Division I American football players. *J. Aust. Strength Cond.* **2013**, *21*, 115–117.
13. Amoako-Sakyi, D.; Obiri-Yeboah, D.; Ofosu, A.; Kusi, K.A.; Osei, K.; Adade, R.; Aniakwaa-Bonsu, E.; Quansah, R.; Arko-Mensah, J.; Amoah, B.Y.; et al. Preponderance of vaccine-preventable diseases hotspots in northern Ghana: A spatial and space-time clustering analysis from 2010 to 2014. *BMC Public Health* **2022**, *22*, 1899. [CrossRef]
14. Henríquez, M.; Peña-González, I.; Albaladejo-García, C.; Sadarangani, K.P.; Reina, R. Sex differences in change of direction deficit and asymmetries in footballers with cerebral palsy. *Scand. J. Med. Sci. Sports* **2023**, *33*, 1519–1530. [CrossRef]
15. Dos'Santos, T.; Thomas, C.; Jones, P.A.; Comfort, P. Assessing Asymmetries in Change of Direction Speed Performance: Application of Change of Direction Deficit. *J. Strength Cond. Res.* **2019**, *33*, 2953–2961. [CrossRef]
16. Nimphius, S.; Callaghan, S.J.; Spiteri, T.; Lockie, R.G. Change of Direction Deficit: A More Isolated Measure of Change of Direction Performance Than Total 505 Time. *J. Strength Cond. Res.* **2016**, *30*, 3024–3032. [CrossRef]
17. Bishop, C.; Weldon, A.; Hughes, J.; Brazier, J.; Loturco, I.; Turner, A.; Read, P. Seasonal Variation of Physical Performance and Inter-limb Asymmetry in Professional Cricket Athletes. *J. Strength Cond. Res.* **2021**, *35*, 941–948. [CrossRef]
18. Lockie, R.G.; Callaghan, S.J.; Berry, S.P.; Cooke, E.R.A.; Jordan, C.A.; Luczo, T.M.; Jeffriess, M.D. Relationship Between Unilateral Jumping Ability and Asymmetry on Multidirectional Speed in Team-Sport Athletes. *J. Strength Cond. Res.* **2014**, *28*, 3557–3566. [CrossRef]
19. Spiteri, T.; Binetti, M.; Scanlan, A.T.; Dalbo, V.J.; Dolci, F.; Specos, C. Physical Determinants of Division 1 Collegiate Basketball, Women's National Basketball League, and Women's National Basketball Association Athletes: With Reference to Lower-Body Sidedness. *J. Strength Cond. Res.* **2019**, *33*, 159–166. [CrossRef]
20. Gonzalo-Skok, O.; Tous-Fajardo, J.; Suarez-Arrones, L.; Arjol-Serrano, J.L.; Casajús, J.A.; Mendez-Villanueva, A. Single-Leg Power Output and Between-Limbs Imbalances in Team-Sport Players: Unilateral Versus Bilateral Combined Resistance Training. *Int. J. Sports Physiol. Perform.* **2017**, *12*, 106–114. [CrossRef]
21. Taylor, J.B.; Nguyen, A.-D.; Westbrook, A.E.; Trzeciak, A.; Ford, K.R. Women's College Volleyball Players Exhibit Asymmetries During Double-Leg Jump Landing Tasks. *J. Sport Rehabil.* **2023**, *32*, 85–90. [CrossRef] [PubMed]
22. Haugen, T.A.; Breitschädel, F.; Seiler, S. Sprint mechanical variables in elite athletes: Are force-velocity profiles sport specific or individual? *PLoS ONE* **2019**, *14*, e0215551. [CrossRef] [PubMed]
23. Samozino, P.; Rabita, G.; Dorel, S.; Slawinski, J.; Peyrot, N.; Saez De Villarreal, E.; Morin, J.B. A simple method for measuring power, force, velocity properties, and mechanical effectiveness in sprint running. *Scand. J. Med. Sci. Sports* **2016**, *26*, 648–658. [CrossRef]
24. Haugen, T.A.; Breitschädel, F.; Samozino, P. Power-Force-Velocity Profiling of Sprinting Athletes: Methodological and Practical Considerations When Using Timing Gates. *J. Strength Cond. Res.* **2020**, *34*, 1769–1773. [CrossRef]
25. Barrera-Domínguez, F.J.; Almagro, B.J.; Sáez De Villarreal, E.; Molina-López, J. Effect of individualised strength and plyometric training on the physical performance of basketball players. *Eur. J. Sport Sci.* **2023**, *23*, 2379–2388. [CrossRef]
26. Lindberg, K.; Lohne-Seiler, H.; Fosstveit, S.H.; Sibayan, E.E.; Fjeller, J.S.; Løvold, S.; Kolnes, T.; Vårvik, F.T.; Berntsen, S.; Paulsen, G.; et al. Effectiveness of individualized training based on force–velocity profiling on physical function in older men. *Scand. J. Med. Sci. Sports* **2022**, *32*, 1013–1025. [CrossRef]
27. Baena-Raya, A.; Rodríguez-Pérez, M.A.; Jiménez-Reyes, P.; Soriano-Maldonado, A. Maximizing Acceleration and Change of Direction in Sport: A Case Series to Illustrate How the Force-Velocity Profile Provides Additional Information to That Derived from Linear Sprint Time. *Int. J. Environ. Res. Public Health* **2021**, *18*, 6140. [CrossRef]

28. Zhang, Q.; Pommerell, F.; Owen, A.; Trama, R.; Martin, C.; Hautier, C.A. Running patterns and force-velocity sprinting profiles in elite training young soccer players: A cross-sectional study. *Eur. J. Sport Sci.* **2021**, *21*, 1718–1726. [CrossRef]
29. Morin, J.-B.; Samozino, P. Interpreting Power-Force-Velocity Profiles for Individualized and Specific Training. *Int. J. Sports Physiol. Perform.* **2016**, *11*, 267–272. [CrossRef]
30. Pleša, J.; Kozinc, Ž.; Šarabon, N. The Association Between Force-Velocity Relationship in Countermovement Jump and Sprint With Approach Jump, Linear Acceleration and Change of Direction Ability in Volleyball Players. *Front. Physiol.* **2021**, *12*, 763711. [CrossRef]
31. Baena-Raya, A.; Soriano-Maldonado, A.; Conceição, F.; Jiménez-Reyes, P.; Rodríguez-Pérez, M.A. Association of the vertical and horizontal force-velocity profile and acceleration with change of direction ability in various sports. *Eur. J. Sport Sci.* **2021**, *21*, 1659–1667. [CrossRef] [PubMed]
32. Andr', A.; Baena-Raya, A.; Jiménez, P.; Jiménez-Reyes, J.; Salinas Romea, E.; Soriano-Maldonado, A.; Rodríguez-Pérez, M.A. Gender-Specific Association of the Sprint Mechanical Properties With Change of Direction Performance in Basketball. *J. Strength Cond. Res.* **2021**, *36*, 2868–2874.
33. Samozino, P.; Peyrot, N.; Edouard, P.; Nagahara, R.; Jimenez-Reyes, P.; Vanwanseele, B.; Morin, J.B. Optimal mechanical force-velocity profile for sprint acceleration performance. *Scand. J. Med. Sci. Sports* **2022**, *32*, 559–575. [CrossRef] [PubMed]
34. Asimakidis, N.D.; Dalamitros, A.A.; Ribeiro, J.; Lola, A.C.; Manou, V. Maturation Stage Does Not Affect Change of Direction Asymmetries in Young Soccer Players. *J. Strength Cond. Res.* **2022**, *36*, 3440–3445. [CrossRef] [PubMed]
35. Bishop, C.R.P.; Lake, J.; Chavda, S.; Turner, A. Interlimb asymmetries: Understanding how to calculate differences from bilateral and unilateral tests. *Strength Cond. J.* **2019**, *40*, 1–6. [CrossRef]
36. Pralle, R.S.; Weigel, K.W.; White, H.M. Predicting blood β-hydroxybutyrate using milk Fourier transform infrared spectrum, milk composition, and producer-reported variables with multiple linear regression, partial least squares regression, and artificial neural network. *J. Dairy Sci.* **2018**, *101*, 4378–4387. [CrossRef]
37. Liu, C.; Zhang, X.; Nguyen, T.T.; Liu, J.; Wu, T.; Lee, E.; Tu, X.M. Partial least squares regression and principal component analysis: Similarity and differences between two popular variable reduction approaches. *Gen. Psychiatry* **2022**, *35*, e100662. [CrossRef]
38. Efron, B. Bootstrap Methods: Another Look at the Jackknife. *Ann. Stat.* **1979**, *7*, 1–26. [CrossRef]
39. Aguirre-Urreta, M.I.; Rönkkö, M. Statistical Inference with PLSc Using Bootstrap Confidence Intervals Statistical Inference with PLSC Using Bootstrap Confidence Intervals 1. *Source: MIS Q.* **2018**, *42*, 1001–1020.
40. Efron, B.; Narasimhan, B. The automatic construction of bootstrap confidence intervals. *J. Comput. Graph. Stat. A Jt. Publ. Am. Stat. Assoc. Inst. Math. Stat. Interface Found. N. Am.* **2020**, *29*, 608. [CrossRef]
41. Chowdhury, S.; Lin, Y.; Liaw, B.; Kerby, L. Evaluation of Tree Based Regression over Multiple Linear Regression for Non-normally Distributed Data in Battery Performance. In Proceedings of the 2022 International Conference on Intelligent Data Science Technologies and Applications (IDSTA), San Antonio, TX, USA, 5–7 September 2022.
42. Morozova, O.; Levina, O.; Uusküla, A.; Heimer, R. Comparison of subset selection methods in linear regression in the context of health-related quality of life and substance abuse in Russia. *BMC Med. Res. Methodol.* **2015**, *15*, 71. [CrossRef] [PubMed]
43. Miller, C.L.; Strachan, S.M. Understanding the role of mother guilt and self-compassion in health behaviors in mothers with young children. *Women Health* **2020**, *60*, 763–775. [CrossRef] [PubMed]
44. Varley, I.; Sale, C.; Greeves, J.P.; Morris, J.G.; Sunderland, C.; Saward, C. Relationship between Football-Specific Training Characteristics and Tibial Bone Adaptation in Male Academy Football Players. *Sports* **2023**, *11*, 86. [CrossRef] [PubMed]
45. Kabir, A.; Rahman, M.J.; Shamim, A.A.; Klemm, R.D.W.; Labrique, A.B.; Rashid, M.; Christian, P.; West, K.P. Identifying maternal and infant factors associated with newborn size in rural Bangladesh by partial least squares (PLS) regression analysis. *PLoS ONE* **2017**, *12*, e0189677. [CrossRef]
46. Sassi, R.H.; Dardouri, W.; Yahmed, M.H.; Gmada, N.; Mahfoudhi, M.E.; Gharbi, Z. Relative and Absolute Reliability of a Modified Agility T-test and Its Relationship With Vertical Jump and Straight Sprint. *J. Strength Cond. Res.* **2009**, *23*, 1644–1651. [CrossRef]
47. Spiteri, T.; Newton, R.U.; Binetti, M.; Hart, N.H.; Sheppard, J.M.; Nimphius, S. Mechanical Determinants of Faster Change of Direction and Agility Performance in Female Basketball Athletes. *J. Strength Cond. Res.* **2015**, *29*, 2205–2214. [CrossRef]
48. Dos'Santos, T.; McBurnie, A.; Thomas, C.; Comfort, P.; Jones, P.A. Biomechanical Determinants of the Modified and Traditional 505 Change of Direction Speed Test. *J. Strength Cond. Res.* **2020**, *34*, 1285–1296. [CrossRef]
49. Cavagna, G.A.; Komarek, L.; Mazzoleni, S. The mechanics of sprint running. *J. Physiol.* **1971**, *217*, 709–721. [CrossRef]
50. Rabita, G.; Dorel, S.; Slawinski, J.; Sàez-De-Villarreal, E.; Couturier, A.; Samozino, P.; Morin, J.B. Sprint mechanics in world-class athletes: A new insight into the limits of human locomotion. *Scand. J. Med. Sci. Sports* **2015**, *25*, 583–594. [CrossRef]
51. Fernandez-Fernandez, J.; Herrero-Molleda, A.; Álvarez-Dacal, F.; Hernandez-Davó, J.L.; Granacher, U. The Impact of Sex and Biological Maturation on Physical Fitness in Adolescent Badminton Players. *Sports* **2023**, *11*, 191. [CrossRef]
52. Freitas, T.T.; Alcaraz, P.E.; Calleja-González, J.; Arruda, A.F.S.; Guerriero, A.; Kobal, R.; Reis, V.P.; Pereira, L.A.; Loturco, I. Differences in Change of Direction Speed and Deficit Between Male and Female National Rugby Sevens Players. *J. Strength Cond. Res.* **2021**, *35*, 3170–3176. [CrossRef] [PubMed]
53. Nuzzo, J.L. Sex differences in skeletal muscle fiber types: A meta-analysis. *Clin. Anat.* **2024**, *37*, 81–91. [CrossRef] [PubMed]
54. Herda, T.J.; Trevino, M.A.; Sterczala, A.J.; Miller, J.D.; Wray, M.E.; Dimmick, H.L.; Gallagher, P.M.; Fry, A.C. Muscular strength and power are correlated with motor unit action potential amplitudes, but not myosin heavy chain isoforms in sedentary males and females. *J. Biomech.* **2019**, *86*, 251–255. [CrossRef] [PubMed]

55. Greska, E.K.; Cortes, N.; Ringleb, S.I.; Onate, J.A.; Van Lunen, B.L. Biomechanical differences related to leg dominance were not found during a cutting task. *Scand. J. Med. Sci. Sports* **2017**, *27*, 1328–1336. [CrossRef]
56. Kozinc, Ž.; Marković, G.; Hadžić, V.; Šarabon, N. Relationship between force-velocity-power profiles and inter-limb asymmetries obtained during unilateral vertical jumping and singe-joint isokinetic tasks. *J. Sports Sci.* **2021**, *39*, 248–258. [CrossRef]
57. Sigward, S.M.; Pollard, C.D.; Havens, K.L.; Powers, C.M. Influence of Sex and Maturation on Knee Mechanics during Side-Step Cutting. *Med. Sci. Sports Exerc.* **2012**, *44*, 1497–1503. [CrossRef]

life

MDPI

Article

Effects of Inertial Flywheel Training vs. Accentuated Eccentric Loading Training on Strength, Power, and Speed in Well-Trained Male College Sprinters

Zhongzhong Hu [1], Yuhang Liu [2], Keke Huang [1], Hao Huang [1], Yu Zhang [1] and Xiaoyi Yuan [2,*]

1 School of Sports Science, Wenzhou Medical University, Wenzhou 325035, China; bsu2014hzz@163.com (Z.H.); huangkekerunning@163.com (K.H.); huanghaomedical@163.com (H.H.); wmu_zhangyu@163.com (Y.Z.)
2 China Athletics College, Beijing Sport University, Beijing 100084, China; 15031958573@163.com
* Correspondence: bsu_yuan@bsu.edu.cn

Abstract: This study aimed to evaluate and compare the effects of inertial flywheel training and accentuated eccentric loading training on the neuromuscular performance of well-trained male college sprinters. Fourteen sprinters were recruited and randomly assigned to either the flywheel training (FWT, $n = 7$) group or the accentuated eccentric loading training (AELT, $n = 7$) group. The FWT group completed four sets of 2 + 7 repetitions of flywheel squats, whereas the AELT group performed four sets of seven repetitions of barbell squats (concentric/eccentric: 80%/120% 1RM). Both groups underwent an eight-week squat training program, with two sessions per week. A two-way repeated ANOVA analysis was used to find differences between the two groups and between the two testing times (pre-test vs. post-test). The results indicated significant improvements in all measured variables for the FWT group: 1RM (5.0%, ES = 1.28), CMJ (13.3%, ES = 5.42), SJ (6.0%, ES = 2.94), EUR (6.5%, ES = 4.42), SLJ (2.9%, ES = 1.77), and 30 m sprint (−3.4%, ES = −2.80); and for the AELT group: 1RM (6.3%, ES = 2.53), CMJ (7.4%, ES = 3.44), SJ (6.4%, ES = 2.21), SLJ (2.2%, ES = 1.20), and 30 m sprint (−3.0%, ES = −1.84), with the exception of EUR (0.9%, ES = 0.63, $p = 0.134$), showing no significant difference. In addition, no significant interaction effects between group and time were observed for 1RM back squat, SJ, SLJ, and 30 m sprint ($p > 0.05$). Conversely, a significant interaction effect between group and time was observed for both CMJ and EUR ($p < 0.001$); post hoc analysis revealed that the improvements in CMJ and EUR were significantly greater in the FWT group compared to the AELT group ($p < 0.001$). These findings indicate that both FWT and AELT are effective at enhancing lower-body strength, power, and speed in well-trained male college sprinters, with FWT being particularly more effective in promoting elastic energy storage and the full utilization of the stretch–shortening cycle.

Keywords: resistance training; flywheel; accentuated eccentric loading; performance

Citation: Hu, Z.; Liu, Y.; Huang, K.; Huang, H.; Zhang, Y.; Yuan, X. Effects of Inertial Flywheel Training vs. Accentuated Eccentric Loading Training on Strength, Power, and Speed in Well-Trained Male College Sprinters. *Life* **2024**, *14*, 1081. https://doi.org/10.3390/life14091081

Academic Editor: Panagiotis Georgianos

Received: 12 July 2024
Revised: 24 August 2024
Accepted: 26 August 2024
Published: 29 August 2024

1. Introduction

A sprint is a track and field event that requires running over a short distance at maximum speed [1], highlighting the athlete's explosive power, acceleration, optimal running technique, and neuromuscular system's ability to rapidly alternate between states of excitation and inhibition [2–6]. Modern sprint training systems place a significant emphasis on strength training [7], as increasing muscle strength and power can greatly enhance mechanical efficiency, muscular coordination, motor unit recruitment patterns, and muscle and tendon stiffness [7,8]. These improvements are crucial for building specific strength in key muscle groups essential for competitive success. Running at maximal speed poses high neurophysiological and biomechanical demands on the sprinter's body [9,10]. Eccentric strength may directly enhance lower-limb stiffness and reactive strength by preventing excessive muscle lengthening under high stretch loads, and indirectly by increasing force production during a subsequent quasi-isometric action through residual force enhancement [11]. Additionally, prolonged and intense eccentric strain from overload may preferentially recruit high-threshold motor units [12]. This recruitment enables the generation of greater maximal force with lower metabolic cost for the same amount of work, maximizing muscle strength gains, regional hypertrophy, and remodeling of muscle architecture [13]. Several studies have shown that eccentric training can accelerate or optimize improvements in maximal muscular strength, power development, optimal muscle length for strength gains, and coordination during eccentric movements [14–17]. Therefore, the systematic inclusion of eccentric-based training protocols in strength and conditioning programs can enhance sprinter performance and help prevent injuries.

Numerous studies have confirmed that the Nordic hamstring exercise effectively improves the eccentric hamstring strength of sprinters [18–20]. However, during sprinting, particularly at maximum speed, the hamstring muscles must generate significant eccentric force in a short amount of time to decelerate knee extension, resist the powerful concentric contraction of the quadriceps [21], and capitalize on reflex potentiation [22]. This maximizes the utilization of elastic structures within the stretch–shortening cycle (SSC) [23]. In other words, the slow movement demands of the Nordic hamstring exercise do not align with the hamstring muscle action patterns required during high-speed running. Consequently, trainers and practitioners are exploring alternative training methods to enhance the strength and force production of eccentric muscle actions and to increase the specificity and efficiency of the SSC. Eccentric overload training (EOT) refers to the training condition where the load intensity applied during the eccentric phase exceeds that of the concentric phase. This approach allows the body to withstand greater loads, disrupt conventional neural adaptations, and activate a larger number of motor units [24–26]. EOT can reduce muscle resistance at the weakest point of motion, increase resistance where strength is greater, and more closely match human strength curves, allowing muscles to operate over a wider range [27,28]. Various methods are employed to achieve eccentric overload, including manual force application, elastic bands, computer-driven devices, weight releasers, and inertial flywheel devices. Among these, flywheels and weight releasers are the most common due to their minimal disruption to exercise mechanics [29].

Specifically, inertial flywheel devices, which harness the inertia of a rotating wheel and the subsequent stored kinetic energy, offer a higher eccentric load compared to traditional weight-training methods [30]. Eccentric overload training using flywheel devices has been shown to be effective in improving jumping and linear sprinting performance in soccer players [31]. Similarly, Murton et al. reported that flywheel and traditional resistance training are equally effective in enhancing lower-body strength and power in male academy rugby union players [32]. Flywheel training (FWT) leads to a significant decrease in muscle oxygen saturation and a longer reoxygenation time, thereby imposing greater physiological stress than traditional strength training [33]. The primary reason FWT induces neural adaptations, such as increased peak power output, muscle cross-sectional area, and tendon stiffness, is the improvement in motor unit recruitment, rate coding (firing frequency), synchronous motor unit activity, and reduced neuromuscular inhibition [12,34]. It is important to note that, despite the advantages and positive aspects discussed earlier, there are some disadvantages associated with the use of inertial flywheel devices, including high costs and difficulties in precisely controlling and prescribing the adequate training load [35,36]. These challenges may hinder or, at the very least, reduce the likelihood of coaches incorporating isoinertial resistance training into their daily practices. The currently popular accentuated eccentric loading training (AELT), which uses only a barbell and weight releasers—adjustable hooks that detach at the lift's lowest point to achieve accentuated eccentric overload—has also been shown to improve maximum strength, power, and sprint ability [29,37–40]. The primary reason is that AELT can significantly increase the cross-sectional area of type IIx muscle fibers and promote shifts towards faster myosin heavy chain isoforms, which are associated with enhancements in force and power production [34,38]. AELT not only allows for precise quantification of training load, but also makes it more affordable; however, it often requires third-party assistance. FWT and AELT produce eccentric overload in different ways, each with its own advantages and disadvantages. This study represents the first attempt to compare the effects of FWT and AELT using weight releasers on the neuromuscular performance of sprinters.

2. Materials and Methods

2.1. Experimental Approach to the Problem

Resistance-trained academy sprinters were recruited to elucidate the effects of AELT vs. FWT on strength, power, and speed. Subjects were randomly assigned to complete either AELT or FWT protocols within their resistance-training program based on a computer-generated sequence (www.randomizer.org, accessed on 15 April 2023). The primary difference between groups was the way to produce eccentric overload. All other elements of the resistance-training program such as exercise selection, sets, repetitions, and rest intervals and frequency were matched between groups. In addition, all subjects were recruited from the same academy program; therefore, the weekly schedule and training load were approximately equivalent across all subjects for the duration of the study. Both AEL and TRT groups completed two training sessions weekly for eight weeks. Dependent variables including muscle strength, power, and speed were assessed before and after the intervention.

2.2. Subjects

Our study employed a randomized two-group design with repeated measures. We calculated the required sample size using G*Power 3.1.9.7 Software (Düsseldorf, Germany), aiming for a power of 90% ($1 - \beta = 0.90$), an alpha error rate of 0.05, and an effect size of 0.95. This effect size was derived from a preliminary experiment on the countermovement jump, a reliable measure of lower-limb power in sprinters [5]. Accordingly, a minimum of 12 subjects was necessary. We successfully enrolled 14 elite academy sprinters (age = 23.5 ± 0.5 years; height = 181.0 ± 2.3 cm; body mass = 77.0 ± 2.6 kg; back squat 1RM = 148.2 ± 10.7 kg; back squat to 1RM ratio = 1.92 ± 0.28, Table 1). To minimize potential biases, subjects were required to meet specific inclusion criteria: (1) absence of lower-extremity musculoskeletal injuries in the past year; (2) well-trained collegiate male sprinters with a minimum of 4 years of training experience; (3) consistent engagement in lower-body resistance training at least twice per week for the six months preceding the study; (4) ability to squat at least 1.5 times their body weight to a depth where the thigh is parallel to the ground. During the program, their typical training volume included five sessions, each lasting 90 min. The weekly training regimen was structured as follows: two sessions focused on speed training, one session dedicated to specific physical fitness exercises and aerobic recovery, and two sessions concentrated on strength training using weight releasers or flywheel devices. Although the participants were highly experienced with free-weight and weight-stack machines, none of them had prior experience with weight releasers or flywheel devices during back squat exercises. The Research Ethics Committee of Beijing Sport University approved the study protocol (Registration number 2023215H), ensuring all procedures adhered to the Declaration of Helsinki standards. Prior to participation, all participants provided written informed consent, acknowledging the study's nature, benefits, and risks.

Table 1. Basic characteristics of sprinters included in the analysis.

	FWT Group (*n* = 7)	AELT Group (*n* = 7)	Difference (*p*-Value)
Age (year)	23.2 ± 0.6	23.8 ± 0.6	0.749
Height (cm)	181.5 ± 3.7	180.6 ± 3.2	0.654
Body weight (kg)	77.4 ± 3.6	76.7 ± 2.4	0.833
Training age (year)	5.1 ± 0.9	5.3 ± 0.6	0.784
1RM back squat (kg)	147.3 ± 9.4	149.1 ± 11.5	0.481
PB time for 100 m (s)	10.87 ± 0.06	10.90 ± 0.06	0.247

Values are mean ± SD and *p*-value of the differences. FWT = flywheel training; AELT = accentuated eccentric loading training; 1RM = one-repetition maximum; PB = personal best.

2.3. Procedures

The experimental process of this study is depicted in Figure 1. Spanning 11 weeks, subjects were advised to refrain from any additional resistance training throughout the study period. The initial week comprised four sessions. The first session was dedicated to familiarizing subjects with the testing procedures for strength, sprinting, and jumping abilities. The second session involved administering back squat 1RM and 30 m linear sprint tests, while the third session focused on countermovement jump (CMJ), squat jump (SJ), and standing long jump (SLJ) tests. Before each test, a standardized warm-up routine was performed, which included 3 min of light jogging, 5 min of dynamic stretching, and 5 min of lower-body strength exercises, such as multi-directional lunges, bodyweight squats, and squat jumps. A rest period of three minutes was allotted between the final practice trial and the commencement of the tests. Subjects performed two practice trials for each test, exerting maximum effort, with at least two minutes of rest between trials. In the fourth session, the flywheel's moment of inertia and the mass of discs for subjects in the FWT group were adjusted based on the mean movement velocity during the concentric phase. The following week included two familiarization sessions to ensure subjects fully comprehended the proper techniques and acclimated to the use of flywheel or weight releasers, aiming to optimize training adaptations. The intervention spanned eight weeks, with the FWT and AELT groups undergoing training sessions twice weekly, maintaining a recovery period of at least 48 h between sessions. Post-training assessments were conducted in two sessions: the first evaluated back squat 1RM and 30 m sprint, and the second measured CMJ, SJ, and SLJ performances. Measurements and data analysis were carried out at the indoor track and field and the scientific research facility of Beijing Sport University.

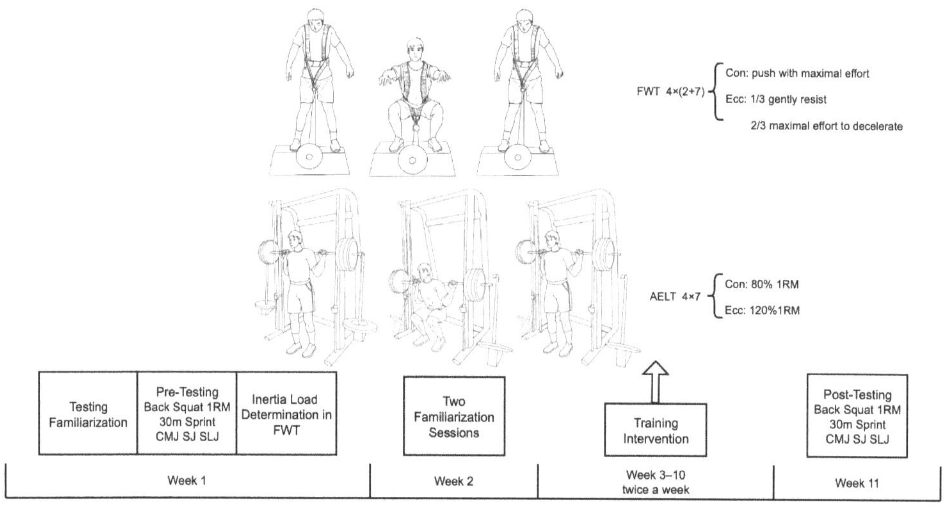

Figure 1. Schematic diagram of the experimental process.

2.4. Inertial Load Determination

The experimental groups (FWT and AELT) underwent identical training regimens, maintaining the same number of sessions, repetitions, sets, and rest intervals to ensure equivolume training protocols. Traditionally, the intensity of resistance training using gravity-dependent devices has been prescribed as a percentage of the maximum strength (1RM). However, with flywheel technology, determining the 1RM is impractical due to the absence of a maximal load limit. Despite peak power frequently serving as the primary parameter [35] to assess flywheel exercise intensity (e.g., during squats), recent studies advocate for velocity monitoring (mean and peak) over power as a more accurate measure [41,42]. Consequently, we adopted the mean concentric velocity as the criterion for setting the inertial load on the flywheel. Subjects were advised against shoulder shrugging at full hip extension and disallowed ankle extension. They were instructed to execute the concentric phase rapidly and to modulate the inertial force in the initial third of the eccentric action, subsequently exerting maximal effort to halt the movement at the motion's end range [43]. Subjects performed 2 + 7 repetitions of the flywheel squat; it should be noted that the first two repetitions, aimed at 'increasing momentum', were not included in the data analysis [40]. A linear transducer (GymAware Power Tool; Kinetic Performance Technologies, Canberra, Australia) was affixed to the chest strap during flywheel squats or positioned 50 mm away from the left hand for barbell squats. The GymAware system is renowned for its accuracy. It uses an extremely accurate LPT linear displacement sensor with a sampling rate of 100 Hz, a distance resolution of 0.8 mm, and an angle sensor resolution of 0.1 degrees, which can avoid the impact of non-vertical movements. It has served as the benchmark for velocity monitoring devices' validation in numerous studies [44]. Immediate velocity performance feedback was provided post-repetition to motivate subjects to maintain maximal velocity throughout. Rest intervals between sets were standardized to three minutes. The inertial load for flywheel squats was adjusted based on the subjects' mean concentric velocity during 7 repetitions of 80% 1RM barbell squat. Prior research has documented that the mean concentric velocities while squatting with a load equivalent to 80% 1RM were 0.55 m/s [45] and 0.62 m/s [46]. Our findings align closely with these figures, recording a similar velocity of 0.58 ± 0.04 m/s. And subjects in the FWT group used an inertial load of 0.05~0.112 kg/m^2, which was consistent with Brien et al. [47] reporting that moderate or large inertial loads maximized eccentric overload.

2.5. Training Program

Subjects engaged in resistance-training protocols for eight consecutive weeks, with sessions held twice weekly, spaced at least 48 h apart. The FWT group performed squats using a flywheel device (Desmotec, D.11 Full, Biella, Italy), executing 4 sets of 2 + 7 repetitions (the first and second repetitions were used to increase the velocity of the weighted disc and were excluded from the data analysis). Conversely, the AELT group conducted squats on a Smith machine equipped with weight releasers, completing 4 sets of 7 repetitions. Both groups took a 3 min rest between sets. FWT subjects were instructed to exert maximal effort throughout the entire concentric phase, from 70° knee flexion to near full extension. Initially, they were to gently resist during the first third of the eccentric phase, then apply maximal braking force to halt the movement at approximately 70° knee flexion [48]. The AELT group performed back squats using barbells equipped with weight releasers (Titan Fitness, Memphis, TN, USA) to induce eccentric overload, setting the concentric load at 80% of their 1RM and the eccentric load at 120% of their 1RM, as this approach was based on prior research [28,49] that training with high loads can be more effective in improving muscle strength and recruitment than low-load training. The height of the weight releasers was adjusted to each participant's lowest squat depth, ensuring disengagement just before the barbell reached the lowest squat position. AELT subjects received strong verbal encouragement to perform the concentric phase explosively in each session. To achieve eccentric overload, two training supervisors reloaded the weight releasers onto the barbell between each repetition.

Back Squat One-Repetition Maximum. Dynamic strength was assessed using a well-established one-repetition maximum (1RM) back squat protocol [50]. Subjects achieved their 1RM within 3 to 4 maximal efforts, following a standardized warm-up tailored to their self-reported 1RM back squat. Starting with feet shoulder-width apart and the barbell positioned on the upper back at shoulder level, with knees and hips fully extended, subjects were directed to lower themselves until their thighs were parallel to the floor, then rise back to a standing position at maximum speed. The squat depth was verified by two supervisors positioned to the side of the power rack. Details such as foot placement, rack height, and safety bar adjustments were documented for consistency in future sessions. The successfully recorded 1RM was then utilized to determine the load for experimental conditions.

Countermovement Jump and Squat Jump. The heights of the countermovement jump (CMJ) and squat jump (SJ) were measured using a Chronojump contact mat (Chronojump, Barcelona, Spain), which has demonstrated excellent reliability (ICC 0.998 for the SJ and 0.997 for the CMJ) compared to the Force Platform [51]. Subjects executed two attempts of each jump type, aiming for maximum height with hands placed on the hips. For the SJ, subjects began from a static position with knees bent at 90°, eliminating the eccentric phase. The CMJ was initiated from a standing position, seamlessly transitioning between the eccentric and concentric phases without pausing, and the knee flexion angle was chosen by the participant. A 30 s active recovery period was provided between attempts. The highest values recorded for both SJ and CMJ were used for subsequent analysis. The eccentric utilization ratio (EUR) was calculated as CMJ divided by SJ.

Standing Long Jump. Each participant positioned themselves on the starting line, with legs aligned parallel and feet placed shoulder-width apart. They then bent their knees and positioned their arms behind their body. Following this, they executed a forceful leg extension, propelled their arms forward, and performed a maximal jump for distance. The jump distance was measured in centimeters by the same individual. Subjects completed two attempts of the standing long jump, with 1 min rest intervals between jumps. The best attempt was documented for statistical analysis.

30 m Sprint. The 30 m sprint time was accurately measured using timing gates with error-correction processing algorithms (Smartspeed pro, Fusion sport, Milton, Australia). This test involved a maximal effort sprint in a straight line across two timing gates spaced 30 m apart, starting from a standing position. Subjects positioned themselves 30 cm behind the starting photocell, which was 80 cm above the ground, placing their preferred foot forward with the toe touching the line marked on the ground. The timing commenced as subjects activated the electronic sensors and ceased as they passed through the sensor plane again. Each participant undertook two trials, interspersed with 3 min active recovery periods. All evaluations were conducted on an indoor plastic track, and subjects were required to wear running shoes. The quickest trial was selected for statistical analysis.

Mechanical Variables. Measures of mean concentric power (MCP) and velocity (MCV) were collected in the execution of all repetitions by a valid and reliable linear position transducer (LPT) (GymAware, Kinetic Performance Technology, Canberra, Australia). Through the LPT, velocity data were calculated from the displacement of the bar or strap relative to time, and the power was calculated by the multiplication of force by velocity. To ensure the data's authenticity and reliability, we confirmed the position of the linear sensor before performing both the flywheel squat and Smith machine squat, verifying that the angles were identical. The angle between the cable and the ground, consistently measured by Dartfish 2022 pro software (Fribourg, Switzerland) at 44 degrees, is illustrated in Figure 2 Based on the application of the Pythagorean theorem in previous research [46], the velocity of the cable (Vc), as recorded by the linear sensor, can be utilized to determine the vertical velocity (Vb) during barbell lifting. Specifically, Vb is calculated as Vc multiplied by the sine of α, which is $Vb = Vc \times \sin(44°) = 0.695$ Vc. Data obtained from GymAware were transmitted via Bluetooth to a tablet (iPad, Apple Inc., Cupertino, CA, USA) using the GymAware v2.1.1 app. We chose the last training session to compare the mechanical effects of each training protocol, which were assessed by perceptual maintenance and decline of velocity and power with the following equation, as proposed by Tufano et al. [52]:

$$\text{Percent decline} = [(\text{repetition}_{\text{lowest}} - \text{repetition}_{\text{highest}})/\text{repetition}_{\text{highest}}] \times 100$$

$$\text{Maintenance} = 100 - [(\text{repetition}_{\text{highest}} - \text{repetition}_{\text{mean}})/\text{repetition}_{\text{highest}} \times 100]$$

Figure 2. The linear transducer showing a deviation of 44° from the vertical during a lift.

2.6. Statistical Analysis

Statistical analyses were conducted using IBM SPSS statistics software (version 26.0, IBM, Chicago, IL, USA). The results are expressed as the mean ± standard deviation. The Shapiro–Wilk test and Levene's test were utilized to assess data normality and equality of variances, respectively. A two-way repeated-measures ANOVA was applied to examine the effects of the group (FWT vs. AELT) and time (Pre vs. Post). Additionally, within-group comparisons of training variables were performed using paired *t*-tests, and an independent sample *t*-test was used in between-group comparisons of mechanical variables. The effect size (ES) for training was calculated for paired variables, following Cohen's classification [53], where ES of 0.2, 0.5, 0.8, and 1.3 indicate small, moderate, large, and very large effects, respectively. The threshold for statistical significance was set at $p < 0.05$.

3. Results

Subjects from both groups successfully completed all the designated training sessions. Table 2 presents the results for the 1RM back squat, vertical jump, eccentric utilization ratio, standing long jump, and 30 m sprint time.

Table 2. Between-group and within-group differences in selected variables with % of improvement and Cohen's effect size (d).

	FWT (n = 7)				AELT (n = 7)						
Variables	Pre	Post	Δ%	Cohen's d	Pre	Post	Δ%	Cohen's d	p (Time)	p (Group)	p (Time × Group)
1RM (kg)	147.29 ± 15.41	154.71 ± 16.10 **	5.0	1.28	149.14 ± 14.06	158.57 ± 13.36 **	6.3	2.53	<0.001	0.283	0.079
CMJ (cm)	50.52 ± 2.19	57.22 ± 2.28 **##	13.3	5.42	50.10 ± 2.86	53.83 ± 2.27 **	7.4	3.44	<0.001	0.006	<0.001
SJ (cm)	47.22 ± 1.71	50.07 ± 1.93 **	6.0	2.94	46.41 ± 1.36	49.39 ± 1.34 **	6.4	2.21	<0.001	0.257	0.339
EUR	1.07 ± 0.02	1.14 ± 0.01 **##	6.5	4.42	1.08 ± 0.02	1.09 ± 0.01	0.9	0.63	<0.001	0.019	<0.001
SLJ (m)	2.73 ± 0.08	2.81 ± 0.05 **	2.9	1.77	2.71 ± 0.07	2.77 ± 0.05 **	2.2	1.20	<0.001	0.332	0.136
S30 (s)	4.09 ± 0.05	3.95 ± 0.05 **	−3.4	−2.80	4.06 ± 0.07	3.94 ± 0.06 **	−3.0	−1.84	<0.001	0.562	0.321

FWT = flywheel training; AELT = accentuated eccentric loading training. One-repetition maximum = 1RM, countermovement jump = CMJ, squat jump = SJ, eccentric utilization ratio (EUR) = CMJ/SJ, standing long jump = SLJ, linear sprint 30 m = S30. Data are presented as mean ± SD, significant Cohen's d, and p-values. Δ% = percentage of improvement, ** = within-group statistical significance ($p < 0.01$), ## = between-group statistical significance ($p < 0.01$).

3.1. Back Squat 1RM

No significant main effects for group (F_group = 1.263; $p = 0.283$) or the interaction between group and time (F_group × time = 3.698; $p = 0.079$) were observed on 1RM back squat values. However, a significant main effect for time was noted (F_time = 262.717; $p < 0.001$). Post hoc analysis revealed a significant increase in 1RM among both FWT and AELT subjects following the training intervention, compared to their baseline values (FWT: $p < 0.001$, ES = 1.28; AELT: $p < 0.001$, ES = 2.53). Furthermore, 1RM increased by 5.0% in the FWT group and by 6.3% in the AELT group post-intervention.

3.2. Vertical Jump and Eccentric Utilization Ratio

Significant main effects of group, time, and the interaction between group and time on CMJ height were observed (F_group = 11.181, $p = 0.006$; F_time = 428.404, $p < 0.001$; F_group × time = 34.833, $p < 0.001$, respectively). Post hoc analysis demonstrated a significant increase in CMJ height for both FWT and AELT subjects post-training compared to the baseline (FWT: $p < 0.001$, ES = 5.42; AELT: $p < 0.001$, ES = 3.44). However, FWT subjects achieved significantly higher CMJ height than AELT subjects post-training ($p < 0.001$). Additionally, CMJ height in the FWT group increased by 13.3%, while the AELT group experienced a 7.4% increase post-intervention. No significant main effects for group (F_group = 1.416; $p = 0.257$) or the interaction between group and time (F_group × time = 0.994; $p = 0.339$) were observed for SJ height. However, a significant main effect for time was noted (F_time = 1956.791; $p < 0.001$). Post hoc analysis revealed a significant increase in SJ height among both FWT and AELT subjects following the training intervention, compared to their baseline values (FWT: $p < 0.001$, ES = 2.94; AELT: $p < 0.001$, ES = 2.21). Furthermore, SJ height increased by 6.0% in the FWT group, and by 6.4% in the AELT group post-intervention.

Significant main effects of group, time, and the interaction between group and time on EUR values were observed (F_group = 7.317, $p = 0.0019$; F_time = 63.473, $p < 0.001$; F_group × time = 36.528, $p < 0.001$, respectively). Post hoc analysis demonstrated a significant increase in EUR values for FWT subjects post-training compared to the baseline ($p < 0.001$, ES = 4.42). In contrast, the EUR values of AELT subjects did not show a significant increase post-training compared to the baseline ($p = 0.199$, ES = 0.63). Additionally, post hoc analysis revealed that the EUR values after the training period were significantly higher in the FWT group compared to the AELT group ($p < 0.001$). EUR values in the FWT group increased by 6.5%, while the AELT group experienced a 0.9% increase post-intervention.

3.3. Standing Long Jump

No significant main effects for group (F_group = 1.023; $p = 0.332$) or the interaction between group and time (F_group × time = 2.557; $p = 0.136$) were observed for SLJ distance. However, a significant main effect for time was noted (F_time = 130.435; $p < 0.001$). Post hoc analysis revealed a significant increase in SLJ distance among both FWT and AELT subjects following the training intervention, compared to their baseline values (FWT: $p < 0.001$, ES = 1.77; AELT: $p < 0.001$, ES = 1.20). Furthermore, SLJ distance increased by 2.9% in the FWT group, while the AELT group experienced a 2.2% increase post-intervention.

3.4. 30 m Sprint

No significant main effects for group (F_group = 0.356; $p = 0.562$) or the interaction between group and time (F_group × time = 1.069; $p = 0.321$) were observed for S30 time. However, a significant main effect for time was noted (F_time = 251.406; $p < 0.001$). Post hoc analysis revealed a significant increase in S30 time among both FWT and AELT subjects following the training intervention, compared to their baseline values (FWT: $p < 0.001$, ES = −2.80; AELT: $p < 0.001$, ES = −1.84). Furthermore, S30 decreased by 3.4% in the FWT group, while the AELT group experienced a 3.0% decrease post-intervention.

3.5. Mechanical Variables

The independent sample *t*-test did not detect significant differences between FWT and AELT regarding the mean concentric velocity (0.58 ± 0.04 m/s and 0.56 ± 0.04 m/s for FWT and AELT, respectively; $p = 0.411$; Figure 3) and peak concentric velocity (1.06 ± 0.13 m/s and 1.03 ± 0.12 m/s for FWT and AELT, respectively; $p = 0.852$). Similarly, differences between groups regarding mean concentric velocity decline (−20.3 ± 7.7% and −22.6± 6.9% for FWT and AELT, respectively; $p = 0.330$) and maintenance (90.6 ± 3.4% and 90.3 ± 3.1% for FWT and AELT, respectively; $p = 0.476$) were not significant. However, mean eccentric velocity was significantly higher in FWT (0.62 ± 0.06 m/s) versus AELT (0.50 ± 0.04 m/s; $p < 0.001$). There were significant differences between FWT and AELT concerning the mean concentric power (1532.69 ± 87.00 W and 1450.84 ± 79.70 W for FWT and AELT, respectively; $p = 0.03$; Figure 4) and peak concentric power (2325.32 ± 295.26 W and 2115.66 ± 243.80 W for FWT and AELT, respectively; $p = 0.02$). However, differences between groups regarding mean concentric power decline (−18.1 ± 5.5% and −17.0 ± 6.6% for FWT and AELT, respectively; $p = 0.852$) and maintenance (90.5 ± 2.9% and 91.1 ± 3.3% for FWT and AELT, respectively; $p = 0.757$) were not significant.

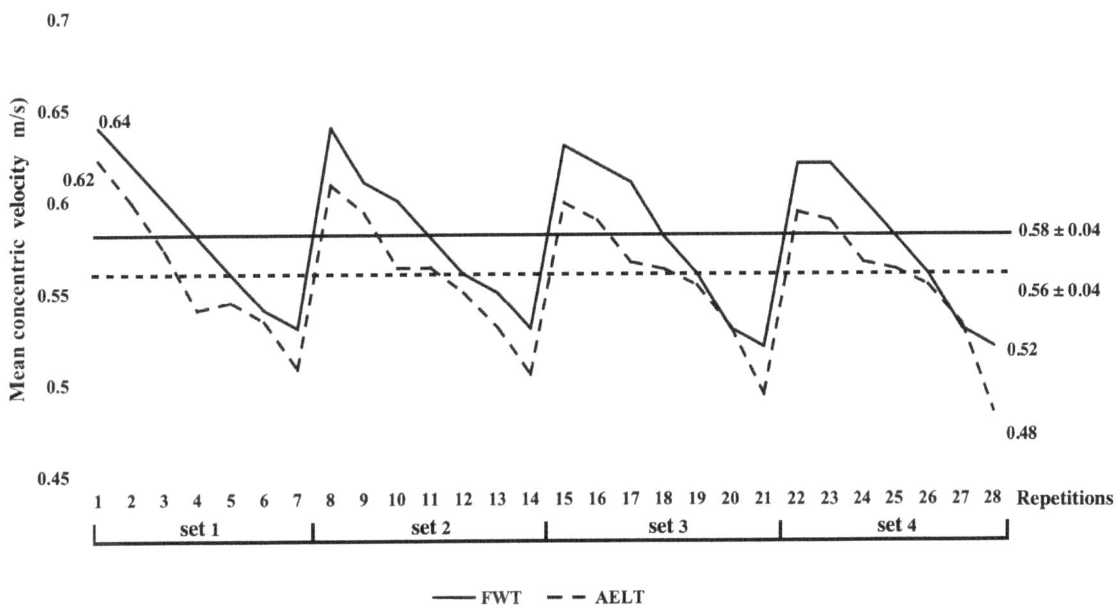

Figure 3. Mean concentric velocity of each repetition flywheel training and accentuated eccentric loading training during the last training session.

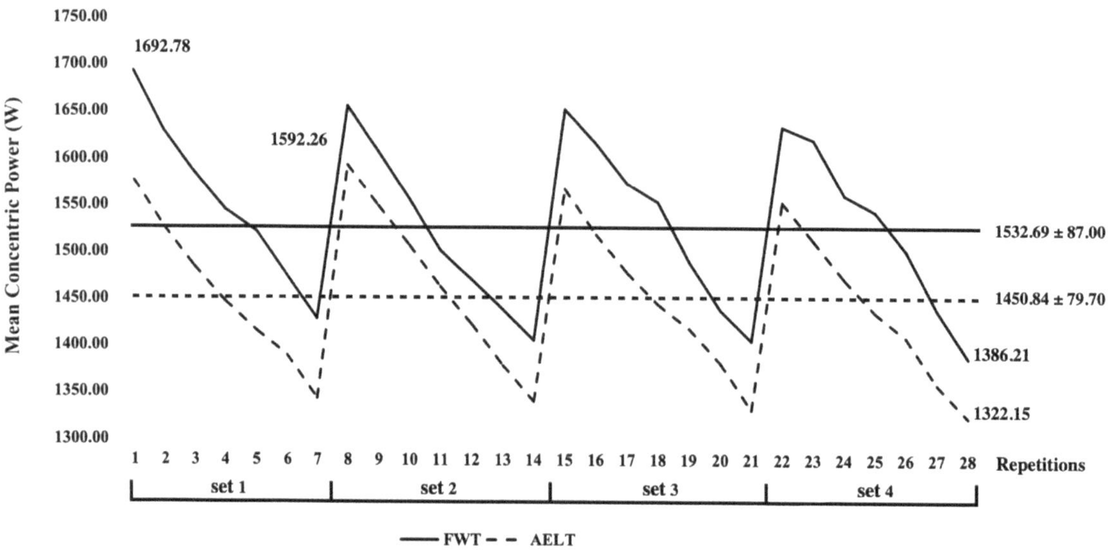

Figure 4. Mean concentric power of each repetition flywheel training and accentuated eccentric loading training during the last training session.

4. Discussion

This study aimed to explore the impact of an eight-week FWT versus AELT on the neuromuscular performance of well-trained male college sprinters. Our findings suggest that both FWT and AELT significantly enhanced lower-body strength, power, and linear sprint capabilities. Notably, the countermovement jump (CMJ) and eccentric utilization ratio (EUR) observed in FWT subjects were superior to those in the AELT group.

The mean concentric velocity (MCV) in the FWT and AELT groups experienced a notable decline between repetitions. However, after a 3 min rest period, the initial repetition of the subsequent set could achieve a relatively high velocity. It is evident that the MCV in each repetition for the AELT group was consistently lower than that observed in the FWT group. It is important to note mean concentric velocity maintenance—a variable highlighting the comparison between the highest velocity of a session and the session's mean velocity. Our findings indicate no significant difference in MCV maintenance between the two groups. The velocity–time profile observed in the FWT group aligns with known characteristics of flywheel training, where the inertial force of the flywheel offers unrestricted resistance from the onset, promoting maximum or near-maximum muscle activation [54]. As repetitions increase, so does fatigue accumulation, leading to a decrease in mean velocity. Additional effort needed to slowly unrack heavy weights (120% of 1RM) during AELT may reduce metabolite clearance in working muscles, potentially impairing the maintenance of mechanical performance.

The mean concentric power (MCP) in the FWT and AELT groups exhibited a significant decline between repetitions. Simultaneously, it is observed that the MCP for each repetition in the AELT group was lower than in the FWT group. However, following a 3 min rest period, the first repetition of the subsequent set was able to attain a relatively high power level. This pattern suggests that longer rest intervals might facilitate better restoration of energy substrates. However, this interpretation remains speculative, as the restoration of energy substrates was not directly assessed in this study.

Our study's findings reveal no significant differences in maximal squat strength gains between the two training protocols. This observation aligns with our previous statement regarding the absence of any difference in mean and peak concentric velocity between the two groups. We observed a notable 5% increase in 1RM squat strength after eight weeks (two sessions per week) of FWT, aligning with Coratella et al. [55], who reported a 7% enhancement in 1RM following an identical period of FWT. The scientific literature acknowledges that FWT induces various morphological and neural adaptations, including enhanced motor unit recruitment, increased firing frequency [56], improved synchronization of motor unit activity [12], and reduced neuromuscular inhibition [34], all of which are critical for strength development. To elicit the optimal training response, subjects are required to concentrate on generating (near) maximal effort in each flywheel squat, as well as on the timing and technique of the braking force during the eccentric phase. This approach ensures (near) maximal muscle activation and enhances the workout's intensity. Maroto-Izquierdo et al. [49] found significant strength improvements following ten weeks of eccentric overload training with supramaximal loads (120% 1RM). Similarly, Walker found that AELT resulted in greater increases in maximum force production, work capacity, and muscle activation, but did not lead to muscle hypertrophy in strength-trained individuals [38]. A potential explanation for the observed increase in maximal strength could be the enhanced neural stimulation of the muscle, attributed to the more substantial stretch of intrafusal muscle fibers (muscle spindles) under increased eccentric loads. This stretch prompts the intrafusal fibers to activate their associated γ motor neurons, which in turn signal the brain to either recruit more α motor neurons or to accelerate their firing rate. Consequently, this leads to a more forceful contraction of the extrafusal muscle fibers, as described by Dietz et al. [57]. Essentially, applying a heavier-loaded eccentric contraction neurologically primes the brain for a more potent concentric contraction, effectively "tricking" it into preparing for increased demands.

Our results demonstrated significantly greater enhancements in the countermovement jump (CMJ) and squat jump (SJ) across both groups, with FWT showing more pronounced improvements in the CMJ and EUR. These findings align with previous research [55,58–60]. Eccentric overload training, whether utilizing flywheel devices or weight releasers, has shown significant increases in type IIx fibers and shifts towards faster myosin heavy chain isoforms, leading to enhanced force and power production [29,61]. FWT has been shown to acutely enhance the rate of muscle lengthening, as evidenced in previous studies [62,63]. This enhancement may lead to temporary storage of additional elastic energy, subsequently available for use during concentric muscle actions. The rationale is based on the higher force generation capabilities and selective recruitment of high-threshold motor units during eccentric muscle actions, which can potentially elicit neuromuscular responses that lead to the desired adaptations [64,65]. Furthermore, the rapid transition between the eccentric and concentric phases during the flywheel squat enhances the stretch of the musculotendinous complex, thereby triggering a myotatic reflex that amplifies the subsequent concentric contraction, as demonstrated by Pecci et al. [66]. A comparison of mean eccentric velocity between groups revealed that the use of weight releasers induces trainers to adopt slower eccentric pacing strategies for managing supramaximal loads, which could reduce the full potential utilization of the stretch–shortening cycle (SSC) [65]. Regarding concentric power output, for both mean and peak power, the FWT group exhibited higher values compared to the AELT group. This discrepancy is likely due to the extended amortization phase in AELT, which may restrict the utilization of the SSC for concentric potentiation [67]. EUR has been established as a reliable indicator of SSC performance across various sports and training stages [68,69]. The present study corroborates this perspective, demonstrating that FWT significantly enhances EUR and CMJ performance in comparison to AELT. Essentially, an improvement in EUR attributable to FWT indicates superior elastic energy storage and utilization within the SSC, which contributes to increased CMJ height [70].

Significant improvements in 30 m sprint times were observed in both groups (FWT: 3.4%, AELT: 3.0%). According to the above research results for 1RM and CMJ, it can be seen that these two have a positive effect on increasing sprint speed. Indeed, it has been consistently demonstrated that improvements in back squat strength positively transfer to sprinting speed [2]. Moreover, the storage and return of energy within the elastic structures of the lower limb play an increasingly important role at higher sprinting speeds up to maximum velocity [4,23,71]. The outcomes in our study align with Petre et al. [72], who reported a 2.4% enhancement in sprint performance after 6–10 weeks of FWT (1–3 sessions per week) among well-trained athletes. Similarly, Cabanillas et al. [73] found that flywheel squat training positively influenced force production in the horizontal vector, leading to improved 30 m sprint times. Douglas et al. [50] also noted that AELT is an effective strategy for boosting sprint ability in rugby players. Previous research [74] has validated the efficacy of eccentric training protocols in increasing tendon stiffness, which significantly enhances sprint performance [23]. Overall, our study provides strong evidence supporting the notion that eccentrically emphasized resistance exercise can potentially contribute to superior neuromuscular adaptation [75], such as improved sprint ability.

Our findings demonstrate that both FWT and AELT are beneficial for enhancing standing long jump (SLJ) performance, with increases of 2.9% for FWT and 2.2% for AELT, but no significant differences in the two training modalities were detected. Increasing the maximum strength of the lower limbs has been conclusively shown to enhance their explosive power [76]. Correspondingly, the aforementioned study reported improvements in maximum squat strength for both experimental groups, which were directly linked to enhancements in the standing broad jump performance. Previous research has shown that flywheel squat exercises have acute enhancement in SLJ ability [77,78], while our study reveals a long-term positive impact of FWT on SLJ performance. Previous studies have confirmed that flywheel squat training significantly contributes to improvements in horizontal jumping, acceleration, and linear sprinting [73,79], all crucial for sprinters. The impact of accentuated eccentric load (AEL) on the SLJ has not been previously investigated. It is likely that AEL training (AELT) enhances SLJ performance due to increased muscle lengthening rates and the potentiation of concentric actions through the activation of higher threshold motor units, as well as the preservation of elastic energy [80].

The study has several limitations that warrant attention. First, the findings, derived from well-trained male college sprinters, may only be generalizable to comparable cohorts and competitive levels. Second, a methodological consideration needs to be acknowledged. Despite efforts to standardize training volume and intensity across groups, achieving an exactly equal workload proved challenging. Although it is feasible to quantify the training volume of accentuated eccentric loading, such quantification is not viable with flywheel devices. Future studies should conduct multi-center, large-sample, and long-term randomized controlled trials to provide more reliable evidence-based support for resistance strength training in sprinters. Additionally, it is crucial to explore the impact of diverse tempo strategies in FWT, including variations in movement velocity during the concentric and eccentric phases, and the eccentric-to-concentric load ratio in AELT on training outcomes. Another important research direction is examining the distinct effects of training with different equipment, such as vertical flywheel devices, horizontal flywheel devices, and seated leg curl flywheel devices.

5. Conclusions

The results of this experimental training study indicate that eight weeks of either FWT or AELT significantly enhanced back squat strength and linear sprint performance in well-trained male college sprinters. Additionally, both FWT and AELT led to significant improvements in jumping abilities, with FWT being more effective in promoting elastic energy storage and full utilization of the stretch–shortening cycle. Consequently, both methods are valuable for integrating into training regimens to facilitate strength and power adaptations. It is recommended to incorporate FWT and AELT into sports periodization to diversify training stimuli for well-trained athletes.

Author Contributions: Conceptualization, Z.H.; Methodology, K.H.; Software, H.H.; Investigation, Y.L.; Resources, Y.L.; Data curation, Y.Z.; Writing—original draft, Z.H.; Writing—review & editing, X.Y.; Project administration, X.Y.; Funding acquisition, X.Y. All authors have read and agreed to the published version of the manuscript.

Funding: This research was funded by National Key R&D Program of China, grant number 2019YFF0302005. And The APC was funded by Xiaoyi Yuan.

Institutional Review Board Statement: The study was conducted in accordance with the Declaration of Helsinki, and approved by the Research Ethics Committee of Beijing Sport University (protocol code 2023215H and 20 September 2023).

Informed Consent Statement: Written informed consent has been obtained from the patients to publish this paper.

Data Availability Statement: The original contributions presented in the study are included in the article, further inquiries can be directed to the corresponding author.

Acknowledgments: We would like to express our gratitude to the volunteers for their time, enthusiasm, and feedback, and to the China Athletics College of Beijing Sport University for providing experimental materials.

Conflicts of Interest: The authors declare no conflict of interest.

References

1. Mero, A.; Komi, P.V. Force-, EMG-, and elasticity-velocity relationships at submaximal, maximal and supramaximal running speeds in sprinters. *Eur. J. Appl. Physiol. Occup. Physiol.* **1986**, *55*, 553–561. [CrossRef]
2. Seitz, L.B.; Reyes, A.; Tran, T.T.; Saez de Villarreal, E.; Haff, G.G. Increases in Lower-Body Strength Transfer Positively to Sprint Performance: A Systematic Review with Meta-Analysis. *Sports Med.* **2014**, *44*, 1693–1702. [CrossRef] [PubMed]
3. Alexander, M.J. The relationship between muscle strength and sprint kinematics in elite sprinters. *Can. J. Sport Sci. J. Can. Des Sci. Du Sport* **1989**, *14*, 148–157.
4. Kale, M.; Asci, A.; Bayrak, C.; Acikada, C. Relationships among Jumping performances and sprint parameters during maximum speed phase in sprinters. *J. Strength Cond. Res.* **2009**, *23*, 2272–2279. [CrossRef] [PubMed]
5. Smirniotou, A.; Katsikas, C.; Paradisis, G.; Argeitaki, P.; Zacharogiannis, E.; Tziortzis, S. Strength-power parameters as predictors of sprinting performance. *J. Sports Med. Phys. Fit.* **2008**, *48*, 447–454.
6. Dowson, M.N.; Nevill, M.E.; Lakomy, H.K.; Nevill, A.M.; Hazeldine, R.J. Modelling the relationship between isokinetic muscle strength and sprint running performance. *J. Sports Sci.* **1998**, *16*, 257–265. [CrossRef]
7. Haugen, T.; Seiler, S.; Sandbakk, O.; Tonnessen, E. The Training and Development of Elite Sprint Performance: An Integration of Scientific and Best Practice Literature. *Sports Med. -Open* **2019**, *5*, 1–16. [CrossRef]
8. Young, W.; Mc Lean, B.; Ardagna, J. Relationship between strength qualities and sprinting performance. *J. Sports Med. Phys. Fit.* **1995**, *35*, 13–19.
9. Rabita, G.; Dorel, S.; Slawinski, J.; Saez-de-Villarreal, E.; Couturier, A.; Samozino, P.; Morin, J.B. Sprint mechanics in world-class athletes: A new insight into the limits of human locomotion. *Scand. J. Med. Sci. Sports* **2015**, *25*, 583–594. [CrossRef]
10. Nagano, Y.; Higashihara, A.; Takahashi, K.; Fukubayashi, T. Mechanics of the muscles crossing the hip joint during sprint running. *J. Sports Sci.* **2014**, *32*, 1722–1728. [CrossRef]
11. Fukutani, A.; Misaki, J.; Isaka, T. Both the elongation of attached crossbridges and residual force enhancement contribute to joint torque enhancement by the stretch-shortening cycle. *R. Soc. Open Sci.* **2017**, *4*, 161036. [CrossRef] [PubMed]
12. Buonsenso, A.; Centorbi, M.; Iuliano, E.; Di Martino, G.; Della Valle, C.; Fiorilli, G.; Calcagno, G.; di Cagno, A. A Systematic Review of Flywheel Training Effectiveness and Application on Sport Specific Performances. *Sports* **2023**, *11*, 76. [CrossRef] [PubMed]
13. Franchi, M.V.; Reeves, N.D.; Narici, M.V. Skeletal Muscle Remodeling in Response to Eccentric vs. Concentric Loading: Morphological, Molecular, and Metabolic Adaptations. *Front. Physiol.* **2017**, *8*, 447. [CrossRef]
14. Seger, J.Y.; Thorstensson, A. Effects of eccentric versus concentric training on thigh muscle strength and EMG. *Int. J. Sports Med.* **2005**, *26*, 45–52. [CrossRef] [PubMed]
15. Mike, J.; Kerksick, C.M.; Kravitz, L. How to Incorporate Eccentric Training Into a Resistance Training Program. *Strength Cond. J.* **2015**, *37*, 5–17. [CrossRef]
16. Stasinopoulos, D.; Stasinopoulos, L. Comparison of effects of eccentric training, eccentric-concentric training, and eccentric-concentric training combined with isometric contraction in the treatment of lateral elbow tendinopathy. *J. Hand Ther.* **2017**, *30*, 13–19. [CrossRef]
17. Harris-Love, M.O.; Gollie, J.M.; Keogh, J.W.L. Eccentric Exercise: Adaptations and Applications for Health and Performance. *J. Funct. Morphol. Kinesiol.* **2021**, *6*, 96. [CrossRef]
18. Alt, T.; Severin, J.; Komnik, I.; Nodler, Y.T.; Benker, R.; Knicker, A.J.; Brueggemann, G.-P.; Strueder, H.K. Nordic Hamstring Exercise training induces improved lower-limb swing phase mechanics and sustained strength preservation in sprinters. *Scand. J. Med. Sci. Sports* **2021**, *31*, 826–838. [CrossRef]
19. Sugiura, Y.; Sakuma, K.; Sakuraba, K.; Sato, Y. Prevention of Hamstring Injuries in Collegiate Sprinters. *Orthop. J. Sports Med.* **2017**, *5*, 2325967116681524. [CrossRef]

20. Alt, T.; Nodler, Y.T.; Severin, J.; Knicker, A.J.; Strueder, H.K. Velocity-specific and time-dependent adaptations following a standardized Nordic Hamstring Exercise training. *Scand. J. Med. Sci. Sports* **2018**, *28*, 65–76. [CrossRef]
21. Sugiura, Y.; Saito, T.; Sakuraba, K.; Sakuma, K.; Suzuki, E. Strength deficits identified with concentric action of the hip extensors and eccentric action of the hamstrings predispose to hamstring injury in elite sprinters. *J. Orthop. Sports Phys. Ther.* **2008**, *38*, 457–464. [CrossRef]
22. Flanagan, E.P.; Comyns, T.M. The Use of Contact Time and the Reactive Strength Index to Optimize Fast Stretch-Shortening Cycle Training. *Strength Cond. J.* **2008**, *30*, 32–38. [CrossRef]
23. Douglas, J.; Pearson, S.; Ross, A.; McGuigan, M. Reactive and eccentric strength contribute to stiffness regulation during maximum velocity sprinting in team sport athletes and highly trained sprinters. *J. Sports Sci.* **2020**, *38*, 29–37. [CrossRef]
24. Hortobágyi, T.; Devita, P.; Money, J.; Barrier, J. Effects of standard and eccentric overload strength training in young women. *Med. Sci. Sports Exerc.* **2001**, *33*, 1206–1212. [CrossRef]
25. Gonzalo-Skok, O.; Tous-Fajardo, J.; Valero-Campo, C.; Berzosa, C.; Vanessa Bataller, A.; Luis Arjol-Serrano, J.; Moras, G.; Mendez-Villanueva, A. Eccentric-Overload Training in Team-Sport Functional Performance: Constant Bilateral Vertical Versus Variable Unilateral Multidirectional Movements. *Int. J. Sports Physiol. Perform.* **2017**, *12*, 951–958. [CrossRef]
26. Liu, R.; Liu, J.; Clarke, C.V.; An, R. Effect of eccentric overload training on change of direction speed performance: A systematic review and meta-analysis. *J. Sports Sci.* **2020**, *38*, 2579–2587. [CrossRef]
27. Kingma, J.J.; de Knikker, R.; Wittink, H.M.; Takken, T. Eccentric overload training in patients with chronic Achilles tendinopathy: A systematic review. *Br. J. Sports Med.* **2007**, *41*, 30916. [CrossRef]
28. Schoenfeld, B.J.; Grgic, J. Eccentric Overload Training: A Viable Strategy to Enhance Muscle Hypertrophy? *Strength Cond. J.* **2018**, *40*, 78–81. [CrossRef]
29. Wagle, J.P.; Taber, C.B.; Cunanan, A.J.; Bingham, G.E.; Carroll, K.M.; DeWeese, B.H.; Sato, K.; Stone, M.H. Accentuated Eccentric Loading for Training and Performance: A Review. *Sports Med.* **2017**, *47*, 2473–2495. [CrossRef]
30. Vicens-Bordas, J.; Esteve, E.; Fort-Vanmeerhaeghe, A.; Bandholm, T.; Thorborg, K. Is inertial flywheel resistance training superior to gravity-dependent resistance training in improving muscle strength? A systematic review with meta-analyses. *J. Sci. Med. Sport* **2018**, *21*, 75–83. [CrossRef]
31. de Hoyo, M.; Pozzo, M.; Sanudo, B.; Carrasco, L.; Gonzalo-Skok, O.; Dominguez-Cob, S.; Moran-Camacho, E. Effects of a 10-Week In-Season Eccentric-Overload Training Program on Muscle-Injury Prevention and Performance in Junior Elite Soccer Players. *Int. J. Sports Physiol. Perform.* **2015**, *10*, 46–52. [CrossRef]
32. Murton, J.; Eager, R.; Drury, B. Comparison of flywheel versus traditional resistance training in elite academy male Rugby union players. *Res. Sports Med.* **2023**, *31*, 214–227. [CrossRef]
33. Timon, R.; Ponce-Gonzalez, J.G.; Gonzalez-Montesinos, J.L.; Olcina, G.; Perez-Perez, A.; Castro-Pinero, J. Inertial flywheel resistance training and muscle oxygen saturation. *J. Sports Med. Phys. Fit.* **2018**, *58*, 1618–1624. [CrossRef]
34. Friedmann-Bette, B.; Bauer, T.; Kinscherf, R.; Vorwald, S.; Klute, K.; Bischoff, D.; Mueller, H.; Weber, M.-A.; Metz, J.; Kauczor, H.-U.; et al. Effects of strength training with eccentric overload on muscle adaptation in male athletes. *Eur. J. Appl. Physiol.* **2010**, *108*, 821–836. [CrossRef]
35. Beato, M.; Dello Iacono, A. Implementing Flywheel (Isoinertial) Exercise in Strength Training: Current Evidence, Practical Recommendations, and Future Directions. *Front. Physiol.* **2020**, *11*, 569. [CrossRef]
36. Beato, M.; McErlain-Naylor, S.A.; Halperin, I.; Dello Iacono, A. Current Evidence and Practical Applications of Flywheel Eccentric Overload Exercises as Postactivation Potentiation Protocols: A Brief Review. *Int. J. Sports Physiol. Perform.* **2020**, *15*, 154–161. [CrossRef]
37. Carter, C.R. Kinetic, Kinematic, and Metabolic Response of Low Volume Accentuated Eccentric Loading in Collegiate Weightlifters. Ph.D. Thesis, East Tennessee State University, Johnson City, TN, USA, 2013.
38. Walker, S.; Blazevich, A.J.; Haff, G.; Tufano, J.J.; Newton, R.U.; Hakkinen, K. Greater Strength Gains after Training with Accentuated Eccentric than Traditional Isoinertial Loads in Already Strength-Trained Men. *Front. Physiol.* **2016**, *7*, 149. [CrossRef]
39. Wagle, J.P.; Taber, C.B.; Carroll, K.M.; Cunanan, A.J.; Sams, M.L.; Wetmore, A.; Bingham, G.E.; DeWeese, B.H.; Sato, K.; Stuart, C.A.; et al. Repetition-to-Repetition Differences Using Cluster and Accentuated Eccentric Loading in the Back Squat. *Sports* **2018**, *6*, 59. [CrossRef]
40. Suchomel, T.J.; Wagle, J.P.; Douglas, J.; Taber, C.B.; Harden, M.; Haff, G.G.; Stone, M.H. Implementing Eccentric Resistance Training-Part 1: A Brief Review of Existing Methods. *J. Funct. Morphol. Kinesiol.* **2019**, *4*, 38. [CrossRef]
41. Carroll, K.M.; Wagle, J.P.; Sato, K.; Taber, C.B.; Yoshida, N.; Bingham, G.E.; Stone, M.H. Characterising overload in inertial flywheel devices for use in exercise training. *Sports Biomech.* **2019**, *18*, 390–401. [CrossRef]
42. Maroto-Izquierdo, S.; Raya-Gonzalez, J.; Hernandez-Davo, J.L.; Beato, M. Load Quantification and Testing Using Flywheel Devices in Sports. *Front. Physiol.* **2021**, *12*, 739399. [CrossRef]
43. Berg, H.E.; Tesch, P.A. Force and power characteristics of a resistive exercise device for use in space. *Acta Astronaut.* **1998**, *42*, 219–230. [CrossRef] [PubMed]
44. Janicijevic, D.; García-Ramos, A.; Lamas-Cepero, J.L.; García-Pinillos, F.; Marcos-Blanco, A.; Rojas, F.J.; Weakley, J.; Perez-Castilla, A. Comparison of the two most commonly used gold-standard velocity monitoring devices (GymAware and T-Force) to assess lifting velocity during the free-weight barbell back squat exercise. *Proc. Inst. Mech. Eng. Part P J. Sports Eng. Technol.* **2023**, *237*, 205–212. [CrossRef]

45. Weakley, J.; Mann, B.; Banyard, H.; McLaren, H.S.; Scott, T.; Garcia-Ramos, A. Velocity-Based Training: From Theory to Application. *Strength Cond. J.* **2021**, *43*, 31–49. [CrossRef]
46. Balsalobre-Fernandez, C.; Torres-Ronda, L. The Implementation of Velocity-Based Training Paradigm for Team Sports: Framework, Technologies, Practical Recommendations and Challenges. *Sports* **2021**, *9*, 47. [CrossRef] [PubMed]
47. Brien, J.O.; Browne, D.; Earls, D.; Lodge, C. The effects of varying inertial loadings on power variables in the flywheel romanian deadlift exercise. *Biol. Sport* **2022**, *39*, 499–503. [CrossRef] [PubMed]
48. Maroto-Izquierdo, S.; Garcia-Lopez, D.; de Paz, J.A. Functional and Muscle-Size Effects of Flywheel Resistance Training with Eccentric-Overload in Professional Handball Players. *J. Hum. Kinet.* **2017**, *60*, 133–143. [CrossRef]
49. Maroto-Izquierdo, S.; Martin-Rivera, F.; Nosaka, K.; Beato, M.; Gonzalez-Gallego, J.; de Paz, J.A. Effects of submaximal and supramaximal accentuated eccentric loading on mass and function. *Front. Physiol.* **2023**, *14*, 1176835. [CrossRef]
50. Douglas, J.; Pearson, S.; Ross, A.; McGuigan, M. Effects of Accentuated Eccentric Loading on Muscle Properties, Strength, Power, and Speed in Resistance-Trained Rugby Players. *J Strength Cond. Res* **2018**, *32*, 2750–2761. [CrossRef]
51. Loturco, I.; Pereira, L.A.; Kobal, R.; Kitamura, K.; Abad, C.C.C.; Marques, G.; Guerriero, A.; Moraes, J.E.; Nakamura, F.Y. Validity and usability of a new system for measuring and monitoring variations in vertical jump performance. *J. Strength Cond. Res.* **2017**, *31*, 2579–2585. [CrossRef]
52. Tufano, J.J.; Conlon, J.A.; Nimphius, S.; Brown, L.E.; Seitz, L.B.; Williamson, B.D.; Haff, G.G. Maintenance of Velocity and Power With Cluster Sets During High-Volume Back Squats. *Int. J. Sports Physiol. Perform.* **2016**, *11*, 885–892. [CrossRef] [PubMed]
53. Cohen, J. *Statistical Power Analysis for the Behavioral Sciences*; Academic Press: Cambridge, MA, USA, 2013.
54. Beato, M.; de Keijzer, K.L.; Munoz-Lopez, A.; Raya-Gonzalez, J.; Pozzo, M.; Alkner, B.A.; Dello Iacono, A.; Vicens-Bordas, J.; Coratella, G.; Maroto-Izquierdo, S.; et al. Current Guidelines for the Implementation of Flywheel Resistance Training Technology in Sports: A Consensus Statement. *Sports Med.* **2024**, *54*, 541–556. [CrossRef] [PubMed]
55. Coratella, G.; Beato, M.; Ce, E.; Scurati, R.; Milanese, C.; Schena, F.; Esposito, F. Effects of in-season enhanced negative work-based vs traditional weight training on change of direction and hamstrings-to-quadriceps ratio in soccer players. *Biol. Sport* **2019**, *36*, 241–248. [CrossRef]
56. Maroto-Izquierdo, S.; Fernandez-Gonzalo, R.; Magdi, H.R.; Manzano-Rodriguez, S.; Gonzalez-Gallego, J.; de Paz, J.A. Comparison of the musculoskeletal effects of different iso-inertial resistance training modalities: Flywheel vs. electric-motor. *Eur. J. Sport Sci.* **2019**, *19*, 1184–1194. [CrossRef]
57. Dietz, V.; Schmidtbleicher, D.; Noth, J. Neuronal mechanisms of human locomotion. *J. Neurophysiol.* **1979**, *42*, 1212–1222. [CrossRef]
58. Madruga-Parera, M.; Bishop, C.; Fort-Vanmeerhaeghe, A.; Beato, M.; Gonzalo-Skok, O.; Romero-Rodriguez, D. Effects of 8 Weeks of Isoinertial vs. Cable-Resistance Training on Motor Skills Performance and Interlimb Asymmetries. *J. Strength Cond. Res.* **2022**, *36*, 1200–1208. [CrossRef] [PubMed]
59. Puustinen, J.; Venojarvi, M.; Haverinen, M.; Lundberg, T.R. Effects of Flywheel vs. Traditional Resistance Training on Neuromuscular Performance of Elite Ice Hockey Players. *J. Strength Cond. Res.* **2023**, *37*, 136–140. [CrossRef]
60. Stojanović, M.D.M.; Mikić, M.; Drid, P.; Calleja-González, J.; Maksimović, N.; Belegišanin, B.; Sekulović, V. Greater Power but Not Strength Gains Using Flywheel Versus Equivolumed Traditional Strength Training in Junior Basketball Players. *Int. J. Environ. Res. Public Health* **2021**, *18*, 1181. [CrossRef]
61. Munger, C.N.; Jones, B.C.; Halloran, I.J.; Eggleston, G.G.; Post, P.G.; Brown, L.E.; Berning, J.M. Short-Term Effects of Eccentric Overload Versus Traditional Back Squat Training on Strength and Power. *Int. J. Kinesiol. Sports Sci.* **2022**, *10*, 1–8. [CrossRef]
62. Shimizu, T.; Tsuchiya, Y.; Ueda, H.; Izumi, S.; Ochi, E. Effects of single and repeated bouts of flywheel exercise on jump performance and muscle damage in athletes and non athletes. *Bmc Sports Sci. Med. Rehabil.* **2024**, *16*, 9. [CrossRef]
63. Arsenis, S.; Gioftsidou, A.; Smilios, I.; Malliou, P.; Chatzinikolaou, A.O.; Gkogkos, D.; Beneka, A. The effects of flywheel training on the strength and power of lower limbs. *Sport Sci.* **2019**, *12*, 8–19.
64. Walker, S.; Trezise, J.; Haff, G.G.; Newton, R.U.; Hakkinen, K.; Blazevich, A.J. Increased fascicle length but not patellar tendon stiffness after accentuated eccentric-load strength training in already-trained men. *Eur. J. Appl. Physiol.* **2020**, *120*, 2371–2382. [CrossRef] [PubMed]
65. Merrigan, J.J.; Tufano, J.J.; Falzone, M.; Jones, M.T. Effectiveness of Accentuated Eccentric Loading: Contingent on Concentric Load. *Int. J. Sports Physiol. Perform.* **2021**, *16*, 66–72. [CrossRef] [PubMed]
66. Pecci, J.; Munoz-Lopez, A.; Jones, P.A.; Sanudo, B. Effects of 6 weeks in-season flywheel squat resistance training on strength, vertical jump, change of direction and sprint performance in professional female soccer players. *Biol. Sport* **2023**, *40*, 521–529. [CrossRef] [PubMed]
67. Su, E.Y.-S.; Carroll, T.J.; Farris, D.J.; Lichtwark, G.A. Musculoskeletal simulations to examine the effects of accentuated eccentric loading (AEL) on jump height. *Peerj* **2023**, *11*, 14687. [CrossRef]
68. Kozinc, Z.; Plesa, J.; Sarabon, N. Questionable Utility of the Eccentric Utilization Ratio in Relation to the Performance of Volleyball Players. *Int. J. Environ. Res. Public Health* **2021**, *18*, 11754. [CrossRef]
69. McGuigan, M.R.; Doyle, T.L.; Newton, M.; Edwards, D.J.; Nimphius, S.; Newton, R.U. Eccentric utilization ratio: Effect of sport and phase of training. *J. Strength Cond. Res.* **2006**, *20*, 992–995. [CrossRef]
70. Weng, Y.; Liu, H.; Ruan, T.; Yang, W.; Wei, H.; Cui, Y.; Ho, I.M.K.; Li, Q. Effects of flywheel resistance training on the running economy of young male well-trained distance runners. *Front. Physiol.* **2022**, *13*, 1060640. [CrossRef]

71. Haugen, T.; McGhie, D.; Ettema, G. Sprint running: From fundamental mechanics to practice-a review. *Eur. J. Appl. Physiol.* **2019**, *119*, 1273–1287. [CrossRef]
72. Petre, H.; Wernstal, F.; Mattsson, C.M. Effects of Flywheel Training on Strength-Related Variables: A Meta-analysis. *Sports Med. Open* **2018**, *4*, 55. [CrossRef]
73. Cabanillas, R.; Serna, J.; Muñoz-Arroyave, V.; Ramos, J.A.E. Effect of eccentric overload through isoinertial technology in basketball players. *Rev. Bras. Cineantropometria Desempenho Hum.* **2020**, *22*, e59831. [CrossRef]
74. Malliaras, P.; Kamal, B.; Nowell, A.; Farley, T.; Dhamu, H.; Simpson, V.; Morrissey, D.; Langberg, H.; Maffulli, N.; Reeves, N.D. Patellar tendon adaptation in relation to load-intensity and contraction type. *J. Biomech.* **2013**, *46*, 1893–1899. [CrossRef]
75. Suchomel, T.J.; Wagle, J.P.; Douglas, J.; Taber, C.B.; Harden, M.; Haff, G.G.; Stone, M.H. Implementing Eccentric Resistance Training-Part 2: Practical Recommendations. *J. Funct. Morphol. Kinesiol.* **2019**, *4*, 55. [CrossRef]
76. Nakai, Y.; Usumoto, Y.; Takeshita, Y. The Effects of Regional Muscle Strength and Mass on Standing Long Jump Performance. *Muscles* **2024**, *3*, 60–70. [CrossRef]
77. Beato, M.; Bigby, A.E.J.; De Keijzer, K.L.; Nakamura, F.Y.; Coratella, G.; McErlain-Naylor, S.A. Post-activation potentiation effect of eccentric overload and traditional weightlifting exercise on jumping and sprinting performance in male athletes. *PLoS ONE* **2019**, *14*, 222466. [CrossRef]
78. de Keijzer, K.L.; McErlain-Naylor, S.A.; Dello Iacono, A.; Beato, M. Effect of Volume on Eccentric Overload-Induced Postactivation Potentiation of Jumps. *Int. J. Sports Physiol. Perform.* **2020**, *15*, 976–981. [CrossRef]
79. Abade, E.; Silva, N.; Ferreira, R.; Baptista, J.; Goncalves, B.; Osorio, S.; Viana, J. Effects of Adding Vertical or Horizontal Force-Vector Exercises to In-season General Strength Training on Jumping and Sprinting Performance of Youth Football Players. *J. Strength Cond. Res.* **2021**, *35*, 2769–2774. [CrossRef]
80. Toien, T.; Haglo, H.P.; Unhjem, R.; Hoff, T.; Wang, E. Maximal strength training: The impact of eccentric overload. *J. Neurophysiol.* **2018**, *120*, 2868–2876. [CrossRef]

life

MDPI

Article

Acute Effects of Transcranial Direct Current Stimulation Combined with High-Load Resistance Exercises on Repetitive Vertical Jump Performance and EEG Characteristics in Healthy Men

Yuping Zhou [1,2,†], Haiting Zhai [3,4,†] and Hongwen Wei [1,5,*]

1 School of Strength and Conditioning Training, Beijing Sport University, Beijing 100084, China; xihuyuping@163.com
2 Department of Public Education, Zhejiang College of Construction, Hangzhou 311231, China
3 School of Basic Sciences for Aviation, Naval Aviation University, Yantai 264001, China; minghuhaiting@163.com
4 School of Sports Coaching, Beijing Sport University, Beijing 100084, China
5 Key Laboratory for Performance Training & Recovery of General Administration of Sport, Beijing Sport University, Beijing 100084, China
* Correspondence: weihw@bsu.edu.cn
† These authors contributed equally to this work.

Abstract: Background: Transcranial direct current stimulation (tDCS) is a non-invasive technique known to enhance athletic performance metrics such as vertical jump and lower limb strength. However, it remains unclear whether combining tDCS with the post-activation effects of high-load resistance training can further improve lower limb performance. Objective: This study investigated the synergistic effects of tDCS and high-load resistance training, using electroencephalography to explore changes in the motor cortex and vertical jump dynamics. Methods: Four experiments were conducted involving 29 participants. Each experiment included tDCS, high-load resistance training, tDCS combined with high-load resistance training, and a control condition. During the tDCS session, participants received 20 min of central stimulation using a Halo Sport 2 headset, while the high-load resistance training session comprised five repetitions of a 90% one-repetition maximum weighted half squat. No intervention was administered in the control group. Electroencephalography tests were conducted before and after each intervention, along with the vertical jump test. Results: The combination of tDCS and high-load resistance training significantly increased jump height ($p < 0.05$) compared to tDCS or high-load resistance training alone. As for electroencephalography power, tDCS combined with high-load resistance training significantly impacted the percentage of α-wave power in the frontal lobe area (F3) of the left hemisphere ($F = 6.33$, $p < 0.05$). In the temporal lobe area (T3) of the left hemisphere, tDCS combined with high-load resistance training showed a significant interaction effect ($F = 6.33$, $p < 0.05$). For β-wave power, tDCS showed a significant main effect in the frontal pole area (Fp1) of the left hemisphere ($F = 17.65$, $p < 0.01$). In the frontal lobe area (F3) of the left hemisphere, tDCS combined with high-load resistance training showed a significant interaction effect ($F = 7.53$, $p < 0.05$). The tDCS combined with high-load resistance training intervention also resulted in higher β-wave power in the parietal lobe area (P4) and the temporal lobe area (T4) ($p < 0.05$). Conclusions: The findings suggest that combining transcranial direct current stimulation (tDCS) and high-load resistance training significantly enhances vertical jump performance compared to either intervention alone. This improvement is associated with changes in the α-wave and β-wave power in specific brain regions, such as the frontal and temporal lobes. Further research is needed to explore the mechanisms and long-term effects of this combined intervention.

Keywords: transcranial direct current stimulation; resistance exercises; repetitive vertical jump; electroencephalography

Citation: Zhou, Y.; Zhai, H.; Wei, H. Acute Effects of Transcranial Direct Current Stimulation Combined with High-Load Resistance Exercises on Repetitive Vertical Jump Performance and EEG Characteristics in Healthy Men. *Life* **2024**, *14*, 1106. https://doi.org/10.3390/life14091106

Academic Editor: Ines Drenjančević

Received: 11 July 2024
Revised: 6 August 2024
Accepted: 22 August 2024
Published: 3 September 2024

1. Introduction

Athletic performance enhancement is driven by diverse factors and a range of training methods designed to foster improvement. However, when advancements stall, some individuals resort to illicit substances to boost brain excitability and overall performance [1]. Consequently, numerous sports scientists are now concentrating on how the central nervous system, particularly the brain, influences athletic performance [2,3]. Given the continuous push to extend physical boundaries in modern competitive sports, emphasizing the importance of training both the body and the brain was crucial. [4]. Exploring the link between neuromodulation techniques and athletic performance, understanding the physiological mechanisms behind performance enhancement, and effectively applying these insights in sports training represented future trends [5].

Transcranial direct current stimulation (tDCS) is a non-invasive technique for brain neuromodulation. It involves applying weak, constant, low-intensity direct currents (1–2 mA) to the scalp to modulate the activity of cortical neurons [6]. tDCS is distinguished by its non-invasive nature, efficiency, user-friendliness, and portability. tDCS has found widespread applications in neurorehabilitation [7,8], the treatment of psychiatric disorders [9,10], skill learning enhancement [11], and memory improvement [12]. Recent studies have also shown that tDCS can effectively enhance athletic performance, improving metrics such as vertical jump, lower limb strength, explosive power, and speed [13–15]. Specifically, the vertical jump is the most commonly used measure in research. tDCS has been shown to positively influence vertical jump performance, enhancing peak power, reducing ground contact time, and increasing jump height [16–18].

Recently, researchers have introduced the portable Halo Sport 2 headphones, which utilize the principles of transcranial direct current stimulation (tDCS). These headphones contain electrodes that deliver electrical currents to the wearer's brain, stimulating the motor cortex to enhance neural pathways and increase the firing rate of neural networks [19]. A systematic review and meta-analysis revealed that by combining transcranial direct current stimulation (tDCS) with resistance training, researchers have further enhanced strength capabilities [20]. However, research into the combined effects of tDCS and strength training remains inconclusive. Some studies suggest that combining tDCS with resistance training can significantly enhance muscle strength and lower limb explosive power [21–24], while others report no significant benefits compared to a sham stimulation combined with resistance training [25–27]. These discrepancies could stem from differences in tDCS parameters such as the current intensity, targeted brain areas, duration of stimulation, and electrode size [28]. Additionally, research has demonstrated significant increases in lower limb power output after acute high-load resistance training ($\geq$80% one-repetition maximum, 1RM), marked by shorter ground contact times, higher jumps, and extended airborne phases during vertical jumps [2]. The synergistic potential of high-load resistance training combined with tDCS to enhance lower limb power has not yet been explored.

If combining high-load resistance training with tDCS can significantly enhance lower limb power output, what are the potential mechanisms behind this effect? There are currently no studies reporting on this. Therefore, this study hypothesizes that the combination of resistance training and tDCS can significantly improve repeated jump performance, and that this effect is related to changes in electroencephalography (EEG) characteristics.

2. Methods

2.1. Participants

Participants were selected based on the following inclusion criteria: (1) Experience in resistance training, with lower extremity strength training at least 3 times/week; (2) male; (3) aged 19–23; and (4) proficiency in large-load half squats and continuous vertical jumps. Additionally, participants were required to have no history of lower extremity or brain injuries or psychiatric illnesses within the last 3 months. Exclusion criteria included individuals with: (1) skin diseases or allergies, (2) fractures or joint injuries in the past six months, (3) no strength training experience, and (4) psychiatric conditions or brain injuries. Thirty-four volunteers initially participated in this study, but only 29 met the experimental requirements and were randomly assigned to one of four experimental conditions: control condition (CON), high-load resistance training (HRT) condition, tDCS intervention (tDCS) condition, and the combination of tDCS and high-load resistance training (tDCS + HRT) condition. Each participant experienced all four conditions in a randomized order. Participants information is shown in Table 1.

Table 1. Basic information on participants.

Number of Participants (n)	Age (yr)	Height (cm)	Body Weight (kg)	1RM Back squat/kg
29	21.8 ± 2.13	178.1 ± 6.15	77.1 ± 7.70	2.15 ± 0.20

2.2. Human Ethics and Consent to Participate Declarations

This study strictly adhered to the principles of the Declaration of Helsinki and was approved by the Institutional Review Board (IRB) of Beijing Sport University Ethics Committee (approval number: 2020120H). The review ensured that the research design complied with international and national ethical standards. Prior to the commencement of the experiment, the research team conducted detailed information sessions to ensure all participants fully understood the purpose, expected benefits, potential risks, and their rights within the study. Each participant signed a written informed consent form, which detailed the research procedures, the handling of personal data, and privacy protection measures. To respect participants' rights, this study ensured that all participants were aware they had the right to withdraw their participation at any stage without providing a reason, and that their decision to withdraw would not affect any services or treatments they were receiving or might receive in the future. The research team committed to immediately removing any information related to a participant's withdrawal from the study records, and such data would not be used in the analysis.

This study adheres to CONSORT guidelines for reporting clinical trials.

2.3. Experimental Design

This study employed a single-blind, randomized, crossover design with repeated measures. Each of the 29 participants was randomly assigned to complete four different experimental conditions (CON, HRT, tDCS, and tDCS + HRT) in separate study visits. The order of conditions was counterbalanced to control for potential order effects. The experiments were conducted at the Sports Science Research Centre of Beijing Sport University, China, with each participant visiting the laboratory five times. On the first day, participants had their weight and height measured after emptying their bladders. Subsequently, according to the NSCA testing procedure [29], participants were assessed for half-squat 1RM, performed repeated vertical jumps, and underwent baseline testing. Participants were randomly allocated to complete four tests in a counterbalanced order, with each set scheduled 5 days apart and conducted at the same time of day to control for circadian rhythm influences. Each participant followed a warm-up regimen in every test, consisting of 4 min of jogging, 4 min of dynamic stretching, a 20 m sprint, practice of the vertical jump technique, and 2 min of slow walking. All groups followed a predetermined experimental protocol, as illustrated in Figure 1. Participants in the CON group underwent an EEG test, a warm-up, another EEG test during an 8 min rest, and then a continuous vertical jump test. The tDCS group had an EEG test, a 10 min warm-up, tDCS stimulation with the Halo Sport 2 headset, an 8 min rest with an EEG test, and then a vertical jump test. The HRT group did an EEG test, a 10 min warm-up, a 2 min rest, five 90% 1RM half squats with 1 min intervals, an 8 min rest with an EEG test, and then a vertical jump test. The tDCS + HRT group had an EEG test, a warm-up, tDCS stimulation with the Halo Sport 2 headset, a 2 min rest, five 90% 1RM half squats with 1 min intervals, an 8 min rest with an EEG test, and then a vertical jump test. Given previous research [30,31] indicating the optimal post-activation effect window is 4–10 min, and considering the time required for EEG testing, we chose to complete the EEG test and the vertical jump test within an 8 min rest period.

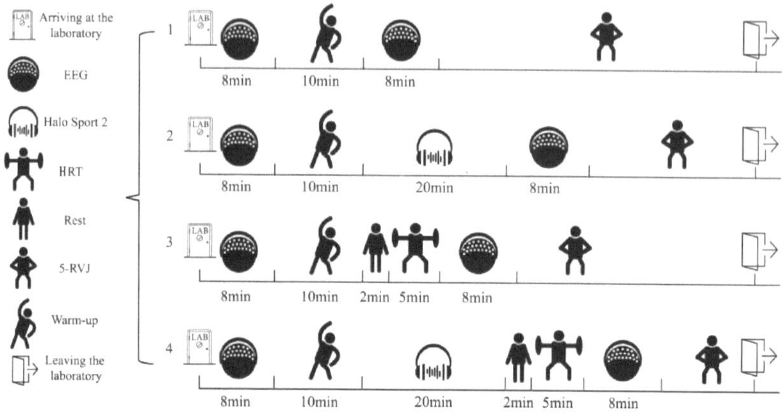

Figure 1. Flow chart of the experiment.

Participants were instructed to refrain from strenuous exercise and avoid alcohol or caffeine 24 h before the experiment. Participants were encouraged to maintain a positive mental state and ensure adequate sleep. The intervention and vertical jump tests were conducted in a controlled environment at an independent training facility to minimize external interference.

2.4. *Intervention*

2.4.1. Transcranial Direct Current Stimulation

The Halo Sport 2 headset (HS002K, Halo Neuroscience, Francisco, CA, USA) (Figure 2) delivers a stimulation current of 2.2 mA, with an output frequency ranging from 0 to 625 Hz over a duration of 20 min. The electrodes were positioned over the motor cortex, and the same stimulation parameters were used for all participants to ensure consistency across the study. Previous research [32–34] supports the safety and efficacy of this stimulation mode. The electrodes extend across the top of the head from one ear to the other, effectively targeting both hemispheres of the motor cortex. Typically, the anode electrodes are positioned over the top of the motor cortex, while the cathode electrodes are placed at other locations on the head, such as the supraorbital area, to complete the circuit. The current intensity is adjustable via the Halo application, accessible on devices such as iPhones or iPads. During the experiment, participants reclined in a relaxed state in a chair with the Halo Sport 2 headset properly positioned on their heads. Over 30 s, the current was gradually increased to 2.0 mA and maintained for 20 min, following protocols from the existing literature demonstrating improved performance in multiple studies.

Figure 2. Halo Sport 2 headset.

Participants had the option to terminate or withdraw from the experiment in case of any discomfort, including head tingling, burning sensations, allergic reactions, significant adverse bodily reactions, nausea, or vomiting.

2.4.2. Lower Extremity High-Load Resistance Training

Following the NSCA testing procedure [29], the participants' half-squat 1RM was measured with the aim of reaching a 1RM value within five attempts. Post-activation potentiation (PAP) is a phenomenon where the force output of a muscle is temporarily enhanced following high-intensity conditioning activity. Performing sets of five reps at 90% of 1RM can induce PAP, which is believed to enhance neural drive to the muscles, increase motor unit recruitment, and improve muscle fiber synchronization [30,31]. This study utilized 90% of 1RM weight for strength induction, effectively enhancing muscle strength and performance in subsequent explosive activities.

Weight-bearing exercises primarily consisted of half squats and deep squats to induce the HRT effect. The weight-bearing half squat was selected for its capacity to increase gluteus maximus muscle force, alleviate knee joint stress, and mitigate the risk of injury [35]. Thus, the weight-bearing half squat was used to induce the HRT effect in this experiment.

2.4.3. Outcome Measurement Tools

Data collection was conducted by an occupational physiology team with over 3 years of experience, capturing experimental intervention and demographic data for all participants. Importantly, the experimenter remained blinded to the group affiliation of the participants.

To accurately reflect changes in reactive strength without excessively increasing participant fatigue and to ensure data reliability and validity, the Repeat the Vertical Jump 5 times (5-RVT) test was chosen as an indicator of lower limb explosive power [36]. Lower limb explosive strength was evaluated through using the Smart Jump apparatus (9281CA, Kistler Instruments, Hook, UK), known for its high reliability [37]. The test parameters included contact time, flight time (FT), height (H), reaction force index (RSI), and peak power output (PPO).

Participants were instructed to maintain crossed arms to prevent arm swing interference, focus on a green signal light for initiating the jump, and aim to achieve a fast, high jump with minimal ground contact time. A total of 11 jumps were performed, with only five jumps being counted, excluding the initial pre-jump, and only records with a touchdown time of less than 0.5 s were considered.

2.4.4. Electroencephalography

EEG signals were recorded using an EEG test system (Nation9128W, Shanghai Nuocheng Company, Shanghai, China). Electrode placement follows the international standard 10–20 system, positioned on the sagittal plane formed by the connection between the root of the nose and the occipital protuberance [38]. Sixteen electrodes were evenly distributed, with reference electrodes on the bilateral earlobes. The electrodes were placed at Fp1, Fp2, F3, F4, C3, C4, P3, P4, O1, O2, F7, F8, T3, T4, T5, and T6. All electrodes were evenly distributed across all skull positions as shown in the Figure 3A,B. Signals were sampled at 512 Hz, filtered (0.5–30 Hz), and cleaned using independent component analysis (ICA) to remove artifacts. The cleaned signals were segmented into 2 s epochs, and power spectral density (PSD) was calculated using fast fourier transform (FFT). Alpha (8–13 Hz) and beta (13–30 Hz) band powers were averaged for each electrode. Power changes were calculated as percentages relative to pre-intervention values.

The primary outcome measures included power changes in the alpha and beta frequency bands. Alpha waves (8–13 Hz) are associated with relaxation and reduced cortical activation, while beta waves (13–30 Hz) are linked to active thinking and focus [38]. Changes in these bands provide insights into cortical excitability and neural efficiency following the interventions. Increased alpha power may indicate enhanced neural synchrony and relaxation, whereas increased beta power suggests improved cognitive processing and motor control. By analyzing these changes, we aimed to understand the neural mechanisms underlying the observed improvements in physical performance.

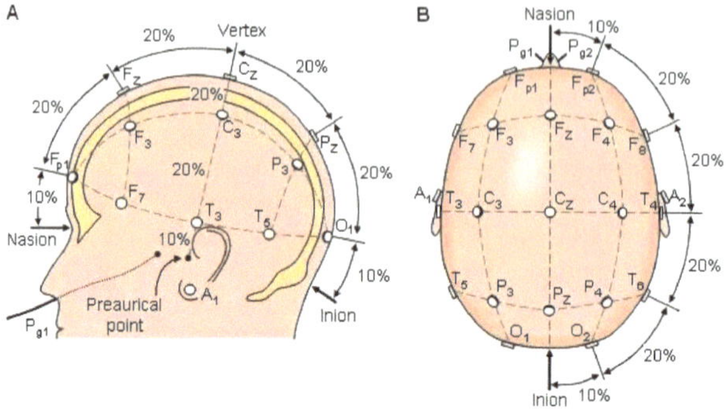

Figure 3. EEG electrode placement diagram. (**A**) Lateral view, (**B**) Top view.

EEG acquisition was conducted in a comfortable, quiet room. Participants were seated in chairs wearing electrode caps and were instructed to remain still.

2.5. Statistical Analysis

Statistical analyses were performed using SPSS (version 20.0, IBM Corp, Armonk, NY, USA). Data were presented as mean ± SD. Paired *t*-tests were used to examine potential differences in basic participant characteristics. A two-way ANOVA (Analysis of Variance) was employed to assess the impact of tDCS and high-load resistance training on continuous repeated vertical jump and EEG signals. The "F" in the results refers to the F-statistic from the ANOVA, which indicates whether there are significant differences between groups. Post hoc comparisons were conducted using the least significant difference method, with a significance level set at $p < 0.05$. Additionally, effect sizes were calculated using Cohen's d ("d"), which measures the standardized difference between two means, providing an indication of the magnitude of the intervention effects.

3. Results

Analysis of the baseline data showed no significant differences between groups regarding demographic characteristics (Table 2). Additionally, no adverse events were reported.

Table 2. Effect of different intervention methods on repeated vertical jump.

	CON	tDCS	HRT	tDCS + HRT
CT (s)	0.5 ± 0.03 ▲	0.4 ± 0.13 *▲	0.4 ± 0.32 ▲	0.3 ± 0.24 *
FT (s)	0.5 ± 0.04	0.5 ± 0.02	0.5 ± 0.03	0.5 ± 0.05
JH (cm)	36.5 ± 3.04 ▲	39.6 ± 3.19 *▲	40.1 ± 3.03 *▲	42.7 ± 3.01 *
RSI	2.0 ± 0.07	2.2 ± 0.07	2.3 ± 0.05	2.4 ± 0.03
PPO (w/kg)	41.5 ± 4.01	45.0 ± 3.36 *	41.6 ± 3.57	45.8 ± 3.53 *

*: significant difference compared to CON ($p < 0.05$). ▲: significant difference compared to tDCS + HRT ($p < 0.05$). Contact time: CT; flight time: FT; jump height: JH; reaction force index: RSI; peak power output: PPO.

3.1. Outcome Measures

3.1.1. Repeat Vertical Jumps

Contact Time (CT)

A significant interaction effect was observed for the tDCS + HRT intervention (F = 5.33, $p < 0.05$). Significant simple effects were found for both the tDCS (F = 14.90, $p < 0.001$) and HRT (F = 4.90, $p < 0.05$) interventions. Post-intervention, the CT values for the tDCS + HRT condition (0.3 ± 0.24 s) were significantly lower compared to the tDCS (0.4 ± 0.13 s) and HRT (0.4 ± 0.32 s) interventions alone, as well as the CON (0.5 ± 0.03 s) ($p < 0.05$). The

Cohen's d indicated that the tDCS + HRT intervention had a larger effect size (d = −1.17) compared to the tDCS (d = −0.52) and HRT (d = −0.35) interventions. The HRT intervention also had a moderate effect size compared to the CON (d = −0.44).

Flight Time (FT)

The tDCS + HRT intervention showed a significant interaction effect (F = 6.33, $p < 0.05$). Significant simple effects were observed for both the tDCS (F = 15.10, $p < 0.001$) and HRT (F = 6.81, $p < 0.05$) interventions. FT values remained consistent across all conditions: CON (0.5 ± 0.04 s), tDCS (0.5 ± 0.02 s), HRT (0.5 ± 0.03 s), and tDCS + HRT (0.5 ± 0.05 s). The SRM for FT showed no significant differences across the interventions (all d ≈ 0).

Jump Height (JH)

The analysis revealed a significant interaction effect for the tDCS + HRT intervention (F = 17.12, $p < 0.001$). Significant simple effects were found for both the tDCS (F = 16.85, $p < 0.001$) and HRT (F = 4.78, $p < 0.05$) interventions. Post-intervention, JH values were significantly higher for the tDCS + HRT condition (42.7 ± 3.01 cm) compared to the tDCS (39.6 ± 3.19 cm) and HRT (40.1 ± 3.03) interventions alone, as well as the control group (36.5 ± 3.04 cm) ($p < 0.05$). The SRM indicated a larger effect size for tDCS + HRT (d = 2.05) compared to tDCS (d = 1.0) and HRT (d = 0.86). The HRT intervention also showed a significant effect size compared to the CON (d = 1.18).

Reactive Strength Index (RSI)

The tDCS + HRT intervention demonstrated a significant interaction effect (F = 9.33, $p < 0.001$). Significant simple effects were found for both the tDCS (F = 20.30, $p < 0.001$) and HRT (F = 8.33, $p < 0.01$) interventions. RSI values were highest in thetDCS + HRTgroup (2.4 ± 0.03), followed by HRT (2.3 ± 0.05), tDCS (2.2 ± 0.07), and control (2.0 ± 0.07). The SRM showed a significantly larger effect for tDCS + HRT (d = 7.41) compared to tDCS (d = 2.86) and HRT (d = 4.21).

Peak Power Output (PPO)

A significant interaction effect was identified for the tDCS + HRT intervention (F = 13.01, $p < 0.001$). Significant simple effects were observed for both the tDCS (F = 10.84, $p < 0.001$) and HRT (F = 4.66, $p < 0.05$) interventions. PPO values were higher in the tDCS + HRT group (45.8 ± 3.53 w/kg) compared to the tDCS (45.0 ± 3.36 w/kg), HRT (41.6 ± 3.57 w/kg), and control (41.5 ± 4.01 w/kg) groups. The SRM indicated a larger effect for tDCS + HRT (d = 1.14) compared to tDCS (d = 0.91) and HRT (d = 0.05).

3.1.2. Electroencephalogram Characteristics Percentage of α-Wave EEG Power

An interaction effect of the tDCS + HRT intervention was found to significantly impact the percentage of α-wave EEG power in the frontal lobe area (Fp1) of the left hemisphere (F = 6.33, $p < 0.05$). The Cohen's d indicated that the tDCS +HRT intervention had a larger effect size (d = 3.33) compared to the tDCS (d = 1.70) and HRT (d = 0.78) interventions. Additionally, a significant simple effect of the tDCS intervention was observed (F = 14.90, $p < 0.001$), while the HRT intervention did not show a significant simple effect (F = 0.231, $p > 0.05$). Similar results were seen in the temporal lobe area (T3) of the left hemisphere, with a significant interaction effect of the tDCS + HRT intervention (F = 6.33, $p < 0.05$). The SRM for the tDCS + HRT intervention was 5.50, indicating a much larger effect size compared to the tDCS (d = 4.00) and HRT (d = 0.39) interventions. A significant simple effect of the tDCS intervention (F = 14.71, $p < 0.001$) was observed, and a nonsignificant simple effect of the HRT intervention (F = 0.24, $p > 0.05$). More detailed information is shown in Figure 4.

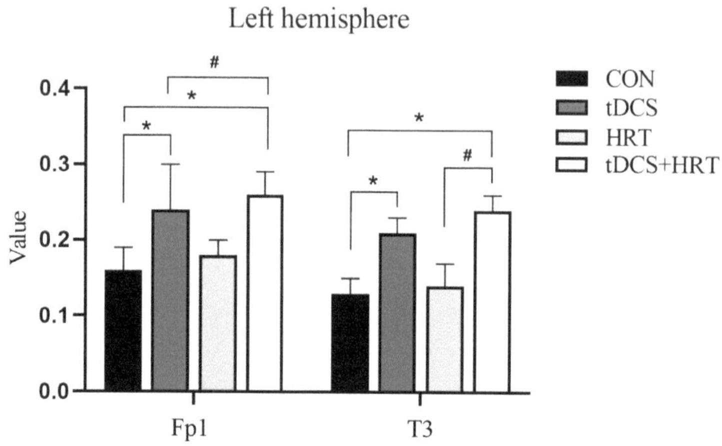

Figure 4. α-wave EEG power. *: Indicates a significant difference compared to CON. #: Indicates a significant difference compared to tDCS + HRT.

3.1.3. Percentage of β-Wave EEG Power

The study found a non-significant interaction effect of the tDCS + HRT intervention on the percentage of β-wave EEG power in the frontal region (Fp1) of the left hemisphere (F = 0.35, $p > 0.05$). However, a significant main effect of the tDCS intervention was observed (F = 17.65, $p < 0.01$), with the Cohen's d indicating a larger effect size for tDCS (d = 9.74) compared to the control group (CON).

Additionally, a significant interaction effect of the tDCS + HRT intervention was found on the percentage of β-wave EEG power in the frontal region (F3) of the left hemisphere (F = 7.53, $p < 0.05$), along with significant simple effects for both the tDCS (F = 11.10, $p < 0.001$) and HRT (F = 5.23, $p < 0.05$) interventions. The SRM showed that the tDCS + HRT intervention had a larger effect size (d = 8.72) compared to both the tDCS (d = 9.51) and HRT (d = 7.04) interventions.

Moreover, the tDCS + HRT intervention resulted in significantly higher levels compared to the tDCS and HRT interventions ($p < 0.05$), while the tDCS intervention showed significantly higher levels than the CON ($p < 0.05$). Furthermore, the percentage of β-wave EEG power in the right hemisphere parietal region (P4) and temporal region (T4) was significantly higher with the tDCS + HRT intervention compared to CON ($p < 0.05$). More detailed information is shown in Figures 5 and 6.

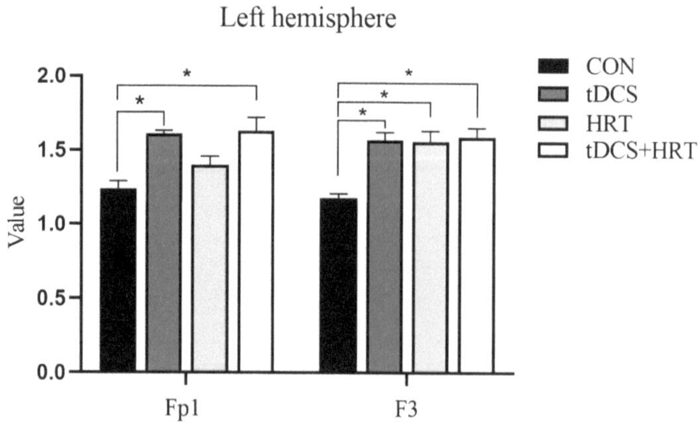

Figure 5. Left hemisphere β-wave EEG power. *: Indicates a significant difference compared to CON.

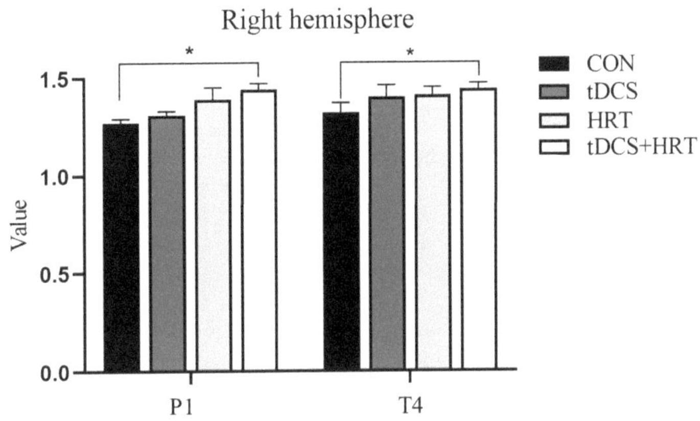

Figure 6. Right hemisphere β-wave EEG power. *: Indicates a significant difference compared to CON.

4. Discussion

4.1. Effects of Different Interventions on Repetition Vertical Jump

This study innovatively employed the portable and easy-to-wear Halo Sport 2 headset, incorporating settings commonly associated with positive results in prior research (stimulation current of 2.2 mA, output frequency ranging from 0 to 625 Hz, and duration of 20 min). Unlike most previous studies that focused solely on the effects of single-session tDCS on athletic performance, our research combined tDCS with resistance training and conducted comparative analyses to explore their combined effects. Notably, our study's innovation lies in integrating tDCS with the post-activation potentiation effect generated by high-load resistance training and investigating its physiological mechanisms through electroencephalography (EEG) changes. This approach not only offers a new perspective on the combination of tDCS and athletic performance but also unveils potential physiological mechanisms, providing a unique and forward-looking contribution to the field.

In this crossover study, we individually investigated the effects of tDCS, HRT, and tDCS + HRT on the 5-RVT and compared these effects within the same participants under control conditions. The results of the experiment demonstrated a significant interaction between HRT and tDCS interventions across various indices such as contact time (CT), flight time (FT), height (H), reactive strength index (RSI), and peak power output (PPO) during the repeated vertical jump. Additionally, both tDCS and HRT showed significant simple effects. These findings suggest that the combination of HRT and tDCS leads to a significant improvement in 5-RVT performance. Consistent with our study findings, Lattari's research involved 10 participants in a randomized, double-blind study under three conditions [39]. Although Lattari et al. [39] conducted countermovement jumps rather than repeated vertical jumps, their analysis indicated significant increases in jump height, flight time, and peak muscular power following stimulation in the anodal condition. Similarly, Grosprêtre et al. [40] showed that an extracephalic anodal montage (anodal at M1, cathodal on the contralateral shoulder) significantly enhanced jump performance, accompanied by increased excitability at both supraspinal and spinal levels.

However, some studies, such as those by Romero-Arenas et al. [41] and Park et al. [42] did not observe any improvements in jump performance due to tDCS, possibly due to factors such as the duration of effects, equipment functionality, stimulation sites, and the training levels of the participants. For instance, Marinus et al. [43] conducted a systematic review and found that the influence of a single tDCS session on physical fitness varies significantly based on these factors. Similarly, Chinzara and Harris [20], in their meta-analysis, highlighted the mixed results of tDCS on physical endurance, muscular strength, and visuomotor skills, emphasizing the need for standardized protocols. Hu et al. [44] also noted, in their systematic review and meta-analysis, the varying effects of tDCS on upper limb muscle strength and endurance, suggesting that differences in protocols could explain these inconsistencies. Furthermore, Savoury et al. [45] discussed methodological issues in enhancing muscle strength and endurance with tDCS, underscoring the importance of consistent and well-controlled experimental designs.

Our study builds on these previous works by incorporating a combined intervention of tDCS and HRT, which has not been extensively explored before. We improved upon existing methodologies by carefully controlling variables such as electrode placement, stimulation intensity, and session duration, as recommended by prior studies. This approach allowed us to better isolate the effects of the combined interventions and understand their synergistic impact on lower limb explosive power.

Future research should investigate the long-term impact of tDCS on jumping performance and offer a more comprehensive analysis of lower limb dynamics in vertical jumping. Additionally, exploring the effects of varying tDCS parameters and their interactions with different training protocols could provide deeper insights into optimizing athletic performance enhancements.

4.2. Physiological Rationale behind Improved Performance

The rationale behind these results can be suggested from a physiological perspective. The data presented in Table 2 revealed a significant increase in vertical jump height following HRT, which may suggest the recruitment of more muscle fibers and enhanced activation of fast-twitch muscle fibers. It was also possible that the activation of muscle contraction was accelerated, leading to increased muscle output after engaging in half-squat exercises with heavy loads. However, these hypotheses require further investigation as we did not directly measure excitability, nerve impulse frequency, or muscle fiber recruitment. Furthermore, the results from Table 2 indicated that RSI and PPO were maximized post-tDCS intervention. These parameters were closely linked to the functioning of the nervous system, suggesting that tDCS might enhance neuromuscular efficiency. Again, further studies are needed to confirm these findings and to explore the underlying mechanisms more thoroughly. Previous research has highlighted that tDCS could expedite the regulation of spinal cord nerves, elevate the speed of muscle contraction, and enhance the coordinated control of lower limb muscles [46,47]. Stimulation of the motor cortex through tDCS augments motor neuron activity and the excitation of tendon organs, resulting in enhanced afferent impulses during muscle contraction. Overall, the enhancements observed in all 5-RVT indices following tDCS combined with HRT intervention signify increased nervous system excitation, frequency and intensity of nerve impulses, enhanced muscle fiber recruitment, heightened activation, and accelerated contraction rate, ultimately leading to an improved 5-RVT performance.

Although the tDCS Halo Sport 2 headset, developed based on the principles, has been shown to positively influence sports performance in sports science studies, there are scholarly inconsistencies. For example, Fortes et al. suggested that tDCS does not improve power and that its biggest influence is the individual's training experience and level [48]. Individuals with no training experience had a more significant effect after being stimulated with tDCS than those with training experience, possibly due to the fact that neuromuscular function is more optimized (e.g., peak power and volume of training) in participants with training experience [49]. Lerner et al. [50] have noted that tDCS produces 'noise' in the motor cortex of the brain when stimulated, thereby attenuating motor performance. One study found that anodal tDCS stimulation of the M1 region decreased participants' pain perception after stimulation of the M1 region, as observed by international 10–20-lead EEG, and helped athletes to reduce fatigue during both endurance and high-load resistance training [51].

The level of activation in the motor cortex may have reflected reflect the underlying mechanisms of enhanced motor performance. Factors influencing brain wave frequency include the duration of neuronal circuit activity, the number of synapses transmitting nerve signals, and the metabolic rate of neuronal substances [52]. An increase in the alpha frequency band suggests that the firing rate of the central nervous system is becoming more coordinated and synchronized, indicating improved stability in EEG changes [53,54]. The study results demonstrated a significant rise in alpha band power following both tDCS + HRT and tDCS interventions, suggesting that tDCS interventions could enhance the organization of the central nervous system in terms of synchronization, desynchronization, and modulation between inhibition and excitation. Furthermore, the 5-RVT results revealed significant increases in CT, RSI, and PPO indices after tDCS and tDCS + HRT interventions, all of which were strongly correlated with nervous system function. This could explain how tDCS led to improved 5-RVT performance.

β-waves played a crucial role in indicating neural activity. β-waves are linked to the level of nerve cell activation in the cerebral cortex [55]. Proper activation suggested the brain was appropriately aroused, while excessive activation may indicate tension [56]. The study's findings revealed that the percentage of power in the β-frequency band after the HRT and tDCS + HRT interventions was significantly higher than that of CON ($p < 0.05$) for each, and the β-frequency band after the tDCS + HRT intervention was higher than that of tDCS ($p < 0.05$). These changes could be attributed to both intervention protocols involving a demanding load resistance exercise (90% 1RM half squat). Research has highlighted that β-wave alterations are particularly responsive to exercise intensity [57].

4.3. Cognitive Functions and Brain Synergy

The prefrontal lobe, which makes up 25% of the entire cerebral hemisphere area, plays a crucial role in higher cognitive functions such as attention, memory, problem solving, and personality development [58–62]. Engaging in cognitive activities has been shown to enhance brain organization [63]. Results from this study indicated that the primary brain areas with the highest EEG power percentage in the α-band were the frontal area (Fp1) and lobe area (T3) in the left cerebral hemisphere. Changes in the β-waves were observed in the central area of the left cerebral hemisphere (C3) and the frontal area (Fp2). Despite variations in left–right brain synergy following stimulation, with a tendency towards dominance in the left side, the frontal regions of both hemispheres exhibited more stable synergy under specific conditions. Brain synergy is a critical mechanism for assessing brain activity, and all human cognitive processes are essentially organized brain functions under the influence of synergy [64]. The Halo Sport 2 delivers stimuli to the M1 area of the primary motor cortex via electrodes integrated into the headset, targeting the primary active brain area identified in this study. The M1 area is responsible for generating nerve impulses that travel down the spinal cord to control human movement execution and enhance cortical excitation [38]. Therefore, the enhanced repeated vertical jump performance observed after tDCS and tDCS + HRT may be linked to changes in EEG frequency bands. However, improvements in motor performance are multifaceted and can also be influenced by individual participants' central cerebral state, emotions, training level, and other factors [65]. The results of this study indicate that combining HRT with tDCS intervention can improve central nervous system function.

5. Conclusions

The study found that tDCS, HRT, and the combination of both significantly improved continuous repeated vertical jump performance. Notably, the combined intervention of tDCS and HRT was found to be the most effective. Specifically, this combined intervention outperformed transcranial direct current stimulation or resistance exercise alone in enhancing repeated vertical jump performance. The positive effects observed were correlated with the impact of transcranial direct current stimulation on brain electrical activity in various regions, resulting in enhanced alpha and beta waves.

5.1. Limitations and Future Directions

In our study, although we found that the combination of tDCS and resistance training significantly enhanced lower limb explosive power, these results may be influenced by the lack of blinding. Therefore, future studies should adopt a double-blind design to enhance the validity of the findings. This can be achieved by using a sham stimulation, where the tDCS device is applied but no current is delivered, ensuring that participants cannot distinguish between the active and sham conditions. Additionally, assessing the success of the blinding process by asking participants whether they believed they received the active or sham treatment can provide further insights into the potential influence of placebo effects.

5.2. Future Applications

The potential of combining transcranial direct current stimulation (tDCS) with high-load resistance training (HRT) for sports training is significant. Future research should focus on how to effectively integrate this combined approach into the training cycles of high-level athletes to optimize their performance. Additionally, it is important to explore the impact of this training method on athlete fatigue. By personalizing these interventions and considering factors such as age, training phase, and specific athletic needs, training outcomes can be further enhanced, providing athletes with scientifically validated and effective training programs.

Author Contributions: Y.Z. and H.W. contributed to the design; Y.Z. and H.W. participated in most of the study steps; H.Z. and Y.Z. prepared the manuscript; H.Z. and Y.Z. assisted in designing the study and helped in the interpretation of the study. All authors have read and agreed to the published version of the manuscript.

Funding: This research was supported by the Special Funds for Basic Scientific Research Services of Chinese Central Universities (2024TNJN003) and the Athletic Sports Specific Program of the China Administration of Sports for Persons with Disabilities (2023-JT&014).

Institutional Review Board Statement: Ethical approval for this study was obtained from Experimental Ethics Committee of Sports Science, Beijing Sport University (approval number: 2020120H, 16 December 2020).

Informed Consent Statement: All participants provided informed consent to participate in the study.

Data Availability Statement: Datasets are available through the corresponding author upon reasonable request.

Acknowledgments: We are grateful to the Sports Science Research Center at Beijing Sport University for providing the necessary facilities and environment that greatly facilitated our research. Furthermore, we acknowledge the collective efforts of all the authors whose contributions have been crucial in bringing this study to fruition.

Conflicts of Interest: The authors declare that the research was conducted in the absence of any commercial or financial relationships that could be construed as a potential conflict of interest.

Abbreviations

1RM	One Repetition Maximum
5-RVT	Repeat the Vertical Jump 5 Times
C3,4	Central Region 3,4
CON	Control Condition
CT	Contact Time
EEG	Electroencephalographic
F7,8	Frontal Electrode 7,8
Fp1,2	Frontal Lobe 1,2
FT	Flight Time
H	Height

NSCA	National Strength and Conditioning Association
O1,2	Occipital Lobe Region 1,2
P3,4	Parietal Region 3,4
PPO	Peak Power Output
RSI	Reaction Force Index
HRT	High-load Resistance Training
T4,7	Temporal Lobe Region 4,7
tDCS	Transcranial Direct Current Stimulation

References

1. Schubert, M.M.; Astorino, T.A. A systematic review of the efficacy of ergogenic aids for improving running performance. *J. Strength Cond. Res.* **2013**, *27*, 1699–1707. [CrossRef] [PubMed]
2. Noakes, D.T. Fatigue is a brain-derived emotion that regulates the exercise behavior to ensure the protection of whole body homeostasis. *Front. Physiol.* **2012**, *3*, 82. [CrossRef]
3. David, T.N. Time to move beyond a brainless exercise physiology: The evidence for complex regulation of human exercise performance. *Appl. Physiol. Nutr. Metab.=Physiol. Appl. Nutr. Et. Metab.* **2011**, *36*, 23–35.
4. Campbell, M.J.; Toth, A.J.; Moran, A.P.; Kowal, M.; Exton, C. eSports: A new window on neurocognitive expertise? *Prog. Brain Res.* **2018**, *240*, 161–174. [CrossRef] [PubMed]
5. Machado, S.; Travassos, B.; Teixeira, D.S.; Rodrigues, F.; Cid, L.; Monteiro, D. Could tDCS Be a Potential Performance-Enhancing Tool for Acute Neurocognitive Modulation in eSports? A Perspective Review. *Int. J. Environ. Res. Public Health* **2021**, *18*, 3678. [CrossRef] [PubMed]
6. Huang, L.; Deng, Y.; Zheng, X.; Liu, Y. Transcranial Direct Current Stimulation With Halo Sport Enhances Repeated Sprint Cycling and Cognitive Performance. *Front. Physiol.* **2019**, *14*, 118. [CrossRef] [PubMed]
7. Jitka, V.; Alireza, G. Non-invasive brain stimulation for improving gait, balance, and lower limbs motor function in stroke. *J. Neuro Eng. Rehabil.* **2022**, *19*, 84. [CrossRef]
8. Hartwigsen, G. Adaptive Plasticity in the Healthy Language Network: Implications for Language Recovery after Stroke. *Neural Plast.* **2016**, *2016*, 9674790. [CrossRef]
9. Brunelin, J.; Mondino, M.; Bation, R.; Palm, U.; Saoud, M.; Poulet, E. Transcranial Direct Current Stimulation for Obsessive-Compulsive Disorder: A Systematic Review. *Brain Sci.* **2018**, *8*, 37. [CrossRef]
10. Palm, U.; Hasan, A.; Strube, W.; Padberg, F. tDCS for the treatment of depression: A comprehensive review. *Eur. Arch. Psychiatry Clin. Neurosci.* **2016**, *266*, 681–694. [CrossRef]
11. Buch, E.R.; Santarnecchi, E.; Antal, A.; Born, J.; Celnik, P.A.; Classen, J.; Gerloff, C.; Hallett, M.; Hummel, F.C.; Nitsche, M.A.; et al. Effects of tDCS on motor learning and memory formation: A consensus and critical position paper. *Clin. Neurophysiol.* **2017**, *128*, 589–603. [CrossRef]
12. Nikolin, S.; Loo, C.K.; Bai, S.; Dokos, S.; Martin, D.M. Focalised stimulation using high definition transcranial direct current stimulation (HD-tDCS) to investigate declarative verbal learning and memory functioning. *NeuroImage* **2015**, *117*, 11–19. [CrossRef] [PubMed]
13. Wu, D.; Qian, L.; Zorowitz, R.D.; Zhang, L.; Qu, Y.; Yuan, Y. Effects on Decreasing Upper-limb Poststroke Muscle Tone Using Transcranial Directcurrent Stimulation: A Randomized Sham-Controlled Study. *Arch. Phys. Med. Rehabil.* **2013**, *94*, 1–8. [CrossRef] [PubMed]
14. Kähkönen, S.; Kesäniemi, M.; Nikouline, V.V.; Karhu, J.; Ollikainen, M.; Holi, M.; Ilmoniemi, R.J. Ethanol Modulates Cortical Activity: Direct Evidence With Combined TMS and EEG. *Neuroimage* **2001**, *14*, 322–328. [CrossRef] [PubMed]
15. Ferreri, F.; Vecchio, F.; Guerra, A.; Miraglia, F.; Ponzo, D.; Vollero, L.; Iannello, G.; Maatta, S.; Mervaala, E.; Rossini, P.M.; et al. Age Related Differences In Functional Synchronization of EEG Activity as Evaluated by Means of TMS-EEG Coregistrations. *Neurosci. Lett.* **2017**, *647*, 141–146. [CrossRef]
16. Bauer, P.; Sansone, P.; Mitter, B.; Makivic, B.; Seitz, L.B.; Tschan, H. Acute Effects of Back Squats on Countermovement Jump Performance Across Multiple Sets of a Contrast Training Protocol in Resistance-Trained Men. *J. Strength Cond. Res.* **2019**, *33*, 995–1000. [CrossRef]
17. Berry, H.R.; Tate, R.J.; Conway, B.A. Transcutaneous Spinal Direct Current Stimulation Induces Lasting Fatigue Resistance and Enhances Explosive Vertical Jump Performance. *PLoS ONE* **2017**, *12*, e0173846. [CrossRef]
18. Yang, D.; Uhm, Y. The Effects of Transcranial Direct Current Stimulation Combined Lumbar Stability Exercise on the Muscle Activity and Jump Performance of the Soccer Player. *Arch. Orthop. Sports Phys. Ther.* **2019**, *15*, 1–10. [CrossRef]
19. Park, I.; Kim, Y.; Kim, S.K. Examining the Effect of Transcranial Direct Current Sstimulation on the Dominant Motor Cortex in the Indirect Measurement of Physical Ability. *J. Men's Health* **2022**, *18*, 69–70. [CrossRef]
20. Chinzara, T.T.; Buckingham, G.; Harris, D.J. Transcranial direct current stimulation and sporting performance: A systematic review and meta-analysis of transcranial direct current stimulation effects on physical endurance, muscular strength and visuomotor skills. *Eur. J. Neurosci.* **2022**, *55*, 468–486. [CrossRef]

21. Vargas, V.Z.; Baptista, A.F.; Pereira, G.O.C.; Pochini, A.C.; Ejnisman, B.; Santos, M.B.; João, S.M.A.; Hazime, F.A. Modulation of Isometric Quadriceps Strength in Soccer Players With Transcranial Direct Current Stimulation: A Crossover Study. *J. Strength Cond. Res.* **2018**, *32*, 1336–1341. [CrossRef]
22. Hazime, F.A.; da Cunha, R.A.; Soliaman, R.R.; Romancini, A.C.B.; Pochini, A.C.; Ejnisman, B.; Baptista, A.F. anodal transcranial direct current stimulation (tDCS) increases isometric strength of shoulder rotators muscles in handball players. *Int. J. Sports Phys. Ther.* **2017**, *12*, 402–407. [PubMed]
23. Frazer, A.; Williams, J.; Spittles, M.; Rantalainen, T.; Kidgell, D. Anodal transcranial direct current stimulation of the motor cortex increases cortical voluntary activation and neural plasticity. *Muscle Nerve* **2016**, *54*, 903–913. [CrossRef] [PubMed]
24. Tanaka, S.; Hanakawa, T.; Honda, M.; Watanabe, K. Enhancement of pinch force in the lower leg by anodal transcranial direct current stimulation. *Exp. Brain Res.* **2009**, *196*, 459–465. [CrossRef] [PubMed]
25. Flood, A.; Waddington, G.; Keegan, R.J.; Thompson, K.G.; Cathcart, S. The effects of elevated pain inhibition on endurance exercise performance. *PeerJ* **2017**, *5*, e3028. [CrossRef]
26. Muthalib, M.; Kan, B.; Nosaka, K.; Perrey, S. Effects of transcranial direct current stimulation of the motor cortex on prefrontal cortex activation during a neuromuscular fatigue task: An fNIRS study. *Adv. Exp. Med. Biol.* **2013**, *789*, 73–79. [CrossRef]
27. Kan, B.; Dundas, J.E.; Nosaka, K. Effect of transcranial direct current stimulation on elbow flexor maximal voluntary isometric strength and endurance. *Appl. Physiol. Nutr. Metab.* **2013**, *38*, 734–739. [CrossRef]
28. Alix-Fages, C.; Romero-Arenas, S.; Castro-Alonso, M.; Colomer-Poveda, D.; Río-Rodriguez, D.; Jerez-Martínez, A.; Fernandez-Del-Olmo, M.; Márquez, G. Short-Term Effects of Anodal Transcranial Direct Current Stimulation on Endurance and Maximal Force Production: A Systematic Review and Meta-Analysis. *J. Clin. Med.* **2019**, *8*, 536. [CrossRef]
29. National Strength and Conditioning Association. *Essentials of Strength Training and Conditioning*, 4th ed.; Human Kinetics: Champaign, IL, USA, 2016.
30. Tillin, N.A.; Bishop, D. Factors modulating post-activation potentiation and its effect on performance of subsequent explosive activities. *Sports Med.* **2009**, *39*, 147–166. [CrossRef]
31. Hodgson, M.; Docherty, D.; Robbins, D. Post-activation potentiation: Underlying physiology and implications for motor performance. *Sports Med.* **2005**, *35*, 585–595. [CrossRef]
32. Pereira, J.B.; Junqué, C.; Bartrés-Faz, D.; Martí, M.J.; Sala-Llonch, R.; Compta, Y.; Falcón, C.; Vendrell, P.; Pascual-Leone, A.; Valls-Solé, J.; et al. Modulation of Verbal Fluency Networks by Transcranial Direct Current Stimulation (tDCS) in Parkinson's Disease. *Brain Stimul.* **2013**, *6*, 16–24. [CrossRef] [PubMed]
33. Nitsche, M.A.; Kuo, M.F.; Karrasch, R.; Wächter, B.; Liebetanz, D.; Paulus, W. Serotonin Affects Transcranial Direct Current-induced Neuroplasticity in Humans. *Biol. Psychiatry* **2009**, *66*, 503–508. [CrossRef]
34. Hordacre, B.; Moezzi, B.; Ridding, M.C. Neuroplasticity and Network Connectivity of the Motor Cortex Following Stroke:A Transcranial Direct Current Stimulation Study. *Hum. Brain Mapp.* **2018**, *39*, 3326–3339. [CrossRef]
35. Muyor, J.M.; Martín-Fuentes, I.; Rodríguez-Ridao, D.; Antequera-Vique, J.A. Electromyographic activity in the gluteus medius, gluteus maximus, biceps femoris, vastus lateralis, vastus medialis and rectus femoris during the Monopodal Squat, Forward Lunge and Lateral Step-Up exercises. *PLoS ONE* **2020**, *15*, e0230841. [CrossRef] [PubMed]
36. Natera, A.O.; Chapman, D.W.; Chapman, N.D.; Keogh, J.W.L. The reliability and validity of repeat power ability assessments and measurement indices in loaded vertical jumps. *PeerJ* **2023**, *11*, e15553. [CrossRef] [PubMed]
37. Bogataj, Š.; Pajek, M.; Hadžić, V.; Andrašić, S.; Padulo, J.; Trajković, N. Validity, Reliability, and Usefulness of My Jump 2 App for Measuring Vertical Jump in Primary School Children. *Int. J. Environ. Res. Public Health* **2020**, *17*, 3708. [CrossRef]
38. Seeck, M.; Koessler, L.; Bast, T.; Leijten, F.; Michel, C.; Baumgartner, C.; He, B.; Beniczky, S. The standardized EEG electrode array of the IFCN. *Clin. Neurophysiol.* **2017**, *128*, 2070–2077. [CrossRef]
39. Lattari, E.; Campos, C.; Lamego, M.K.; Legey, S.; Neto, G.M.; Rocha, N.B.; Oliveira, A.J.; Carpenter, C.S.; Machado, S. Can Transcranial Direct Current Stimulation Improve Muscle Power in Individuals with Advanced Weight-Training Experience? *J. Strength Cond. Res.* **2020**, *34*, 97–103. [CrossRef]
40. Grospretre, S.; Grandperrin, Y.; Nicolier, M.; Gimenez, P.; Vidal, C.; Tio, G.; Haffen, E.; Bennabi, D. Effect of transcranial direct current stimulation on the psychomotor, cognitive, and motor performances of power athletes. *Sci. Rep.* **2021**, *11*, 9731. [CrossRef]
41. Romero-Arenas, S.; Calderón-Nadal, G.; Alix-Fages, C.; Jerez-Martínez, A.; Colomer-Poveda, D.; Márquez, G. Transcranial Direct Current Stimulation Does Not Improve Countermovement Jump Performance in Young Healthy Men. *J. Strength Cond. Res.* **2021**, *35*, 2918–2921. [CrossRef]
42. Park, S.B.; Han, D.H.; Hong, J.; Lee, J.W. Transcranial Direct Current Stimulation of Motor Cortex Enhances Spike Performances of Professional Female Volleyball Players. *J. Mot. Behav.* **2022**, *55*, 18–30. [CrossRef]
43. Marinus, N.; Van Hoornweder, S.; Aarts, M.; Vanbilsen, J.; Hansen, D.; Meesen, R. The influence of a single transcranial direct current stimulation session on physical fitness in healthy subjects: A systematic review. *Exp. Brain Res.* **2023**, *241*, 31–47. [CrossRef] [PubMed]
44. Hu, K.; Chen, Y.; Guo, F.; Wang, X. Effects of transcranial direct current stimulation on upper limb muscle strength and endurance in healthy individuals: A systematic review and meta-analysis. *Front. Physiol.* **2022**, *13*, 834397. [CrossRef]
45. Savoury, R.; Kibele, A.; Behm, D.G. Methodological issues with transcranial direct current stimulation for enhancing muscle strength and endurance: A narrative review. *J. Cogn. Enhanc.* **2022**, *6*, 114–125. [CrossRef]

46. Ahmed, I.; Mustafaoglu, R.; Rossi, S.; Cavdar, F.A.; Agyenkwa, S.K.; Pang, M.Y.C.; Straudi, S. Non-invasive Brain Stimulation Techniques for the Improvement of Upper Limb Motor Function and Performance in Activities of Daily Living After Stroke: A Systematic Review and Network Meta-analysis. *Arch. Phys. Med Rehabil.* **2023**, *104*, 1683–1697. [CrossRef]
47. Grossman, N.; Bono, D.; Dedic, N.; Kodandaramaiah, S.B.; Rudenko, A.; Suk, H.J.; Cassara, A.M.; Neufeld, E.; Kuster, N.; Tsai, L.H.; et al. Noninvasive Deep Brain Stimulation via Temporally Interfering Electric Fields. *Cell* **2017**, *169*, 1029–1041.e16. [CrossRef] [PubMed]
48. Fortes, L.S.; Mazini-Filho, M.; Lima-Júnior, D.; Machado, D.G.S.; Albuquerque, M.R.; Fonseca, F.S.; Ferreira, M.E.C. Transcranial Stimulation Improves Volume and Perceived Exertion but does not Change Power. *Int. J. Sports Med.* **2021**, *42*, 630–637. [CrossRef]
49. Teymoori, H.; Amiri, E.; Tahmasebi, W.; Hoseini, R.; Grospretre, S.; Machado, D.G.D.S. Effect of tDCS targeting the M1 or left DLPFC on physical performance, psychophysiological responses, and cognitive function in repeated all-out cycling: A randomized controlled trial. *J. Neuroeng. Rehabil.* **2023**, *20*, 97. [CrossRef]
50. Lerner, O.; Friedman, J.; Frenkel-Toledo, S. The effect of high-definition transcranial direct current stimulation intensity on motor performance in healthy adults: A randomized controlled trial. *J. Neuroeng. Rehabil.* **2021**, *18*, 103. [CrossRef]
51. Stagg, C.J.; Antal, A.; Nitsche, M.A. Physiology of Transcranial Direct Current Stimulation. *J. ECT* **2018**, *34*, 144–152. [CrossRef]
52. Vitrac, C.; Benoit-Marand, M. Monoaminergic Modulation of Motor Cortex Function. *Front. Neural Circuits* **2017**, *11*, 72. [CrossRef]
53. Chapeton, J.I.; Haque, R.; Wittig JHJr Inati, S.K.; Zaghloul, K.A. Large-Scale Communication in the Human Brain Is Rhythmically Modulated through Alpha Coherence. *Curr. Biol.* **2019**, *29*, 2801–2811.e5. [CrossRef]
54. van Wijk, B.C.; Beek, P.J.; Daffertshofer, A. Neural synchrony within the motor system: What have we learned so far? *Front. Hum. Neurosci.* **2012**, *6*, 252. [CrossRef]
55. Sacca, V.; Maleki, N.; Wen, Y.; Hodges, S.; Kong, J. Modulation Effects of Repeated Transcranial Direct Current Stimulation at the Dorsolateral Prefrontal Cortex: A Pulsed Continuous Arterial Spin Labeling Study. *Brain Sci.* **2023**, *13*, 395. [CrossRef]
56. Lipski, W.J.; Wozny, T.A.; Alhourani, A.; Kondylis, E.D.; Turner, R.S.; Crammond, D.J.; Richardson, R.M. Dynamics of human subthalamic neuron phase-locking to motor and sensory cortical oscillations during movement. *J. Neurophysiol.* **2017**, *118*, 1472–1487. [CrossRef] [PubMed]
57. Engchuan, P.; Wongsuphasawat, K.; Sittiprapaporn, P. Brain electrical activity during bench press weight training exercise. *Asian J. Med. Sci.* **2019**, *10*, 80–85. [CrossRef]
58. Jones, D.T.; Graff-Radford, J. Executive Dysfunction and the Prefrontal Cortex. *Continuum* **2021**, *27*, 1586–1601. [CrossRef] [PubMed]
59. Friedman, N.P.; Robbins, T.W. The role of prefrontal cortex in cognitive control and executive function. *Neuropsychopharmacology* **2022**, *47*, 72–89. [CrossRef]
60. Choi, E.; Shin, S.H.; Ryu, J.K.; Jung, K.I.; Kim, S.Y.; Park, M.H. Commercial video games and cognitive functions: Video game genres and modulating factors of cognitive enhancement. *Behav. Brain Funct.* **2020**, *16*, 2. [CrossRef]
61. Drigas, A.; Karyotaki, M. Executive Functioning and Problem Solving: A Bidirectional Relation. *Int. J. Eng. Pedagog.* **2019**, *9*, 76–98. [CrossRef]
62. Bo, W.; Lei, M.; Tao, S.; Jie, L.T.; Qian, L.; Lin, F.Q.; Ping, W.X. Effects of combined intervention of physical exercise and cognitive training on cognitive function in stroke survivors with vascular cognitive impairment: A randomized controlled trial. *Clin. Rehabil* **2019**, *33*, 54–63. [CrossRef] [PubMed]
63. Moscatelli, F.; Messina, A.; Valenzano, A.; Monda, V.; Salerno, M.; Sessa, F.; La Torre, E.; Tafuri, D.; Scarinci, A.; Perrella, M.; et al. Transcranial magnetic stimulation as a tool to investigate motor cortex excitability in sport. *Brain Sci.* **2021**, *11*, 432. [CrossRef] [PubMed]
64. Cirillo, J. Physical activity, motor performance and skill learning: A focus on primary motor cortex in healthy aging. *Exp. Brain Res.* **2021**, *239*, 3431–3438. [CrossRef]
65. Tyng, C.M.; Amin, H.U.; Saad, M.N.M.; Malik, A.S. The Influences of Emotion on Learning and Memory. *Front. Psychol.* **2017**, *8*, 1454. [CrossRef] [PubMed]

life

MDPI

Article

The Impact of Stroboscopic Visual Conditions on the Performance of Elite Curling Athletes

Tianhe Li [1,†], Chiyue Zhang [2,†], Xiaoyao Wang [1], Xinai Zhang [1], Zhiqiang Wu [1,*] and Yapu Liang [1,*]

[1] School of Strength and Conditioning Training, Beijing Sport University, Beijing 100084, China; 2023240685@bsu.edu.cn (T.L.); 2022240687@bsu.edu.cn (X.W.); 2023240662@bsu.edu.cn (X.Z.)
[2] China Ice Sport College, Beijing Sport University, Beijing 100084, China; zhangchiyue2022@bsu.edu.cn
* Correspondence: 2021992891@bsu.edu.cn (Z.W.); yapu_liang@bsu.edu.cn (Y.L.)
† These authors contributed equally to this work.

Abstract: Background: In elite curling, precise time perception, speed control, and accuracy are critical components of performance. Stroboscopic training enhances visual processing speed, reaction time, motor skill control, and cognitive abilities by challenging the brain to make quick decisions with limited visual information. Purpose: This study aimed to investigate the impact of stroboscopic visual conditions on the key performance aspects of elite athletes in curling to determine whether these effects can be leveraged in long-term training to enhance elite curling performance. Methods: This study involved the participation of 32 national-level male curling athletes (n = 32, age: 19.9 ± 2.2 years, height: 178.0 ± 6.2 cm, body mass: 71.9 ± 10.6 kg, and training age: 2.7 ± 0.9 years). A cross-over controlled experiment was conducted, with participants randomly assigned to either a stroboscopic-first group (n = 16) or a control-first group (n = 16). Each participant completed tests under both stroboscopic and normal visual conditions, including assessments of time perception error, speed control error, and curling accuracy. Paired sample t-tests were employed to analyse performance differences across conditions, and two-factor ANOVA was used to analyse sequence effects. Bonferroni post-hoc tests were used to compare differences if the main effect was significant. Cohen's d was used for two-group comparisons, whereas η_p^2 and Cohen's f were used for comparisons involving three or more groups. Results: under stroboscopic conditions, participants experienced increased errors in time perception ($p < 0.001$, Cohen's d = 1.143), delivery speed control (p = 0.016, Cohen's d = 0.448), and reduced accuracy (p = 0.029, Cohen's d = 0.404). The sequence main effect on speed control error was significant (p = 0.025, η_p^2 = 0.081, Cohen's f = 0.297). Conclusions: Stroboscopic visual conditions negatively impacted cognition (especially time perception) and delivery performance focused on speed control and accuracy in elite curling, highlighting the potential and feasibility of using stroboscopic training to enhance elite curling performance.

Keywords: stroboscopic visual conditions; elite curling athletes; curling performance

Citation: Li, T.; Zhang, C.; Wang, X.; Zhang, X.; Wu, Z.; Liang, Y. The Impact of Stroboscopic Visual Conditions on the Performance of Elite Curling Athletes. *Life* **2024**, *14*, 1184. https://doi.org/10.3390/life14091184

Academic Editor: Herbert Loellgen

Received: 23 August 2024
Revised: 13 September 2024
Accepted: 18 September 2024
Published: 19 September 2024

1. Introduction

The sport of curling originated in Scotland during the 16th century. It has subsequently grown in popularity and is now a prominent feature of the Winter Olympics [1,2]. The game is played on a rectangular sheet of ice, with the primary objective being to slide granite stones towards a circular target area segmented into four concentric circles, known as the "house" [3,4].

The accuracy of stone delivery can be used as a performance indicator in curling. The precision of stone delivery is contingent upon the perception of time and regulation of the stone's velocity. In the sport of curling, there is a strong correlation between time perception and performance [5]. Time perception encompasses the estimation of the timing of actions, as well as anticipation and preparation for the occurrence of events [6]. Furthermore, cognitive abilities, including time perception, can impact performance in time-constrained sports [7], particularly those requiring precise timing and accuracy, such as badminton, ice hockey, and others [7,8]. Additionally, movement can also influence or further enhance an athlete's perception of time [9].

Furthermore, curling athletes must regulate the velocity of the stone in order to guarantee its adherence to the intended trajectory, thereby facilitating the attainment of tactical objectives or the scoring of points. The control of speed is contingent upon the acquisition and processing of information, which in turn gives rise to the execution of highly automated technical movements. Sensory perception (vision, proprioception, balance sense, and hearing) serves as the foundation for the reception of information pertaining to human motion, upon which technical movements are developed [10].

In recent years, the impact of visual training on athletic performance, especially regarding accuracy and cognition, has been the subject of considerable discussion and attention. Various studies have demonstrated that visual attention and occlusion training can significantly enhance perceptual abilities, anticipatory capabilities, and motor skills across a range of sports, including badminton, football, basketball, and motor racing [11–13].

Furthermore, stroboscopic visual training, which employs specific intermittent visual disruption frequencies, achieves visual occlusion in both temporal and spatial dimensions [14,15]. Firstly, stroboscopic training has been demonstrated to enhance visual–motor control and dynamic vision, with the potential to even improve jumping performance [16,17]. Additionally, visual occlusion training has been shown to reinforce athletes' non-visual senses by disrupting visual feedback, thereby enhancing proprioception and vestibular function [18–20]. In the context of curling, it is of paramount importance to facilitate continuous enhancement in temporal perception and judgement, as well as muscle control and stability, in order to achieve speed control and enhance accuracy. This is closely linked with visual feedback, proprioception, and vestibular functions [21–23]. Nevertheless, no studies have substantiated the influence of stroboscopic visual conditions on curling performance, nor have they addressed the potential for visual training, particularly in relation to whether the cognitive and performance aspects of elite curling athletes are affected by stroboscopic visual conditions. Although research has validated the positive effects of stroboscopic training on cognition and performance in other sports, some controversy still exists [24]. Consequently, further investigation into the specific impact of stroboscopic visual conditions on high-level curling athletes is required.

This study aimed to investigate the impact of stroboscopic visual conditions on time perception and judgement, stone delivery speed control, and accuracy in elite curling athletes. This was achieved through a cross-over controlled experiment. In light of these findings, this study will inform the training of elite athletes, directing further research into the effects of stroboscopic training interventions and elite performance. This study investigated the specific impacts of stroboscopic visual conditions on the potential and feasibility of elite curling stroboscopic training, thereby deepening the theoretical understanding of visual training. This research hypothesised that stroboscopic visuals may disrupt the attention, anticipatory timing, and technical movements of elite curling athletes, thereby impairing their ability to perceive and judge time. This, in turn, may negatively impact delivery speed control and accuracy when compared to normal visual conditions. This hypothesis was based on previous studies in this field.

2. Materials and Methods

2.1. Subjects

This study involved the participation of 32 national-level male curling athletes (mean ± SD: male, n = 32, age = 19.9 ± 1.2 years, height = 178.0 ± 6.2 cm, body mass = 71.9 ± 10.6 kg, curling training age = 2.7 ± 0.9 years), as shown in Table 1. Athletes participating in this study were aged between 18 and 22 and engaged in curling-specific training at least seven times per week over five days. On training days, athletes engaged in one or two on-ice training sessions. In order to be eligible for inclusion in this study, participants were required to meet a number of specific criteria. The inclusion criteria required that participants regularly engaged in on-ice technical training, had a minimum of two years' experience in curling, were right-handed, and possessed normal or corrected-to-normal vision. The exclusion criteria included whether participants experienced any sports injuries during the study, exhibited signs of infection or fatigue on the assessment day, had taken medications or drugs (including antidepressants and painkillers) on the assessment day, or consumed caffeine on the assessment day that could affect neuromuscular function during the tests. In addition, all participants were required to complete a medical history questionnaire in order to confirm their good health and the absence of any musculoskeletal, neurological, or other conditions that could affect their ability to perform curling techniques. Participant recruitment began on 1 January 2024 and ended on 7 January 2024. All subjects involved in this study signed an informed consent form. There were no statistically significant differences in these personal characteristics among the groups.

Table 1. Basic information of participants.

Sample Size	Age (Years)	Body Height (cm)	Body Mass (kg)	Training Age (Years)
$n = 32$	19.9 ± 2.2	178.0 ± 6.2	71.9 ± 10.6	2.7 ± 0.9

The data are presented as mean ± standard deviation (SD).

2.2. Ethical Approval

This study was conducted according to the guidelines of the Declaration of Helsinki and approved by the Sports Science Experiment Ethics Committee of Beijing Sport University (2024105H). Before the start of this study, all participants provided their written informed consent to participate after the benefits and potential risks were explained. The study was conducted according to the guidelines of the Declaration of Helsinki and approved by the Sports Science Experiment Ethics Committee of Beijing Sport University (2024105H).

2.3. Experimental Protocol

The principal aim of the initial phase of this investigation was to assess the impact of stroboscopic visuals on the performance of curling athletes, with a particular focus on their ability to control stone delivery speed and precision. A cross-over controlled trial methodology was employed in this study, which recruited a cohort of 32 participants. Subsequently, the participants were randomly allocated to either the stroboscopic priority group (Group A, comprising 16 individuals) or the control priority group (Group B, consisting of 16 individuals). Each participant was subjected to two distinct assessments conducted under disparate visual conditions. In particular, the individuals in Group A were initially evaluated under stroboscopic visual conditions (SVC), after which they underwent a passive interval of 15 min. They then underwent a second assessment under normal visual conditions (NVC). The frequency was set at 0.3 s of transparency, followed by 0.1 s of opacity. In contrast, the sequence was inverted for Group B. The assessments comprised evaluations of time perception in curling and on-ice curling performance, which included both the regulation of delivery speed and the precision of stone placement. The research framework is depicted in Figure 1 and the experimental procedure is delineated in Figure 2.

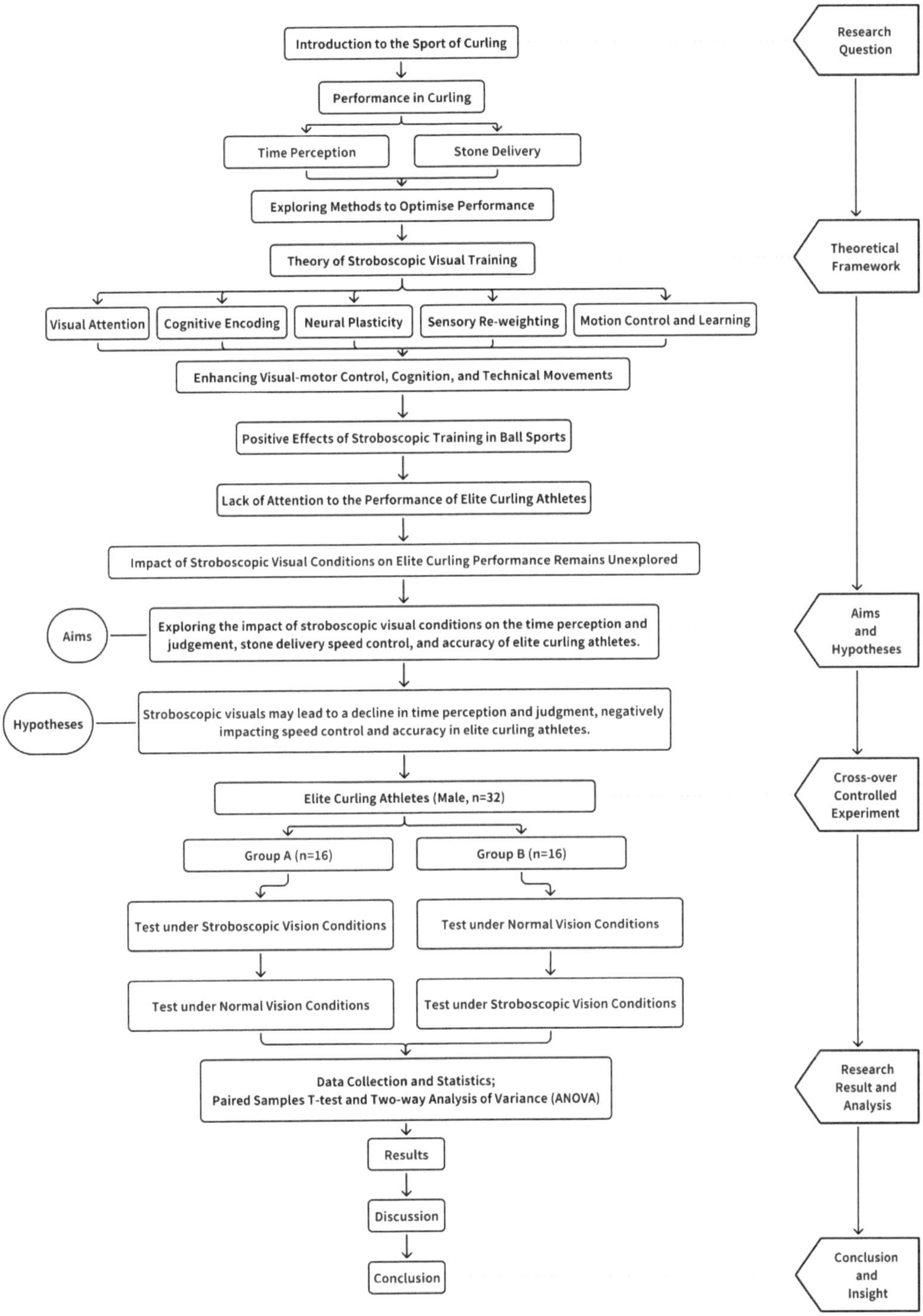

Figure 1. Research Framework. The figure illustrates the theoretical framework and design approach of this study.

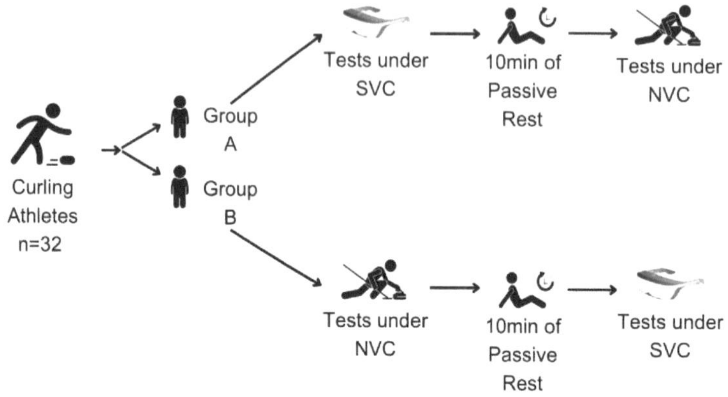

Figure 2. Experimental flowchart. SVC: stroboscopic visual conditions; NVC: normal visual conditions.

2.4. Time Perception and Judgment Test

Prior to the commencement of the time perception and judgment test, the experiment leader provided an explanation of the test content and tasks, followed by a practice session. Once the participants demonstrated proficiency in the experimental task, the practice session was brought to a close and the official test began. The participants were required to view video materials showing a curling stone moving in a defined pattern towards the hog line at the end of the playing area and then entering and covering a specified zone.

Subsequently, participants were required to ascertain the precise moment at which the front of the stone made contact with the T-line (highlighted in red in the video) and then press the button. It was required that each participant complete a series of judgement tasks on 10 randomly selected video materials. The discrepancy between each judgement time and the actual situation was recorded as a judgment error, with a precision of 0.01 s. A smaller average judgement error is indicative of enhanced temporal perception and judgement capabilities within the specified context.

This study employed the 3D modelling software 3D MAX 2020 (Autodesk, San Francisco, CA, USA) to create three-dimensional virtual curling videos as stimulus material. The videos had a resolution of 1920 × 1080 pixels, a frame rate of 30 frames per second, a focal length of 50 mm, and an initial viewing angle of 39.598 degrees. These specifications simulated the horizontal field of view of a human eye at a height of 170 cm. The footage depicted a curling stone in motion, a hog line, and the house. To avoid the possibility of a learning effect among participants due to the repetition of test materials, the curling stone was initiated from different points at varying speeds and with varying trajectories. In total, 30 distinct stimuli were prepared.

2.5. Speed Control Test

Prior to the speed control test, participants engaged in a standardised warm-up comprising 10 min of power cycling and a dynamic stretching routine. Subsequently, the participants proceeded to the ice rink to undertake the speed control test. During the test, the Rock Hawk timing system was set up at the delivery end of the rink with the objective of recording the time it takes for the curling stone to travel from the back line to the hog line, measured in seconds. This duration was employed to represent the range of speed control necessary for executing a slow delivery technique. The test personnel randomly selected four target values from a pre-established range (3.70–4.20 s, with 0.05-s intervals) to be used as benchmarks in the stone delivery tests. The participants commenced their delivery from the hack, with the Rock Hawk system recording the time taken for the front of each stone to travel from the back line to the hog line. The data were then immediately relayed to both the participant and the test personnel via an iPad screen. For each target value, participants delivered the curling stone on three occasions. The test personnel recorded the actual time taken for each delivery alongside the target value, with a precision of 0.01 s. The discrepancy between the exact times and the target values was used to evaluate the efficacy of the speed control mechanism.

2.6. Accuracy Test

Subsequently, an accuracy test was conducted following the speed control test. The delivery of the stone by the athlete is analogous to the delivery of a stone in a curling match, whereby the stone is released before reaching the hog line. A higher score is awarded the closer the stone comes to the centre of the house. Prior to the official commencement of the test, each participant was permitted two trial throws. Each athlete was required to deliver the stone in a clockwise direction on three occasions. During the official test, participants were required to utilise the same stone for each delivery. Furthermore, sweeping was not permitted as a means of adjusting the stone after its release. It was the responsibility of at least one member of the test personnel to ensure that the ice surface was cleaned before each delivery. This procedure was implemented to mitigate the potential impact of debris, such as ice chips, on the experimental results.

The scoring system for the stone's final resting position was as follows: a score of five points was awarded if the stone successfully covered the button, four points if it landed within the one-foot ring, three points within the four-foot ring, two points within the eight-foot ring, and one point within the twelve-foot ring. In the event that the stone rested on the boundary between two rings, the score corresponding to the inner ring was awarded. From an overall perspective, the stone's edge must enter the outer edge of a designated ring.

2.7. Statistical Analysis

Data were summarised and analysed using Excel 2021 and SPSS 27.0 (USA). All results are presented as mean ± standard deviation (M ± SD). Paired sample t-tests were used to compare the differences in test results under different visual conditions. A two-factor analysis of variance (ANOVA) was used to evaluate the effects of stroboscopic visual conditions and test sequence on performance in the time perception and judgment tests, as well as the curling stone delivery speed control and accuracy tests. If the main effect was significant, Bonferroni post-hoc tests were used to compare the differences between different sequences and conditions. The magnitude of the differences (effect sizes) was reported using Cohen's *d* for paired sample t-tests. The effect size of the differences was reported using partial eta squared (η_p^2) and Cohen's *f* for two-way ANOVA. A Cohen's *d* greater than 0.8 was considered large, between 0.8 and 0.5 was categorised as medium, between 0.5 and 0.2 was considered small, and less than 0.2 was deemed insignificant. A η_p^2 greater than 0.14 was considered large, between 0.06 and 0.14 was categorised as medium, between 0.06 and 0.01 was considered small, and less than 0.01 was deemed insignificant. A Cohen's *f* greater than 0.40 was considered large, between 0.40 and 0.25 was categorised as medium, between 0.25 and 0.10 was considered small, and less than 0.10 was deemed insignificant. The significance level was set at $p < 0.05$.

3. Results

Significant differences were observed in time perception and judgment under different visual conditions, with greater errors occurring under stroboscopic conditions ($p < 0.001$, Cohen's $d = 1.143$, large effect size). Regarding performance in curling, the participants showed higher velocity control errors when wearing stroboscopic glasses ($p = 0.016$, Cohen's $d = 0.448$, small effect size). In terms of accuracy, the participants under stroboscopic visual conditions scored significantly lower than under normal visual conditions ($p = 0.029$, Cohen's $d = 0.404$, small effect size). The above results are illustrated in Table 2. The test results under different visual conditions are shown in Figure 3.

Table 2. Comparison of test results under different visual conditions.

Variable	SVC	NVC	Mean Difference	*p*	CI	Cohen's *d*
Time Perception and Judgment Error (s)	0.85 ± 0.15	0.66 ± 0.15	0.19	<0.001 **	0.132~0.254	1.143
Speed Control Error (s)	0.53 ± 0.30	0.38 ± 0.23	0.15	0.016 *	0.029~0.264	0.448
Curling Scoring (points)	7.38 ± 2.97	9.19 ± 3.35	−1.81	0.029 *	−3.431~−0.194	0.404

The data are presented as mean ± SD; SVC: stroboscopic visual conditions; NVC: normal visual conditions; Cohen's *d* greater than 0.8 was considered large, between 0.8 and 0.5 was categorised as medium, between 0.5 and 0.2 was considered small, and less than 0.2 was deemed insignificant; CI: 95% CI; * $p < 0.05$, ** $p < 0.01$ for differences between conditions.

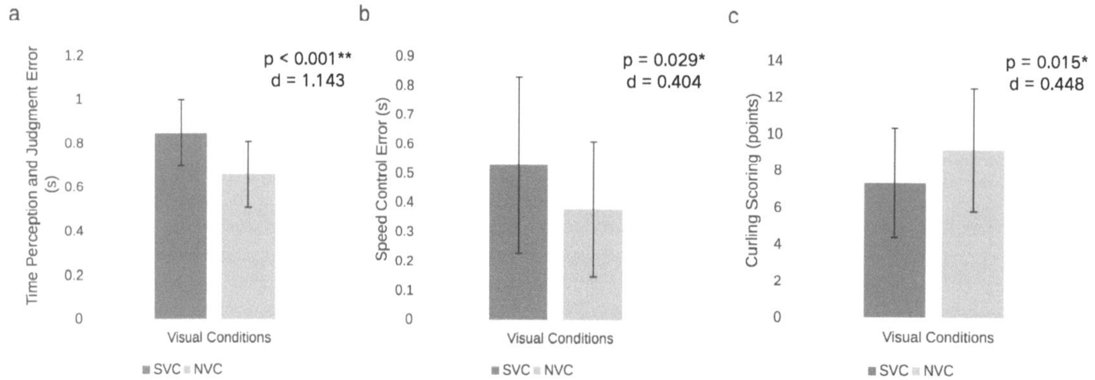

Figure 3. Time perception and stone delivery efficiency under different visual conditions. (**a**) Time perception and judgment error under different visual conditions (s); (**b**) speed control error under different visual conditions (s); (**c**) curling scoring under different visual conditions (points); * $p < 0.05$, ** $p < 0.01$ for differences between conditions. The data are presented as mean ± standard error (SE) in the figure.

The results of the two-factor ANOVA showed that a significant main effect of the sequence was observed only in delivery speed control ($p = 0.002$, $\eta_p^2 = 0.150$, Cohen's $f = 0.420$, large effect size), with no interaction effect ($p = 0.996$, $\eta_p^2 = 0.000$, Cohen's $f = 0.000$, insignificant effect size). Subsequent post-hoc multiple comparisons revealed that participants who prioritised the stroboscopic glasses in testing had significantly higher speed control errors compared to the other group ($p = 0.002$, Cohen's $d = 0.794$, medium effect size). The results of the two-factor ANOVA are shown in Table 3. The post-hoc multiple comparisons are shown in Tables 4 and 5.

Table 3. The impact of test sequence and visual conditions on test results.

Variable	Source	SS	df	MS	*F*	*p*	η_p^2	*f*
Time Perception and Judgment Error (s)	Test Sequence	0.024	1	0.024	1.076	0.304	0.018	0.135
	Visual Conditions	0.599	1	0.599	26.554	<0.001 **	0.307	0.666
	Sequence × Visual Conditions	0.038	1	0.038	1.700	0.197	0.028	0.170
Speed Control Error (s)	Test Sequence	0.687	1	0.687	10.626	0.002 **	0.150	0.420
	Visual Conditions	0.344	1	0.344	5.317	0.025 *	0.081	0.297
	Sequence × Visual Conditions	0.000	1	0.000	0.000	0.996	0.000	0.000
Curling Scoring (points)	Test Sequence	0.028	1	0.028	0.003	0.958	0.000	0.000
	Visual Conditions	61.164	1	61.164	5.999	0.017 *	0.091	0.316
	Sequence × Visual Conditions	8.601	1	8.601	0.844	0.362	0.014	0.119

η_p^2: partial eta squared; η_p^2 greater than 0.14 was considered large, between 0.06 and 0.14 was categorised as medium, between 0.06 and 0.01 was considered small, and less than 0.01 was deemed insignificant; *f*: Cohen's *f* greater than 0.40 was considered large, between 0.40 and 0.25 was categorised as medium, between 0.25 and 0.10 was considered small, and less than 0.10 was deemed insignificant; * $p < 0.05$, ** $p < 0.01$ for differences between conditions.

Table 4. Post-hoc multiple comparisons for the main effect of sequence.

Variable	Sequence A	Sequence B	Mean Difference	F	p	Cohen's d
Speed Control Error (s)	0.56 ± 0.30	0.35 ± 0.21	0.207	1.087	0.002 **	0.794

The data are presented as mean ± SD; Sequence A: SVC first, then NVC; Sequence B: NVC first, then SVC; Cohen's d greater than 0.8 was considered large, between 0.8 and 0.5 was categorised as medium, between 0.5 and 0.2 was considered small, and less than 0.2 was deemed insignificant; ** $p < 0.01$ for differences between conditions.

Table 5. Post-hoc multiple comparisons for main effect between groups.

Variable	SVC	NVC	Mean Difference	F	p	Cohen's d
Time Perception and Judgment Error (s)	0.85 ± 0.15	0.66 ± 0.15	0.193	26.226	<0.001 **	1.280
Speed Control Error (s)	0.53 ± 0.30	0.38 ± 0.23	0.147	4.668	0.035 *	0.540
Curling Scoring (points)	7.38 ± 2.97	9.19 ± 3.35	−1.813	5.253	0.025 *	−0.573

The data are presented as mean ± SD; SVC: stroboscopic visual conditions; NVC: normal visual conditions; Cohen's d greater than 0.8 was considered large, between 0.8 and 0.5 was categorised as medium, between 0.5 and 0.2 was considered small, and less than 0.2 was deemed insignificant; * $p < 0.05$, ** $p < 0.01$ for differences between conditions.

4. Discussion

4.1. Main Findings

The findings revealed that, in comparison to normal visual conditions, the stroboscopic visual conditions resulted in a notable increase in the average error of time perception and judgement among the elite curlers, a discernible rise in the error of stone delivery speed control, and a significant reduction in delivery accuracy. Furthermore, the sequence of testing had an impact on the speed control of elite athletes, although no interaction was observed between the sequence and visual conditions.

4.2. The Impact of Stroboscopic Visual Conditions on Cognition

It has been demonstrated that stroboscopic visual conditions negatively impact the cognitive performance of athletes. In the sport of curling, the ability to "read the ice" is of paramount importance for the successful execution of stone delivery and sweeping techniques [25]. The term refers to the process of understanding the ice conditions by observing the movement of the curling stones on the ice surface, including their slipperiness and curvature. This is considered a crucial cognitive skill in the sport [26]. The available evidence suggests that professional curling athletes possess a distinct advantage in time perception abilities in specific scenarios, characterised by the accuracy of their predictions and judgements about stone movement. This cognitive skill is closely linked to athletes' performance in delivering stones and can, to some extent, predict their overall performance in the sport [5,27]. The presence of on-ice markings, such as the hog line, tee line, and back line, constitutes a significant environmental factor that may exert an influence on athletes' predictions regarding stone movement during a curling match. It is important to note that the cognitive advantages observed in athletes are only evident in cognitive tests that are specifically designed to assess these skills. Consequently, the video stimuli prepared for this study accurately replicated the essential markings of a curling sheet. The experiment demonstrated that under stroboscopic visual interference, athletes exhibited significantly reduced time perception and judgement abilities in specific scenarios compared to those observed under normal visual conditions. This finding emphasises the pronounced impact of stroboscopic visual interference on these cognitive processes. The use of stroboscopic glasses, which impede the input of visual information, presented a more significant challenge for subjects engaged in time perception and judgement tasks. In accordance with prior research, the use of stroboscopic glasses has been observed to impede the processing speed of the central nervous system regarding visuomotor perception while simultaneously reducing visuomotor response speed. This may provide a neurophysiological basis for the observed performance improvements following stroboscopic training [28].

4.3. The Impact of Stroboscopic Visual Conditions on Stone Delivery Performance

Furthermore, stroboscopic visual conditions have an impact on athletes' stone delivery performance, which represents the most critical specialised skill in curling. The initial velocity necessary for stone delivery is attained by exerting a force on the ice surface with the hack. The force and rhythm of the push exerted on the stone directly influence the stone's speed and acceleration during its slide. Curlers regulate the launch speed within a specified range to guarantee that the stone acquires optimal initial velocity before it departs from their control. This enables the stone to reach a predetermined position at the far end of the track, thereby fulfilling the tactical intent as set out in reference [29]. In the majority of cases, team members use a stopwatch to accurately record and assess the speed of the stone. The most common reference point employed by athletes during competition is the time taken for the stone to traverse a fixed distance from the back line to the hog line. Sweepers may utilise this timing to anticipate the need for sweeping and the requisite intensity of said sweeping, whereas the delivering athlete may adjust the stone's speed based on the discrepancy between the target time and the actual time. Fixed distance timing is the most commonly utilised aid in both curling training and competition [30].

The stone delivery speed control test employed in this study was designed to replicate the conditions encountered in competitive and training scenarios, wherein athletes are required to exercise precise control over the sliding speed towards a range of targets. This was achieved through the utilisation of the Rock Hawk timing system, which enabled precise recording over fixed distances. The athlete's ability to control the speed of their delivery is dependent on their capacity to perceive time and speed in specific scenarios, which is influenced by the integration of visual, auditory, and proprioceptive input and feedback [5]. The significant reduction in visual information input under stroboscopic visual interference could also impact the athletes' dynamic balance [31] and time perception abilities, consequently affecting both their control of delivery speed and the accuracy of their deliveries. This aligns with findings from another study involving a specialised football test, where visual feedback obstruction provided by stroboscopic glasses significantly affected performance, particularly accuracy, and had a greater impact on higher-level athletes [32]. In a separate investigation into volleyball performance, the same research team observed comparable outcomes. In this study, athletes demonstrated impaired jumping performance when exposed to stroboscopic conditions, in comparison to their performance under full-field vision. This finding suggests that incorporating stroboscopic conditions into plyometric training may be a potential avenue for further exploration [33].

4.4. The Potential of Stroboscopic Training in Enhancing Curling Performance

Research indicates that stroboscopic training, by limiting the input of visual information, forces athletes to perform tasks under conditions of incomplete visual feedback [28]. This scenario may prompt the brain to enhance the processing capabilities of other sensory information, such as auditory and proprioceptive inputs, thereby compensating for the deficiency in visual information [34]. Notably, under dynamic sporting conditions, the central nervous system integrates sensory information to facilitate motor control [35]. Our study's findings suggest that delivering a curling stone under visual constraints detrimentally affects athletic performance. Evidence indicates that when either visual or proprioceptive input is restricted, the central nervous system recalibrates the contribution of individual sensory inputs [36]. This recalibration is likely more pronounced during stroboscopic training, as opposed to the reduced reliance on sensory inputs observed during steady-state movements [37]. The additional visual load introduced during training induces adaptive modifications in sensory integration, which subsequently enhances motor performance through the refinement or optimisation of sensory integration processes [38]. An increase in stroboscopic visual input may result in a reallocation of cognitive resources during training, thereby enhancing concentration and the speed of information processing. This may lead to faster reaction times and more accurate execution of movements in normal competitive environments [39]. It has been demonstrated that the curling delivery process comprises three distinct phases: the collection of information regarding the duration of the stone's movement, the transformation and processing of motion information, and output of the delivery action [40]. The findings of this study indicate that all three stages are affected by stroboscopic visuals, suggesting that stroboscopic visuals present certain challenges to the input, processing, and output of information. As a result of restricted vision, athletes may be compelled to rely more on other senses and their intrinsic movement perception, which could lead to an overall enhancement of motor skills and cognitive abilities.

4.5. Prospects of this Study

This study paves the way for the potential use of stroboscopic training for elite curling athletes, with subsequent research set to further examine the impact of stroboscopic visual training on high-level curling performance. Further research could examine the effect of varying stroboscopic frequencies, athletes at different skill levels, and gender differences and compare a range of emerging training methods. The efficacy of stroboscopic training in enhancing curling performance can be gauged using a number of indicators employed in this study.

Theoretically, this study emphasises the necessity of comprehending multisensory integration in sports training and its influence on performance, demonstrating how visual information restriction impacts perception–action coupling. These findings have significant implications for the development of theories about the complex interactions between perception and action within the field of sports science. In practical terms, this study serves to bridge the gap between subsequent research and useful application. This study not only advances scientific understanding of the effects of stroboscopic conditions, but also provides a baseline for changes in curling performance under stroboscopic conditions. This will facilitate the design of subsequent studies to verify and optimise the efficacy of stroboscopic training applications.

4.6. Limitations of this Study

It is important to acknowledge the limitations of the present study. The relatively small sample size was primarily due to the limited availability of elite athletes and challenges in participant recruitment. Additionally, the use of a single stroboscopic frequency setting may have restricted the study's ability to fully replicate the competitive environment, including factors such as noise and the presence of opponents. Furthermore, the results of this study may not be directly applicable to the general population of curling athletes, as only the performance of elite male curlers was measured. This study did not include biochemical or bioelectrical testing of the athletes, which future research could explore to assess the impact of stroboscopic visual conditions. Additionally, this study did not involve long-term stroboscopic training interventions, and future research could investigate the effects of such long-term training on athletic performance.

5. Conclusions

Stroboscopic visual interference has a significant impact on the time perception and judgement abilities of curling athletes, as well as their stone delivery performance. In conditions of stroboscopic visual interference, athletes are prone to an increase in errors in time perception and judgement, as well as a significant decline in both the control of delivery speed and accuracy. This suggests that limiting visual feedback by wearing stroboscopic glasses increases the difficulties encountered by athletes during stone delivery, thereby demonstrating the potential of stroboscopic training to improve elite curling performance.

Supplementary Materials: The following supporting information can be downloaded at: https://www.mdpi.com/article/10.3390/life14091184/s1.

Author Contributions: Conceptualization, Y.L. and Z.W.; methodology, Y.L., X.W., C.Z. and T.L.; software, C.Z., T.L. and X.W.; validation, Y.L., C.Z., X.W. and T.L.; formal analysis, Y.L., C.Z., X.W. and T.L.; investigation, Y.L., C.Z., X.Z., X.W. and T.L.; resources, Y.L., C.Z., T.L. and Z.W.; data curation,

C.Z., X.W. and T.L.; writing—original draft preparation, C.Z., X.W. and T.L.; writing—review and editing, Y.L. and Z.W.; visualization, Y.L., C.Z., X.Z., X.W. and T.L.; supervision, Y.L. and Z.W.; project administration, Y.L.; funding acquisition, Y.L. and Z.W. All authors have read and agreed to the published version of the manuscript.

Funding: This research was funded by the Fundamental Research Funds for the Central Universities and the Chinese Universities Scientific Fund, grant numbers 2021QN015 (Y.L.) and 2023GCZX006 (Z.W.).

Institutional Review Board Statement: This study was conducted in accordance with the Declaration of Helsinki and was approved by the Institutional Review Board (or Ethics Committee) of the Sports Science Experiment Ethics Committee of Beijing Sport University (2024105H; 2024.4.10).

Informed Consent Statement: Informed consent was obtained from all subjects involved in this study.

Data Availability Statement: The original contributions presented in this study are included in the article/Supplementary Material, further inquiries can be directed to the corresponding author/s.

Acknowledgments: The authors express their gratitude to China Ice Sports College for providing the venue and human resources. Special thanks to the athletes, coaches, teachers, and all participants from China Ice Sports College who contributed to this research. We also extend our appreciation to the researchers from China Ice Sports College and the School of Strength and Conditioning Training for their crucial theoretical support. Finally, we thank the School of Strength and Conditioning Training and China Ice Sports College for their assistance in organizing and implementing the experiment, as well as their support in the survey and analysis.

Conflicts of Interest: The authors declare no conflicts of interest.

References

1. Tranter, N. Sport and the economy in nineteenth and Early Twentieth Century Scotland: A review of recent literature. *Scott. Hist. Rev.* **1989**, *68*, 53–69.
2. Wieting, S.G.; Lamoureux, D. Curling in Canada. In *Sport and Memory in North America*; Routledge: London, UK, 2019; pp. 140–153.
3. Kim, T.-W.; Lee, S.-C.; Kil, S.-K.; Choi, S.-H.; Song, Y.-G. A case study on curling stone and sweeping effect according to sweeping conditions. *Int. J. Environ. Res. Public Health* **2021**, *18*, 833. [CrossRef] [PubMed]
4. Shi, X.; Wang, Q.; Wang, C.; Wang, R.; Zheng, L.; Qian, C.; Tang, W. An AI-based curling game system for winter olympics. *Research* **2022**. [CrossRef] [PubMed]
5. Song, M.; Zhao, Q.; Du, C.; Zhou, C.; Li, R. The relationship between the accuracy of curling athletes' duration judgment and delivery performance. *PeerJ* **2022**, *10*, e13541. [CrossRef] [PubMed]
6. Thönes, S.; Stocker, K. A standard conceptual framework for the study of subjective time. *Conscious. Cogn.* **2019**, *71*, 114–122. [CrossRef]
7. Behm, D.G.; Carter, T.B. Effect of exercise-related factors on the perception of time. *Front. Physiol.* **2020**, *11*, 770. [CrossRef]
8. Vial, S.; Cochrane, J.; Blazevich, A.J.; Croft, J.L. Using the trajectory of the shuttlecock as a measure of performance accuracy in the badminton short serve. *Int. J. Sports Sci. Coach.* **2019**, *14*, 91–96. [CrossRef]
9. De Kock, R.; Gladhill, K.A.; Ali, M.N.; Joiner, W.M.; Wiener, M. How movements shape the perception of time. *Trends Cogn. Sci.* **2021**, *25*, 950–963. [CrossRef]
10. Hülsdünker, T.; Ostermann, M.; Mierau, A. The speed of neural visual motion perception and processing determines the visuomotor reaction time of young elite table tennis athletes. *Front. Behav. Neurosci.* **2019**, *13*, 165. [CrossRef]
11. Hagemann, N.; Strauss, B.; Cañal-Bruland, R. Training perceptual skill by orienting visual attention. *J. Sport Exerc. Psychol.* **2006**, *28*, 143–158. [CrossRef]
12. Dunton, A.; O'Neill, C.; Coughlan, E.K. The impact of a spatial occlusion training intervention on pass accuracy across a continuum of representative experimental design in football. *Sci. Med. Footb.* **2020**, *4*, 269–277. [CrossRef]
13. Dunton, A.B. The Impact of Spatial Occlusion Training on Complex Motor Skills in Sport. Doctoral Thesis, Cork Institute of Technology, Cork, Ireland, 2020.
14. Appelbaum, L.G.; Lu, Y.; Khanna, R.; Detwiler, K.R. The effects of sports vision training on sensorimotor abilities in collegiate softball athletes. *Athl. Train. Sports Health Care* **2016**, *8*, 154–163.
15. Talwar, S.; Smith, J.W.; Lamberth, J. The Effect of Stroboscopic Vision Training on Blind-folded Straight-line Walking. *Int. J. Exerc. Sci.* **2024**, *17*, 438.
16. Appelbaum, L.G.; Schroeder, J.E.; Cain, M.S.; Mitroff, S.R. Improved visual cognition through stroboscopic training. *Front. Psychol.* **2011**, *2*, 276.
17. Harrison, K.D.; Dakin, C.J.; Beethe, A.Z.; Louder, T. Effects of Stroboscopic Vision on Depth Jump Motor Control: A Biomechanical Analysis. *Bioengineering* **2024**, *11*, 290. [CrossRef]
18. Assländer, L.; Peterka, R.J. Sensory reweighting dynamics in human postural control. *J. Neurophysiol.* **2014**, *111*, 1852–1864.
19. Peterka, R.J. Sensory integration for human balance control. *Handb. Clin. Neurol.* **2018**, *159*, 27–42.

20. Peterka, R.J.; Loughlin, P.J. Dynamic regulation of sensorimotor integration in human postural control. *J. Neurophysiol.* **2004**, *91*, 410–423. [CrossRef] [PubMed]
21. Diener, H.-C.; Dichgans, J. On the role of vestibular, visual and somatosensory information for dynamic postural control in humans. *Prog. Brain Res.* **1988**, *76*, 253–262.
22. Fitzpatrick, R.; McCloskey, D. Proprioceptive, visual and vestibular thresholds for the perception of sway during standing in humans. *J. Physiol.* **1994**, *478*, 173–186. [CrossRef]
23. Gaerlan, M.G.; Alpert, P.T.; Cross, C.; Louis, M.; Kowalski, S. Postural balance in young adults: The role of visual, vestibular and somatosensory systems. *J. Am. Assoc. Nurse Pract.* **2012**, *24*, 375–381.
24. Fortes, L.S.; Faro, H.; Faubert, J.; Freitas-Júnior, C.G.; Lima-Junior, D.d.; Almeida, S.S. Repeated stroboscopic vision training improves anticipation skill without changing perceptual-cognitive skills in soccer players. *Appl. Neuropsychol. Adult* **2023**, *8*, 1–15.
25. Yang, Q.; Li, S.; Li, J.; Zhang, W.; Wang, Q.; Ma, X. Research on Evaluation Indicator of Ice Rink and Curling Stone Motion for the 2022 Beijing Winter Olympic Games Based on Video Recognition Method. *Lubricants* **2023**, *11*, 370. [CrossRef]
26. Ağduman, F.; Bedir, D. From past to present the most mysterious olympic sport: A review of curling sport. *Online J. Recreat. Sports* **2023**, *12*, 314–325.
27. Yilmaz, U.; Buzdagli, Y.; Polat, M.L.; Bakir, Y.; Ozhanci, B.; Alkazan, S.; Ucar, H. Effect of single or combined caffeine and L-Theanine supplementation on shooting and cognitive performance in elite curling athletes: A double-blind, placebo-controlled study. *J. Int. Soc. Sports Nutr.* **2023**, *20*, 2267536.
28. Hülsdünker, T.; Fontaine, G.; Mierau, A. Stroboscopic vision prolongs visual motion perception in the central nervous system. *Scand. J. Med. Sci. Sports* **2023**, *33*, 47–54.
29. Ravanbod, A.; Maw, S.; Hritzuk, E. *An Examination of the Relationships between Curling Rock Rotational and Linear Velocities, and Resulting Displacements, at Low Throwing Speeds*; ISEA: Seoul, Republic of Korea, 2022.
30. Kivi, D.; Auld, T. Temporal methods to estimate the displacement of a curling rock: Comparison between competitive and recreational curlers. In Proceedings of the ISBS-Conference Proceedings Archive, Marquette, MI, USA, 19–23 July 2010.
31. Lee, H.; Han, S.; Page, G.; Bruening, D.A.; Seeley, M.K.; Hopkins, J.T. Effects of balance training with stroboscopic glasses on postural control in chronic ankle instability patients. *Scand. J. Med. Sci. Sports* **2022**, *32*, 576–587.
32. Beavan, A.; Hanke, L.; Spielmann, J.; Skorski, S.; Mayer, J.; Meyer, T.; Fransen, J. The effect of stroboscopic vision on performance in a football specific assessment. *Sci. Med. Footb.* **2021**, *5*, 317–322.
33. Kroll, M.; Preuss, J.; Ness, B.M.; Dolny, M.; Louder, T. Effect of stroboscopic vision on depth jump performance in female NCAA Division I volleyball athletes. *Sports Biomech.* **2023**, *22*, 1016–1026. [CrossRef]
34. Kroll, M. The Effect of Stroboscopic Vision on Reactive Strength in a Sample of Female NCCA Division I Volleyball Athletes. Master's Thesis, University of South Dakota, Vermillion, SD, USA, 2019.
35. Helm, M.; Ritzmann, R.; Gollhofer, A.; Freyler, K. Anticipation modulates neuromechanics of drop jumps in known or unknown ground stiffness. *PLoS ONE* **2019**, *14*, e0211276.
36. Anson, E.; Jeka, J. Sensory Reweighting: A Rehabilitative Mechanism? In *Handbook of Medical Neuropsychology: Applications of Cognitive Neuroscience*; Springer: Berlin/Heidelberg, Germany, 2019; pp. 789–800.
37. Kim, K.-M.; Kim, J.-S.; Oh, J.; Grooms, D.R. Stroboscopic vision as a dynamic sensory reweighting alternative to the sensory organization test. *J. Sport Rehabil.* **2020**, *30*, 166–172. [PubMed]
38. Campos, J.L.; Marusic, U.; Mahoney, J.R. The intersection of cognitive, motor, and sensory processing in aging: Links to functional outcomes, Volume I. *Front. Aging Neurosci.* **2022**, *14*, 1009532.
39. Ellison, P.; Jones, C.; Sparks, S.A.; Murphy, P.N.; Page, R.M.; Carnegie, E.; Marchant, D.C. The effect of stroboscopic visual training on eye–hand coordination. *Sport Sci. Health* **2020**, *16*, 401–410.
40. Muratori, L.M.; Lamberg, E.M.; Quinn, L.; Duff, S.V. Applying principles of motor learning and control to upper extremity rehabilitation. *J. Hand Ther.* **2013**, *26*, 94–103.

life

Article

Construction and Analysis of the Physical Fitness Evaluation Index System for Elite Male Singles Badminton Players: Based on Delphi and AHP Methods

Binyong Ye [1,†], Houwei Zhu [1,†], Zhen Yang [1], Zhanyang He [1], Gongju Liu [2], Huiju Pan [1] and Haiying Guo [2,*]

[1] College of Physical Education and Health Sciences, Zhejiang Normal University, Jinhua 321000, China; binyongye@zjnu.edu.cn (B.Y.); zhuhouwei@zjnu.edu.cn (H.Z.); yz19550596533@163.com (Z.Y.); 2908161568@zjnu.edu.cn (Z.H.); panhuiju@zjnu.cn (H.P.)

[2] Laboratory of Aquatic Sports Science of General Administration of Sports China, Zhejiang College of Sports, Hangzhou 311200, China; liugongju@hotmail.com

* Correspondence: ghy@zjcs.net.cn

† These authors contributed equally to this work.

Abstract: Objective: To construct and validate a physical fitness evaluation index system for elite male singles badminton players. Methods: Utilizing the Delphi method to establish a comprehensive evaluation system, the analytic hierarchy process (AHP) was employed to calculate the influence weights of various indicators. The validity of the comprehensive evaluation system was verified using testing methods. Results: After three rounds of expert selection, the physical fitness evaluation index system for elite male singles badminton players includes three primary indicators, nine secondary indicators, and twenty-one tertiary indicators. Among the primary indicators, specialized physical fitness holds a significant weight in the evaluation with a value of 0.651, whereas body morphology has a smaller weight of 0.077. Among the secondary indicators, specialized agility, strength, and endurance have higher weights of 0.223, 0.217, and 0.210, respectively. Among the tertiary indicators, four-corner ball touch, 400 m × 5 shuttle run, smash-and-rush, and vertical jump height hold higher weights of 0.119, 0.114, 0.104, and 0.096, respectively. The results after randomly selecting ten elite male singles badminton players and applying the evaluation index system demonstrated that this system has high feasibility and validity. It can not only comprehensively assess the physical fitness of athletes but also provide significant practical guidance for enhancing their competitive performance. Conclusions: The evaluation system and weight assignments constructed in this study can scientifically and comprehensively reflect the physical fitness status of athletes. It can guide coaches in formulating targeted training plans and optimizing training outcomes.

Keywords: elite athlete assessment; sports performance indicators; performance measurement; assessment indicators

Citation: Ye, B.; Zhu, H.; Yang, Z.; He, Z.; Liu, G.; Pan, H.; Guo, H. Construction and Analysis of the Physical Fitness Evaluation Index System for Elite Male Singles Badminton Players: Based on Delphi and AHP Methods. *Life* **2024**, *14*, 944. https://doi.org/10.3390/life14080944

Academic Editor: Nicola Smania

Received: 9 July 2024
Revised: 24 July 2024
Accepted: 26 July 2024
Published: 28 July 2024

1. Introduction

As a globally popular sport, badminton has numerous enthusiasts and outstanding athletes in China. Since the change to the 21-point scoring system in 2006, the unpredictability of match results has significantly increased, greatly enhancing the intensity and watchability of the games. Under the new scoring system, the pace of the game is faster, and rallies are more frequent, requiring athletes to start over 500 times and cover a distance of more than 3000 m in a single match [1]. These changes place unprecedented physical demands on athletes. Players need not only exceptional explosive power and speed but also the ability to maintain consistent performance during prolonged high-intensity competition, challenging their endurance, recovery ability, and physical reserves [2,3]. Physical fitness is considered the ability of an individual to perform muscle work satisfactorily, which requires a good level of cardiovascular function, muscular strength, speed, endurance, agility, and other attributes. Physical fitness often becomes the key factor in determining victory, especially in critical stages of competition. The coordinated development of these attributes not only enhances individual athletic performance but also improves adaptability and stress resistance in daily life [4]. The importance of physical fitness is evident not only in sports competitions but also as an indispensable part of our daily lives. Therefore, the systematic and scientific nature of physical training is particularly important in modern badminton [5]. Through scientific physical training, athletes can not only enhance their competitive level and technical stability but also effectively prevent sports injuries, thereby enabling them to stand out in intense competitions and achieve excellent results [6].

As a key competitive sport and a major gold-winning event in China, badminton boasts numerous outstanding players. However, how to evaluate the physical fitness of athletes scientifically and objectively and develop personalized training programs remains an urgent issue to be addressed. An analysis of existing literature reveals current research on physical fitness evaluation systems in badminton focuses mainly on young athletes, talent selection, and certain special groups [7–9]. Some studies have primarily focused on specific physical fitness indicators, such as speed and strength, while neglecting other critical factors [10,11]. Additionally, certain studies have not employed scientific methodologies, such as the Analytic Hierarchy Process, to establish a systematic and objective evaluation indicator system. Instead, they have relied excessively on the researchers' personal experience and intuition, potentially leading to results with a high degree of subjectivity [12]. Research on the physical fitness evaluation system for high-level adult male badminton players has also been lacking, resulting in a failure to meet the needs of practical training fully. Additionally, existing research generally lacks practical application verification of the constructed evaluation systems, thus failing to sufficiently demonstrate their scientific validity and feasibility, making it difficult to ensure that these systems have practical application.

In the absence of practical evidence or recommendations, individual experience becomes the practitioner's alternative. As Minas et al. [13] stated, although the final result of expert consultation may be incorrect, consensus usually provides a better basis than individual judgment. Therefore, in the absence of direct evidence, the Delphi method provides a scientific decision-making tool to ensure scientific validity and reliability [14]. The Delphi method is widely used in establishing evaluation systems and determining indicators. It systematically collects and synthesizes opinions through repeated consultations with a group of independent experts, aiming to achieve consensus and predictions, ultimately forming more objective and comprehensive conclusions [15]. However, while the Delphi method significantly reduces individual bias through expert consensus, its results are often qualitative, difficult to quantify directly, and may carry subjectivity. It is suitable for identifying key issues and forming preliminary plans but may lack precision and objectivity in practical application [16,17]. Some studies have followed the Delphi method with the analytic hierarchy process (AHP) to address these shortcomings and further refine and quantify expert opinions, thereby enhancing the scientific validity and reliability of decisions. However, these studies often do not apply the constructed indicator systems in practice, which may undermine their scientific validity and reliability, ultimately affecting the practical significance and application effectiveness of the indicator systems.

In summary, this study aims to construct a physical fitness evaluation index system for elite male singles badminton players using the Delphi method and the AHP. In the final stage of the study, high-level athletes were selected randomly to apply and validate the established index system. Through the scientific and rational establishment and application of the index system, this study provides a theoretical basis for the comprehensive evaluation of athletes' physical fitness and has important practical significance for improving their competitive performance.

Research Hypothesis: Through several rounds of expert consultation and feedback, the Delphi Method can establish a comprehensive and scientific physical fitness evaluation index system. By employing the Analytic Hierarchy Process (AHP), the weights of various indicators can be calculated to determine their relative importance in the overall evaluation. The application of this comprehensive evaluation system can accurately assess the physical fitness of elite male singles badminton players, thereby guiding coaches in formulating targeted training plans, optimizing training outcomes, and enhancing the competitive performance of athletes.

2. Research Process and Methods

2.1. Research Process

The research process of this study is illustrated in Figure 1. First, the Delphi method was employed to invite experts to evaluate the necessity and importance of the preliminary-drafted indicators. Consultation ceased when expert opinions tended to converge and the establishment of the evaluation index system was completed. Subsequently, a hierarchical structure model was constructed. The AHP was then used to address the qualitative and quantitative analyses of each indicator, providing an in-depth analysis of the complex decision-making problem from the perspectives of the essence of the problem, influencing factors, and their intrinsic relationships. This approach aims to offer concise and clear decision-making solutions. Finally, the completed decision-making plan was applied in practice to verify and support these evaluation system standards further.

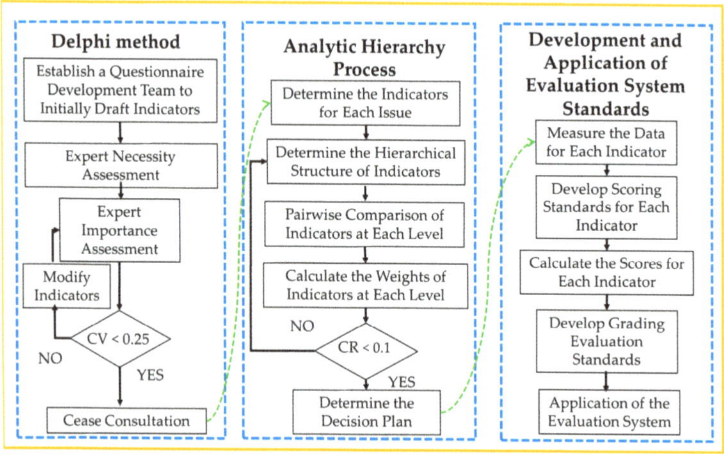

Figure 1. Research Process.

2.2. Research Methods

Preliminary Establishment of Evaluation Indicators

This study systematically analyzed recent literature, works, and policy documents concerning badminton physical training, specialized physical fitness, physical attributes, and the physical fitness measurement and evaluation standards established by the General Administration of Sport of China. Additionally, in-depth discussions and detailed records were conducted with high-level badminton players, coaches, referees, and related personnel regarding influencing factors and other relevant issues. Then, the three basic dimensions were preliminarily identified, including the body morphology, basic physical fitness, and specialized physical fitness of badminton players. Based on these dimensions, a consultation questionnaire was prepared, initially establishing three primary indicators.

A questionnaire preparation team of eight members was formed to ensure the validity, objectivity, comprehensiveness, and comparability of the physical fitness evaluation index system consultation questionnaire. This team included two coaches with over 30 years of coaching experience, two researchers with over 20 years of research experience, one national-level coach with over 10 years of coaching and research experience, two associate professors, and one lecturer. Following the fundamental value orientations of contributing to the development of national sports, guiding scientific training for coaches, and promoting the personal quality improvement of athletes, after multiple discussions, the first round of the consultation questionnaire was determined to include three primary indicators, nine secondary indicators, and twenty-seven tertiary indicators (Figure 2) [9,18–28].

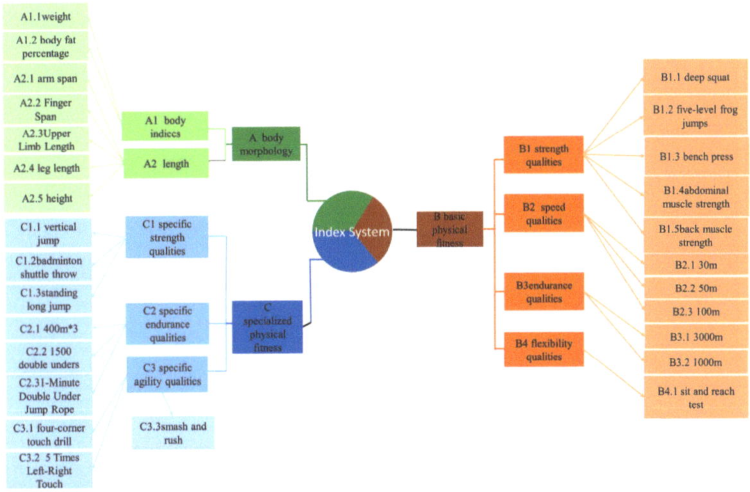

Figure 2. Preliminary indicators of the physical fitness evaluation system for elite male singles badminton players.

2.3. Delphi Expert Consultation Method

2.3.1. Expert Selection Criteria, Expert Positive Coefficient, and Authority Coefficient

This study invited a total of 15 experts from both industry and academia to participate in the Delphi method consultation to ensure the representativeness and comprehensiveness of the expert consultation. The selection of experts was based on the following basic criteria: 1. The expert must have at least ten years of relevant research experience, especially a solid theoretical and practical background in sports science and physical fitness evaluation. 2. The expert must have extensive coaching experience, with a minimum of five years of coaching high-level sports teams and guiding athletes to achieve excellent results in provincial or national competitions. All experts voluntarily participated in the study and committed to completing at least two rounds of consultation. The basic information about the expert group is detailed in Table 1.

Table 1. Summary of the Delphi expert group's basic information.

Number	Name	Position
1	Wang *	National Team Coach
2	Wang *	National Team Coach
3	Xu *	First-line Coach
4	Ni *	First-line Coach
5	Fang *	First-line Coach
6	Chen*	First-line Coach
7	Ding *	First-line Coach
8	Jing *	First-line Coach
9	Lin *	Professor
10	Zhang *	Associate Professor
11	Wang *	Professor
12	Zhu *	Professor
13	Zhang *	Associate Professor
14	Chen *	Professor
15	Zheng *	Professor

Note: The "*" is used to protect the privacy of the consulting experts.

The expert positive coefficient is expressed by the questionnaire recovery rate, reflecting the degree of attention experts pay to this study. Typically, a recovery rate greater than 70% indicates high expert enthusiasm [29]. The expert authority coefficient (Cr) is calculated through statistical analysis based on the expert's judgment basis (Ca) and degree of familiarity (Cs), using the formula Cr = (Ca + Cs)/2. When CS $\geq$ 1, it indicates "Very Familiar"; for 1 > CS $\geq$ 0.8, it denotes "Fairly Familiar"; for 0.8 > CS $\geq$ 0.5, it signifies "Familiar"; for 0.5 > CS $\geq$ 0.3, it represents "Slightly Familiar"; and for 0.3 > CS $\geq$ 0.1, it is described as "Not Familiar". The range of Cr is usually between 0 and 1, with higher values indicating higher authority among the participating experts and greater reliability of the consultation results. Generally, a Cr $\geq$ 0.70 is considered acceptable [7,30]. The values assigned to Ca are detailed in Table 2.

Table 2. Expert judgment basis coefficient assignment table.

Judgment Basis	The Impact of Criteria on Expert Judgment (Ca)		
	Significant Impact	Moderate Impact	Minor Impact
Practical experience	0.5	0.4	0.3
Logical reasoning	0.3	0.2	0.1
Research experience	0.1	0.1	0.1
Intuition	0.1	0.1	0.1
Aggregate	1.0	0.8	0.6

2.3.2. Expert Consultation Process

Step 1: Consultation on the necessity of physical fitness evaluation indicators for elite male singles badminton players. Experts were asked to assess the necessity of three primary indicators, nine secondary indicators, and 27 tertiary indicators through the options "Select", "Delete", and "Modify". Step 2: Incorporate the modifications suggested by the experts in the first round to form a new physical fitness evaluation index system for elite male singles badminton players. Experts were then invited to rate the importance of the new index system on a five-point scale: "Very Important", "Important", "Moderately Important", "Less Important" and "Not Important".

2.3.3. Indicator Screening Criteria

The number of evaluation criteria is not necessarily the more, the better; the focus should be on the significance of each criterion's role in the evaluation. In the initial stage of establishing the evaluation index system, a large amount of redundant information may affect the accuracy of the evaluation results. Thus, to further ensure the accuracy of the evaluation results, a degree of expert opinion coordination was introduced to screen the value of the indicators. The degree of expert opinion coordination reflects the consistency of consulting experts' judgments on various indicators. It is represented by the coefficient of variation (CV, the standard deviation of each indicator divided by its mean value) and the Kendall coordination coefficient (W) [29]. A smaller CV indicates that experts' opinions on a particular indicator tend to be consistent, with a CV < 0.25 being acceptable. The range of W is between 0 and 1, with values closer to 1 indicating better coordination among experts on the ratings of all factors. Generally, a Kendall coordination coefficient < 0.2 indicates poor consistency; 0.2–0.4 indicates moderate consistency; 0.4–0.6 indicates medium consistency; 0.6–0.8 indicates strong consistency; and 0.8–1.0 indicates very strong consistency. This study uses the mean score of experts' importance ratings for each indicator, CV and W, as the screening criteria. An indicator passes the screening, and consultation stops if the mean score > 3.5, CV < 0.25, W is between 0.4 and 0.5, and the asymptotic significance $p < 0.05$ [31–33].

2.4. Analytic Hierarchy Process

The AHP is a commonly used method for quantifying qualitative issues and is widely applied in studies involving the establishment of indicator systems [34,35]. Its application in assigning weights to indicators involves establishing an ordered hierarchy and comprehensively calculating the weight coefficients of the indicators by comparing the relative importance of each indicator within the same level [36,37].

The AHP establishes a hierarchical structure model composed of the overall goal, criteria, and candidates, each of which is independently unaffected by the others. Based on different evaluation factors, candidates are compared pairwise according to the 1–9 relative ranking scale provided by Saaty to form a comparative judgment matrix. The scoring matrices of all experts are combined into a judgment integration matrix using the geometric mean method, and the weight value of each candidate relative to the overall goal is calculated [38,39]. Through this process, the relative importance of each indicator at each level is judged, thus quantifying subjective evaluations and systematically assessing the importance weights of each indicator relative to the overall goal. The specific calculation process can be seen in Figure 3.

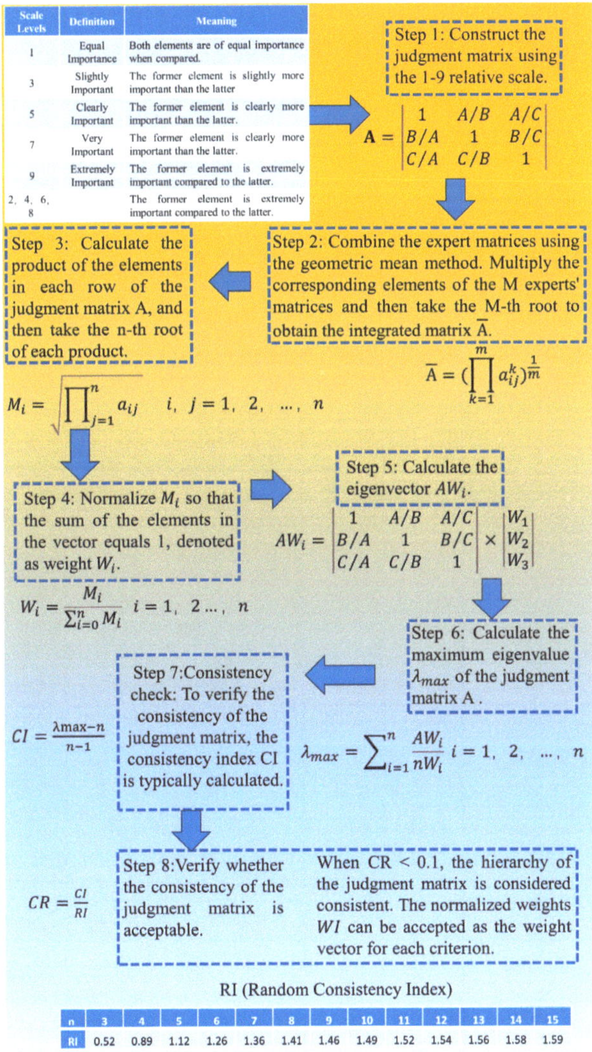

$$\mathbf{A} = \begin{vmatrix} 1 & A/B & A/C \\ B/A & 1 & B/C \\ C/A & C/B & 1 \end{vmatrix}$$

$$\bar{A} = (\prod_{k=1}^{m} a_{ij}^{k})^{\frac{1}{m}}$$

$$M_i = \sqrt[n]{\prod_{j=1}^{n} a_{ij}} \quad i,\ j = 1,\ 2,\ \ldots,\ n$$

$$W_i = \frac{M_i}{\sum_{i=0}^{n} M_i} \quad i = 1,\ 2 \ldots,\ n$$

$$AW_i = \begin{vmatrix} 1 & A/B & A/C \\ B/A & 1 & B/C \\ C/A & C/B & 1 \end{vmatrix} \times \begin{vmatrix} W_1 \\ W_2 \\ W_3 \end{vmatrix}$$

$$\lambda_{max} = \sum_{i=1}^{n} \frac{AW_i}{nW_i} \quad i = 1,\ 2,\ \ldots,\ n$$

$$CI = \frac{\lambda max - n}{n-1}$$

$$CR = \frac{CI}{RI}$$

n	3	4	5	6	7	8	9	10	11	12	13	14	15
RI	0.52	0.89	1.12	1.26	1.36	1.41	1.46	1.49	1.52	1.54	1.56	1.58	1.59

Figure 3. Calculation process of analytic hierarchy process.

Note: Step 1: A (judgment matrix); A, B, C (element); A/B (relative importance of element A compared with element B).

Step 2: $\overline{A}$ (combined matrix); a_{ij}^{k} (element in the i-th row and j-th column of the k-th expert's judgment matrix); m (number of experts).

Step 3: M_i (n-th root of the product of elements in the i-th row); n (dimension of the matrix); a_{ij} (element in the i-th row and j-th column of the judgment matrix).

Step 4: W_i (normalized weight of the i-th element); $\sum_{i=0}^{n} M_i$ (sum of the products of all elements).

Step 5: AW_i (eigenvector); W_i (normalized weight of the i-th element).

Step 6: λ_{max} (maximum eigenvalue of the judgment matrix); AW_i (eigenvector); nW_i (product of the matrix dimension and the normalized weight).

Step 7: CI (consistency Index); λ_{max} (maximum eigenvalue of the judgment matrix); n (dimension of the matrix).

Step 8: CR (consistency ratio); CI (consistency index); RI (random consistency index).

2.5. Testing Method

This study was conducted in two phases. In the first phase, 15 elite male singles badminton players were randomly selected to test predefined indicators. The mean, standard deviation, maximum, and minimum values for each indicator were calculated to establish a general value model for the physical fitness of elite male badminton players. Scoring standards were then developed to assign accurate numerical values to specific evaluation criteria, ensuring that the results are fair, objective, and quantifiable. To verify the independence of the evaluation index system and reduce potential sample bias, the second phase of the study employed a different sample group. An additional 10 players, who were distinct from the initial 15 and varied in their abilities and training backgrounds, were tested. This approach was taken to confirm the universality and effectiveness of the evaluation system.

The inclusion criteria for participants were as follows: age between 18 and 35 years, no injuries in the past year, at least three years of professional training experience, and outstanding performance in various major competitions. This stringent selection process ensured that the participants were representative of elite badminton players, thus providing robust data for validating the evaluation system. All tests were conducted under standardized conditions using calibrated equipment to ensure consistency and safety. Participants received detailed instructions and demonstrations to ensure proper execution of each test. Professional supervisors were present throughout the testing process to oversee procedures and ensure accurate data recording. This study was approved by the Ethics Committee of Zhejiang Normal University (No. ZSRT2024165). Participants were informed about the experimental procedures and provided written informed consent prior to the experiment. The study was conducted in accordance with the principles of the Declaration of Helsinki, with all methods and procedures designed to uphold ethical standards.

3. Results

3.1. Statistical Analysis of Experts' Basic Information

In this study, a survey was conducted with 15 experts from industry and academia. The survey included information on the experts' age, years of work experience, and profession. All 15 experts participated in the first round of the survey. The survey results indicate that these experts possess high authority. The results indicate the expert group has substantial work experience and high educational backgrounds, providing a certain level of representativeness and rationality, which ensures the content authenticity and structural rationality of the constructed evaluation index system.

Regarding expert enthusiasm, this study conducted four rounds of expert consultations, with a 100% questionnaire recovery rate for each round. The questionnaire recovery rate is typically used to calculate the expert positive coefficient, and a recovery rate above 70% indicates high expert enthusiasm [29]. Therefore, the 15 experts consulted in this study showed high levels of support and assistance, giving the survey results high reliability and application value. Additionally, based on the quantitative method of experts' judgment and familiarity, the authority coefficient of the first round of experts was calculated. The values were Ca = 0.892 and Cs = 0.73, with Cr values ranging between 0.7 and 1.0 and a mean value of 0.826, indicating that the experts consulted in this study have high authority, further ensuring the scientific validity and authority of the research results.

3.2. Results of the First Round of Expert Consultation

The purpose of the first round of expert consultation was to assess the necessity of the initially drafted three primary indicators, nine secondary indicators, and 27 tertiary indicators using the options "Select", "Delete" and "Modify", with a consensus threshold set at 70% [40]. The results of the expert consultation indicated that no modifications, deletions, or additions were needed for the primary and secondary indicators. However, for the tertiary indicators, "Finger Span", "Upper Limb Length", "100 m Run" and "5 Times Left-Right Touch" were deleted. Additionally, "Smash and Rush" was modified to "Smash and Net Kill".

3.3. Results of the Second Round of Expert Consultation

Incorporating the modifications suggested by experts in the first round, a new physical fitness evaluation index system for elite male singles badminton players was formed after qualitative screening of the indicators. For the second round, a Likert five-point scale questionnaire was designed to evaluate the importance of the new index system. Experts rated each indicator on a scale of "Very Important", "Important", "Moderately Important", "Less Important" and "Not Important" assigning scores of 5, 4, 3, 2, and 1, respectively. The mean scores, standard deviations, and CVs for the expert ratings were calculated.

All 15 experts returned valid questionnaires, achieving a 100% response rate. Tables 3 and 4 show the average scores for the three primary indicators were above 3.5, with CVs all below 0.25. The Kendall coordination coefficient was W = 0.628; W = 0.628; W = 0.628, and the consistency test showed $p < 0.001$; $p < 0.001$; $p < 0.001$. The results for the secondary indicators revealed the average scores for all 11 secondary indicators were above 3.5, with coefficients of variation below 0.25. The Kendall coordination coefficient W = 0.443; W = 0.443; W = 0.443, and the consistency test showed $p < 0.001$; $p < 0.001$; $p < 0.001$. These results indicate a high level of consistency and agreement among experts for the primary and secondary indicators, meeting the standard requirements.

Table 3. Analysis parameters of indicators at each level in the second round.

Primary Indicators		
Indicators	**M ± SD**	**Cv**
A body morphology	3.93 ± 0.616	0.157
B basic physical fitness	4.36 ± 0.745	0.17
C specialized physical fitness	5.00 ± 0.000	0
Secondary Indicators		
Indicators	**M ± SD**	**Cv**
A1 body indices	4.36 ± 0.497	0.114
A2 length	4.57 ± 0.514	0.112
B1 strength qualities	4.36 ± 0.497	0.114
B2 speed qualities	4.43 ± 0.646	0.146
B3 endurance qualities	4.57 ± 0.514	0.112
B4 flexibility qualities	4.50 ± 0.519	0.115
C1 specific strength qualities	4.43 ± 0.514	0.116
C2 specific endurance qualities	4.64 ± 0.497	0.107
C3 specific agility qualities	4.86 ± 0.363	0.074

Table 3. *Cont.*

Tertiary Indicators		
Indicators	**M ± SD**	**Cv**
A1.1 weight	4.71 ± 0.469	0.1
A1.2 body fat percentage	4.64 ± 0.497	0.107
A2.1 arm span	4.50 ± 0.519	0.115
A2.2 leg length	4.57 ± 0.514	0.112
A2.3 height	4.71 ± 0.469	0.1
B1.1 deep squat	4.64 ± 0.497	0.107
B1.2 Five-Level Frog Jumps	2.50 ± 0.759	0.303
B1.3 bench press	4.79 ± 0.426	0.089
B1.4 abdominal muscle strength	4.71 ± 0.469	0.1
B1.5 back muscle strength	4.79 ± 0.426	0.089
B2.1 30 m	4.64 ± 0.497	0.107
B2.2 50 m	4.50 ± 0.519	0.115
B3.1 3000 m	4.79 ± 0.426	0.089
B3.2 1000 m	4.71 ± 0.469	0.1
B4.1 sit and reach test	4.79 ± 0.426	0.089
C1.1 vertical jump	4.79 ± 0.426	0.089
C1.2 badminton shuttle throw	4.53 ± 0.516	0.115
C1.3 standing long jump	4.60 ± 0.507	0.110
C2.1 400 m×3	4.57 ± 0.514	0.112
C2.2 1500 double unders	4.86 ± 0.363	0.075
C2.3 1-Minute Double Under Jump Rope	2.57 ± 0.646	0.251
C3.1 four-corner touch drill	4.79 ± 0.426	0.089
C3.2 Smash and Net Kill	4.64 ± 0.497	0.107

Table 4. Consistency test statistics of indicators at each level in the second round.

	Kendall (W)	p
Primary Indicators	0.628	<0.001
Secondary Indicators	0.443	<0.001
Tertiary Indicators	0.672	<0.001

However, for the tertiary indicators, the results indicated that the coefficients of variation for B1.2 (Five-Level Frog Jumps) and C2.3 (1-Minute Double Under Jump Rope) were greater than 0.25, reflecting significant disagreement among experts. Consequently, these two indicators were deleted. In this round of expert consultation, some experts suggested that C2.1 (400 m×3) be changed to 400 m×5 to better reflect specificity, A2.1 (Arm Span) be changed to Arm Length, and B1 (Strength Quality) be changed to Relative Strength Quality. These suggestions were adopted. All other indicators met the screening criteria and required no modifications.

3.4. Results of the Third Round of Expert Consultation

Building on the modifications made by experts in the second round, a third round of indicator consultation was conducted. All 15 experts returned valid questionnaires, achieving a 100% response rate. Tables 5 and 6 show the average scores for the three primary indicators in this round were all greater than 3.5, with CVs all less than 0.25. The Kendall coordination coefficient W = 0.588; W = 0.588; W = 0.588, and the consistency test showed $p < 0.001$; $p < 0.001$; $p < 0.001$.

For the nine secondary indicators, the average scores were all greater than 3.5, with CVs all less than 0.25. The Kendall coordination coefficient was W = 0.465; W = 0.465; W = 0.465, and the consistency test showed $p < 0.001$; $p < 0.001$; $p < 0.001$. Based on these standards, the primary and secondary indicators met the requirements.

Table 5. Analysis parameters of indicators at each level in the third round.

Indicators	M ± SD	Cv
Primary Indicators		
Indicators	**M ± SD**	**Cv**
A body morphology	4.40 ± 0.507	0.11
B basic physical fitness	4.60 ± 0.507	0.11
C specialized physical fitness	5.00 ± 0.000	0
Secondary Indicators		
Indicators	**M ± SD**	**Cv**
A1 body indices	4.20 ± 0.676	0.161
A2 length	4.33 ± 0.617	0.142
B1 relative strength quality	4.40 ± 0.507	0.115
B2 speed qualities	4.40 ± 0.632	0.143
B3 endurance qualities	4.40 ± 0.737	0.167
B4 flexibility qualities	4.47 ± 0.516	0.115
C1 specific strength qualities	4.53 ± 0.516	0.114
C2 specific endurance qualities	4.40 ± 0.507	0.115
C3 specific agility qualities	4.60 ± 0.507	0.110
Tertiary Indicators		
Indicators	**M ± SD**	**Cv**
A1.1 weight	4.33 ± 0.816	0.188
A1.2 body fat percentage	4.47 ± 0.640	0.143
A2.1 arm length	4.33 ± 0.816	0.188
A2.2 leg length	4.40 ± 0.737	0.168
A2.3 height	4.47 ± 0.516	0.115
B1.1 deep squat	4.47 ± 0.743	0.166
B1.2 bench press	4.53 ± 0.516	0.113
B1.3 abdominal muscle strength	4.53 ± 0.516	0.113
B1.4 back muscle strength	4.27 ± 0.704	0.164
B2.1 30 m	4.27 ± 0.704	0.164
B2.2 50 m	4.33 ± 0.816	0.188
B3.1 3000 m	4.13 ± 0.743	0.180
B3.2 1000 m	4.33 ± 0.617	0.142
B4.1 sit and reach test	4.73 ± 0.594	0.126
C1.1 vertical jump	4.67 ± 0.488	0.104
C1.2 badminton shuttle throw	4.80 ± 0.414	0.086
C1.3 standing long jump	4.73 ± 0.458	0.097
C2.1 400 m×5	4.20 ± 0.676	0.161
C2.2 1500 double unders	4.33 ± 0.617	0.142
C3.1 four-corner touch drill	4.40 ± 0.507	0.115
C3.2 smash and net kill	4.79 ± 0.426	0.089

Table 6. Consistency test statistics of indicators at each level in the third round.

	Kendall (W)	*p*
Primary Indicators	0.588	<0.001
Secondary Indicators	0.465	<0.001
Tertiary Indicators	0.471	<0.001

For the 21 tertiary indicators in this round, the average scores were all greater than 3.5, with CVs all less than 0.25. The Kendall coordination coefficient was W = 0.471; W = 0.471; W = 0.471, and the consistency test showed $p < 0.001$; $p < 0.001$; $p < 0.001$. All tertiary indicators met the requirements, thus concluding the expert consultation phase. The final determined indicators are shown in Figure 4.

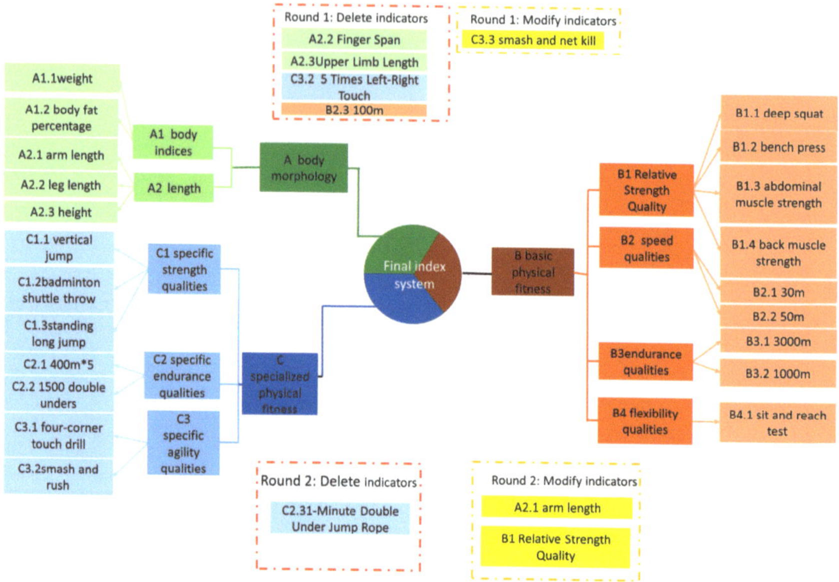

Figure 4. Physical fitness evaluation system for elite male singles badminton players.

3.5. Establishing the Hierarchical Structure Model

In this study, "Physical Fitness Evaluation of Elite Male Singles Badminton Players" is set as the decision-making goal. The hierarchical structure model is constructed with "Body Morphology", "Basic Physical Fitness" and "Specialized Physical Fitness" as the primary indicators, and "Body Index", "Length", "Relative Strength", "Speed Quality", "Endurance Quality", "Flexibility Quality", "Specialized Strength Quality", "Specialized Endurance Quality" and "Specialized Agility Quality" as the secondary indicators. The 21 tertiary indicators serve as the bottom-layer elements. The hierarchical structure model is illustrated in Figure 5.

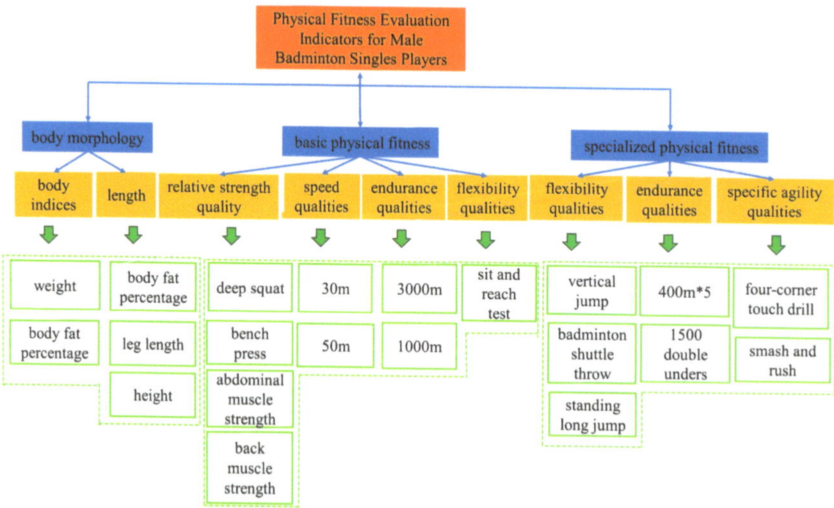

Figure 5. Hierarchical Structure Model.

3.6. Calculation of Indicator System Weights

Figure 6 shows that the weight matrices provided by the experts passed the consistency test. The process involves first calculating the weights of the primary indicators, then the secondary indicators and tertiary indicators within the same level. Subsequently, the global weights of the secondary and tertiary indicators are calculated (Figure 7).

	Indicators	Judgment Aggregation Matrix				N	λmax	CI	RI	CR	Result
Primary Indicators		1	0.293	0.124		3	3.011	0.005	0.52	0.008	Pass
		3.410	1	0.355							
		7.022	2.818	1							
Secondary Indicators	body morphology	1	1.632			≤2	2.000	0.000	0.000	Null	Pass
		0.613	1								
	basic physical fitness	1	0.214	0.373	3.390	4	4.134	0.045	0.89	0.05	Pass
		4.674	1	2.827	6.423						
		2.681	0.354	1	5.548						
		0.295	0.156	0.180	1						
	specialized physical fitness	1	0.393			3	3.008	0.004	0.52	0.008	Pass
		2.544	1								
Tertiary Indicators	body indices	1	0.393			≤2	2.000	0.000	0.000	Null	Pass
		2.544	1								
	length	1	0.551	0.649		3	3.004	0.002	0.52	0.004	Pass
		1.814	1	0.943							
		1.541	1.027	1							
	relative strength quality	1	0.914	2.016	1.842	4	4.023	0.008	0.89	0.009	Pass
		1.094	1	1.840	2.491						
		0.496	0.543	1	1.545						
		0.543	0.401	0.647	1						
	speed qualities	1	0.780			≤2	2.000	0.000	0.000	Null	Pass
		1.282	1								
	endurance qualities	1	0.601			≤2	2.000	0.000	0.000	Null	Pass
		1.663	1								
	flexibility qualities					1					
	specific strength qualities	1	1.011	2.852		3	3.001	0.001	0.52	0.001	Pass
		0.989	1	3.102							
		0.351	0.322	1							
	specific endurance qualities	1	1.340			≤2	2.000	0.000	0.000	Null	Pass
		0.746	1								
	specific agility qualities	1	1.349			≤2	2.000	0.000	0.000	Null	Pass
		1.741	1								

Figure 6. Consistency test results of indicators at each level.

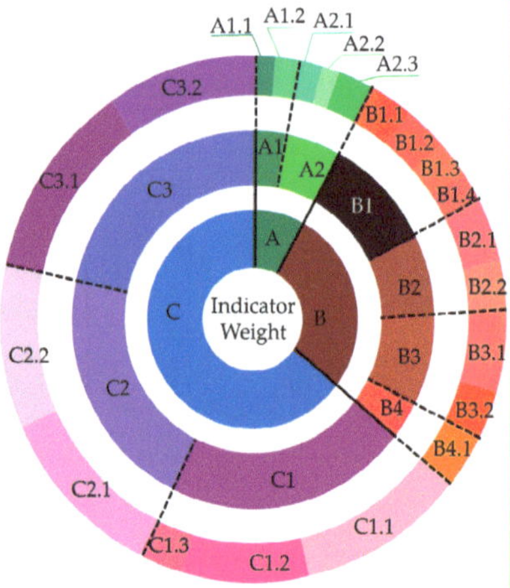

Indicator	Weight
A:body morphology	0.077
A1: body indices	0.371
A2:length	0.629
A1.1:weight	0.313
A1.2:body fat percentage	0.687
A2.1:arm length	0.243
A2.2:leg length	0.300
A2.3:height	0.457

Indicator	Weight
B:basic physical fitness	0.272
B1:relative strength quality	0.379
B2:speed qualities	0.238
B3:endurance qualities	0.280
B4:flexibility qualities	0.103
B1.1:deep squat	0.309
B1.2:bench press	0.322
B1.3: abdominal muscle strength	0.172
B1.4:back muscle strength	0.197
B2.1:30m	0.554
B2.2:50m	0.446
B3.1:3000m	0.594
B3.2:1000m	0.406
B4.1:sit and reach test	1

Indicator	Weight
C:specialized physical fitness	0.651
C1:specific strength qualities	0.334
C2:specific endurance qualities	0.323
C3:specific agility qualities	0.343
C1.1: vertical jump	0.442
C1.2:badminton shuttle throw	0.379
C1.3:standing long jump	0.179
C2.1:400m*5	0.545
C2.2:1500 double unders	0.455
C3.1:four-corner touch drill	0.532
C3.2:smash and net kill	0.467

Figure 7. Weight of each indicator in the physical fitness evaluation system for elite male badminton singles players.

Comprehensive weight calculations and rankings for the indicators in the physical fitness evaluation system of elite male singles badminton players were conducted through the AHP (Figure 8). The results indicate that among the primary indicators, specialized physical fitness holds a dominant weight of 0.651, while Body Morphology has a smaller weight of 0.077. Among the secondary indicators, specialized agility, specialized strength, and specialized endurance have higher weights of 0.223, 0.217, and 0.210, respectively. Among the tertiary indicators, four-corner shuttle run, 400 m×5, smash and net kill, and vertical jump height have higher weights of 0.119, 0.114, 0.104, and 0.096, respectively.

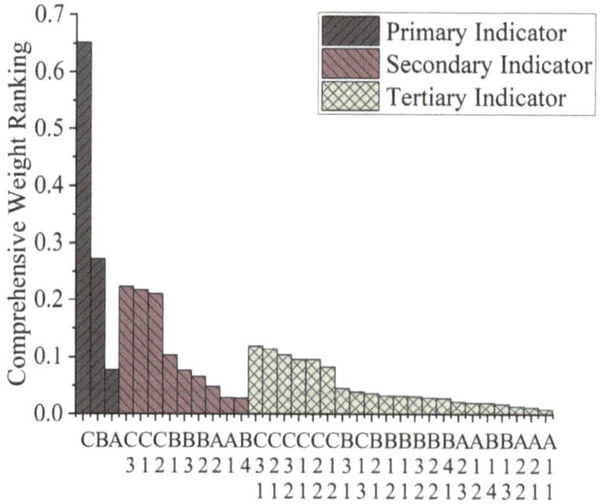

Indicator	Composite Weight
C	0.651
B	0.272
A	0.077
C3	0.223
C1	0.217
C2	0.21
B1	0.103
B3	0.076
B2	0.065
A2	0.048
A1	0.029
B4	0.028
C31	0.119
C21	0.114
C32	0.104
C11	0.096
C22	0.096

Indicator	Composite Weight
C12	0.082
B31	0.045
C13	0.039
B21	0.036
B12	0.033
B11	0.032
B32	0.031
B22	0.029
B41	0.028
A23	0.022
A12	0.02
B14	0.02
B13	0.018
A22	0.014
A21	0.012
A11	0.009

Figure 8. Comprehensive weight ranking of indicators at each level.

These findings further emphasize the primary importance of specialized physical fitness, highlighting its decisive impact on an athlete's performance. The high levels of agility, strength, and endurance suggest that athletes need comprehensive physical abilities to cope with various match situations. For coaches, the results provide clear directions and priorities for training. First, training should be designed based on the indicators to enhance athletes' overall physical fitness. For instance, exercises like the four-corner shuttle run can improve agility and quick-response capabilities. Second, specialized strength training is essential, such as smash and net kill drills, to boost explosive power and hitting strength. Interval training and long-distance running, such as the 400 m×5 exercise, should be implemented to enhance cardiovascular function and endurance. The scientific design and implementation of these training contents will help comprehensively improve athletes' competitive levels, laying a solid foundation for outstanding performance in matches.

3.7. Establishment and Application of the Evaluation System Standards

3.7.1. Development of the Evaluation System Scoring Standards

The development of scoring standards aims to assign accurate numerical values to specific evaluation criteria to obtain fair, objective, and quantifiable evaluation results. The specific steps are as follows:

Data Collection: The average value, standard deviation, maximum value, and minimum value for each indicator were calculated to establish a general value model table for the physical fitness of elite male badminton players (Table 7).

Table 7. General measurement model for physical fitness of elite male badminton singles players.

Test Indicator	Unit	Minimum	Maximum	Average	SD
A1.1 weight	kg	65.00	90.00	72.94	4.07
A1.2 body fat percentage	%	10.00	19.00	13.69	2.82
A2.1 arm length	cm	52.00	67.00	58.66	4.07
A2.2 leg length	cm	102.00	118.00	108.22	4.74
A2.3 height	cm	175.00	193.00	182.00	4.25
B1.1 deep squat	Kg/BW	1.33	2	1.68	0.23
B1.2 bench press	Kg/BW	0.73	1.4	1.01	0.20
B1.3 abdominal muscle strength	min	6.85	7.33	2.36	0.82
B1.4 back muscle strength	min	2.04	3.09	2.24	0.34
B2.1 30 m	s	3.81	4.57	4.13	0.22
B2.2 50 m	s	6.61	8.25	7.31	0.49
B3.1 3000 m	s	611.00	768.00	678.47	47.02
B3.2 1000 m	s	190	270	208	11.67
B4.1 sit and reach test	cm	10.00	26.00	15.25	4.45
C1.1 vertical jump	cm	59.00	71.00	65.75	2.91
C1.2 badminton shuttle throw	cm	595	789	697	63.54
C1.3 standing long jump	cm	269	286	276.06	5.03
C2.1 400 m×5	s	55.00	70.00	66.38	3.38
C2.2 1500 double unders	min	13.13	17.30	15.62	1.13
C3.1 four-corner touch drill	s	29.00	32.68	31.22	0.91
C3.2 smash and net kill	s	30.00	36.11	32.20	1.30

Note: BW = body weight.

This study used the percentile method to divide the collected data into 20 equal parts, with each part corresponding to 5 percentage points (Tables 8 and 9). The percentile method is a widely used statistical analysis technique that effectively describes the value at a specific percentile position in the data distribution. Its main advantage is that it does not require the assumption that the data conform to a normal distribution, thereby eliminating the need for normality testing before setting the standards. Specifically, the percentile method involves sorting the data in ascending order and determining the data value at a specific percentile position to reflect the data distribution. For example, the 50 th percentile indicates that 50% of the data values are less than or equal to this value, and the remaining 50% are greater than or equal to it. This method uses the median and other percentiles as measures of dispersion, allowing for a categorized evaluation of the data and a graded assessment of the athletes' performance levels. The percentile method is extensively used in various fields, both domestically and internationally, due to its simplicity and wide applicability.

This method provides an intuitive description of data distribution and a solid foundation for establishing scientific and reasonable evaluation standards by analyzing data from large sample surveys [41,42]. It should be noted that eight indicators, Body Fat Percentage (A1.2), 30 m (B2.1), 50 m (B2.2), 3000 m (B3.1), 1000 m (B3.2), 400 m×5 (C2.1), Four-Corner Shuttle Run (C3.1), and Smash and Net Kill (C3.2), are in descending order, contrary to other indicators.

Table 8. Performance rating criteria for individual physical fitness indicators of elite male badminton singles players.

Score	A1.1	A1.2	A2.1	A2.2	A2.3	B1.1	B1.2	B1.3	B1.4	B2.1	B2.2
100	100	7	80	132	198	2.3	1.3	7.42	5.58	3.74	6.52
95	98	8	78	130	196.64	2.24	1.26	7.39	5.56	3.76	6.55
90	95	9	76	128	195.28	2.18	1.22	7.36	5.54	3.79	6.58
85	92	10	74	126	193.92	2.12	1.18	7.33	5.52	3.82	6.61
80	89	11	72	124	192.56	2.06	1.14	7.30	5.50	3.85	6.64
75	85	12	70	122	191.2	2	1.1	7.27	5.48	3.88	6.67
70	84	13	68	120	189.84	1.94	1.06	7.24	5.46	3.91	6.70
65	83	14	66	118	188.48	1.88	1.02	7.21	5.44	3.94	6.74
60	81	15	64	116	187.12	1.82	0.98	7.18	5.42	3.97	6.77
55	80	16	62	114	185.76	1.76	0.94	7.15	5.40	4.00	6.80
50	78	17	60	112	184.4	1.7	0.9	7.12	5.38	4.03	6.83
45	75	18	58	110	183.04	1.64	0.86	7.09	5.36	4.06	6.87
40	73	19	56	108	181.68	1.58	0.82	7.06	5.34	4.09	6.90
35	71	20	54	106	180.32	1.52	0.78	7.03	5.32	4.12	6.93
30	68	21	52	104	178.96	1.46	0.74	7.00	5.30	4.15	6.96
25	65	22	50	102	177.6	1.4	0.7	6.93	5.28	4.18	6.99
20	62	23	48	100	176.24	1.34	0.66	6.90	5.26	4.21	7.02
15	60	24	46	98	174.88	1.28	0.62	6.87	5.24	4.25	7.05
10	58	25	44	96	173.52	1.22	0.58	6.85	5.22	4.27	7.08
5	56	26	42	94	172.16	1.16	0.54	6.82	5.20	4.30	7.71
0	53	27	40	92	170.75	1.1	0.5	6.79	5.18	4.33	7.74

Table 9. Performance Rating Criteria for Individual Specialized Physical Fitness Indicators of Elite Male Badminton Singles Players.

Score	B3.1	B3.2	B4.1	C1.1	C1.2	C1.3	C2.1	C2.2	C3.1	C3.2
100	600	180	28	100	950	330	42	27	23	27
95	610	185	27	97	930	325	44	26	23.5	27.5
90	620	190	26	94	910	320	46	25	24	28
85	630	195	25	91	890	315	48	24	24.5	28.5
80	640	200	24	88	870	310	50	23	25	29
75	650	205	23	85	850	305	52	22	25.5	29.5
70	660	210	22	82	830	300	54	21	26	30
65	670	215	21	79	810	295	56	20	26.5	31.5
60	680	220	20	76	790	290	58	19	27	32
55	690	225	19	73	770	285	60	18	27.5	32.5
50	700	230	18	70	750	280	62	17	28	33
45	710	235	17	67	730	275	64	16	28.5	33.5
40	720	240	16	64	710	270	66	15	29	34
35	730	245	15	61	690	265	68	14	29.5	34.5
30	740	250	14	58	670	260	70	13	30	35
25	750	255	13	55	650	255	72	12	31.5	35.5
20	760	260	12	52	630	250	74	11	32	36
15	770	265	11	49	610	245	76	10	32.5	36.5
10	780	270	10	46	590	240	78	9	33	37
5	790	275	9	43	570	235	80	8	33.5	37.5
0	800	280	8	40	550	230	82	7	34	38

The comprehensive score results are obtained by combining individual scores with weights using the following calculation formula: $M = \sum m_i \beta_i$, where m_i represents the score of each indicator, and β_i represents the weight of each indicator. The calculation steps for the comprehensive evaluation results are as follows: Calculation of secondary indicator scores: Multiply the scores of each tertiary indicator by their respective weights and obtain their sum to determine the scores of each secondary indicator. Calculation of primary indicator scores: Multiply the scores of each secondary indicator by their respective weights and obtain their sum to determine the scores of each primary indicator. Calculation of comprehensive evaluation result: Multiply the scores of each primary indicator by their corresponding weights and obtain their sum to obtain the comprehensive evaluation result. The weighted scores for each level of indicators are shown in Figures 9–11.

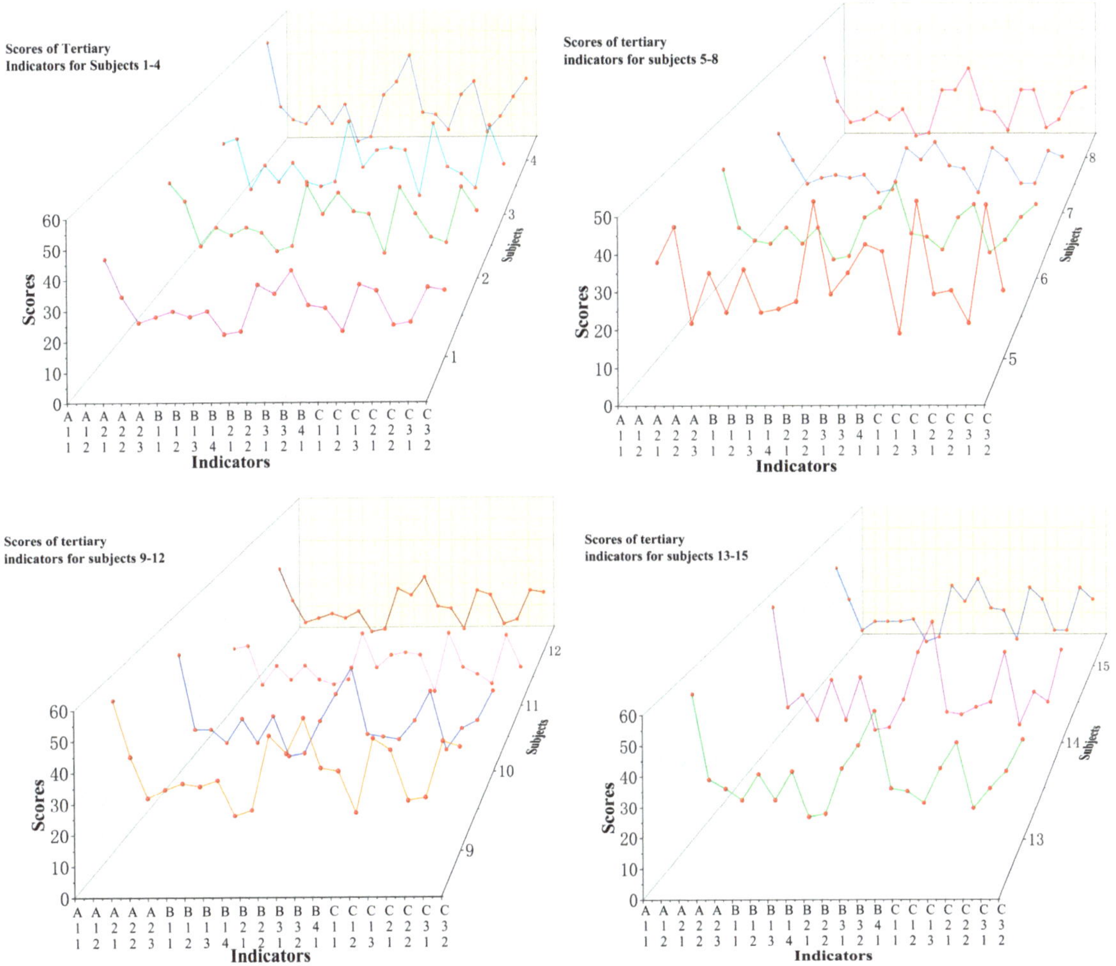

Figure 9. Weighted scores of tertiary indicators for physical fitness of elite male singles badminton players. Note: Different colors of lines represent the scoring conditions of different athletes.

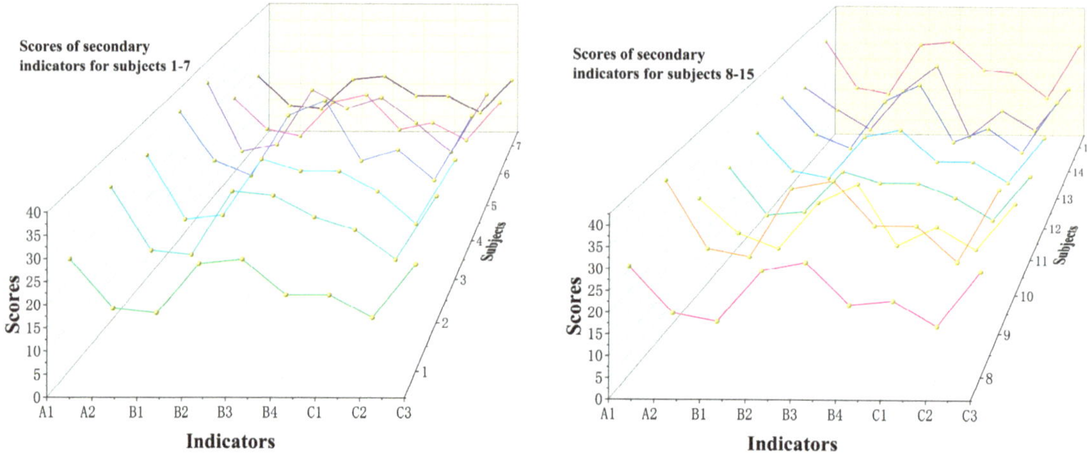

Figure 10. Weighted scores of secondary indicators for physical fitness of elite male singles badminton players. Note: Different colors of lines represent the scoring conditions of different athletes.

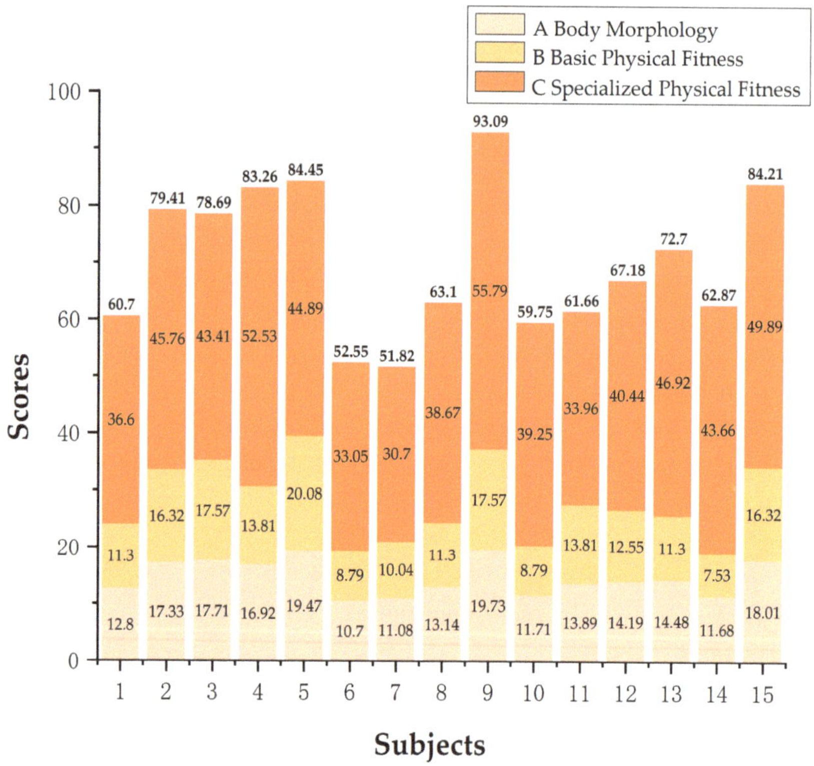

Figure 11. Weighted scores of primary indicators and total scores for physical fitness of elite male singles badminton players.

Based on the data above, a physical fitness level evaluation standard for elite male badminton players was established. The percentile method used in sports measurement and evaluation divides the rating into five levels: excellent, good, average, pass, and poor. The specific steps are as follows: Determine the levels: define the five levels as excellent, good, average, pass, and poor.

Calculate the threshold values: determine the actual measured values corresponding to each level's threshold. Assign scores: assign different scores based on the interval divisions of the levels. List the results: tabulate the results to form a percentile-based rating table. This study uses the common rating division standard: above 90% is excellent, 75%–90% is good, 25%–75% is average, 10%–25% is pass, and below 10% is poor (Table 10). The level evaluation results for each primary indicator are shown in Table 11.

Table 10. Percentile method grading standards.

Evaluation Grade	Standard	Percentage
Excellent	≥90%	10%
Good	75% (incl.)~90%	15%
Average	25% (incl.)~75%	50%
Pass	10%~25%	15%
Poor	≤10%	10%

Table 11. Special physical fitness evaluation standards for elite male badminton singles players.

	Poor	Pass	Average	Good	Excellent
Body Morphology	≤6.42	6.43–12.42	12.43–24.14	24.15–30.5	≥30.55
Basic Physical Fitness	≤8.36	8.37–15.77	15.78–28.94	28.95–40.14	≥40.15
Specialized Physical Fitness	≤14.61	14.62–22.84	22.85–40.75	40.76–54.72	≥54.73

3.7.2. Application of the Physical Fitness Evaluation System

This study randomly selected ten elite male singles badminton players from the China national badminton team to apply the established physical fitness evaluation system for analysis.

The athletes' test data were entered into the physical fitness evaluation system for elite male badminton players, as shown in Table 12. The table shows that out of the 10 athletes, only three achieved an excellent rating in body morphology, four were rated as good, and three as average. In terms of basic physical fitness, five athletes were rated excellent, one good, and four average. For specialized physical fitness, six athletes achieved an excellent rating, one good rating, and three average ratings. Overall, four athletes were rated excellent, two good, and four average.

Table 12. Evaluation grade results for special physical fitness of elite male badminton singles players.

Subjects	Body Morphology	Basic Physical Fitness	Specialized Physical Fitness	Overall Situation
N1	Good	Excellent	Excellent	Good
N2	Excellent	Excellent	Excellent	Excellent
N3	Average	Excellent	Average	Average
N4	Good	Excellent	Excellent	Excellent
N5	Average	Average	Average	Average
N6	Excellent	Average	Excellent	Excellent
N7	Good	Average	Average	Average
N8	Excellent	Good	Good	Good
N9	Average	Excellent	Excellent	Excellent
N10	Good	Average	Excellent	Average

4. Discussion

The physical fitness evaluation index system for elite male badminton players, established through the Delphi method, selects body morphology, basic physical fitness, and specialized physical fitness as primary indicators. This system includes nine secondary indicators and twenty-one tertiary indicators (Figure 6). The aim is to comprehensively understand the physical fitness status of the athletes through systematic and scientific indicator evaluation, enabling the development of more targeted training programs to enhance their competitive performance and match outcomes.

Research has shown that body composition is an important indicator of physical fitness, and an athlete's physique directly impacts their performance. By precisely managing body composition, athletes can not only improve their performance in the short term but also maintain good physical condition and health throughout their athletic careers. [24–26,43]. Different sports require different body morphology characteristics, which are particularly important for the biomechanical properties during physical activity [44,45]. Among the body morphology indicators, weight, body fat percentage, arm length, leg length, and height are included.

Weight and body fat percentage are crucial factors for evaluating an athlete's physical fitness because excessive weight and body fat can significantly negatively impact athletic performance. Previous studies on badminton players have indicated a significant negative correlation between fat content and an athlete's aerobic capacity. A higher body fat percentage can increase the athlete's body temperature during exercise, making them more susceptible to fatigue and thereby reducing aerobic capacity. Additionally, it imposes an extra strain on heart function, which affects the efficiency and endurance of the cardiovascular system [46–49]. Excessive weight also means athletes bear more load during repeated jumps and quick movements, requiring them to have a higher capacity to carry their body weight [26,50]. These adverse factors can prevent athletes from maintaining optimal performance during high-intensity activities, thus affecting their overall performance and competitive level. Phoumsoupha et al. [51] pointed out that top-ranked badminton players are, on average, 5 cm taller than lower-ranked players, indicating that greater height might be an advantage, with taller players performing better. From a sports biomechanics perspective, arm length, leg length, and height provide significant mechanical leverage advantages [52]. Specifically, longer limbs can increase the length of the lever arm, thereby improving torque efficiency, which is crucial for quick movements and effective hitting in badminton. Longer arms and legs help athletes cover the court faster and hit the shuttlecock from higher positions, enhancing their offensive and defensive effectiveness [53]. These physical characteristics provide athletes with a significant performance advantage, allowing them to demonstrate higher efficiency and stability in quick responses and high-intensity movements during matches.

Badminton is considered the most demanding racquet sport in the world [51,54]. Athletes need to make quick movements and change directions as necessary, which undoubtedly requires them to possess excellent strength, speed, endurance, and flexibility [55]. Tiwari et al. [56] found a significant correlation between these qualities and athletes' competitive abilities in studies of sub-elite athletes. Therefore, relative strength, speed, endurance, and flexibility were included in the basic physical fitness indicators. Relative strength was included with four tertiary indicators: deep squat, bench press, abdominal muscles, and back muscles. Deep squats and bench presses are essential indicators for assessing an athlete's strength, reflecting the lower and upper body strength critical for jumping and hitting in badminton [21,57]. Evaluating the abdominal and back muscles helps understand the athlete's core strength, which is crucial for maintaining body stability and power transfer. Quick movements, posture changes, hitting power, and accuracy in badminton require strong core strength and high dynamic balance [28,58].

The speed indicators of the 30- and 50-m runs evaluate short-distance explosive power and acceleration. Frequent quick movements and sudden stops are required in badminton matches, and these short-distance speed tests can accurately reflect the athlete's acceleration and reaction capabilities, enhancing quick movements and positional changes in matches [51]. Badminton matches typically last between 40 min and 1 h [27,59]. The high intensity and frequency of matches require badminton players to have strong aerobic energy capacity [60,61]. The endurance indicators of the 3000- and 1000-m runs are chosen based on the high intensity and duration requirements of badminton matches to assess the athlete's cardiovascular function and endurance level, ensuring they can maintain physical fitness during long, high-intensity matches. Flexibility plays a crucial role in badminton because it allows athletes to move their bodies and limbs to a wide range of motion within the limited space of the court. This flexibility is essential for effective returns and quick turns and affects an athlete's agility, technical performance, and injury prevention [56,62].

For non-rhythmic sports such as badminton, handball, and basketball, traditional physical fitness testing methods using standard bicycles or treadmills are insufficient to accurately predict an athlete's performance in competition [27]. Therefore, this study included specialized physical fitness as one of the primary indicators to ensure the evaluation results accurately reflect the physical demands and performance levels required in actual matches. Within the specialized physical fitness primary indicator, three secondary indicators were included: specialized strength, specialized endurance, and specialized agility. These indicators aim to assess the specific physical fitness requirements of athletes in badminton matches. Specialized Strength: The tertiary indicators of vertical jump, badminton throw, and standing long jump effectively assess the athlete's explosive power and strength application ability, which are crucial for jumping and hitting actions [18,51,57,63]. Specialized Endurance: The 400 m$\times$5 and 1-min double-under jump rope tests evaluate the athlete's ability to sustain high-intensity activity throughout the match, ensuring they can maintain high performance levels during the later stages of the game [51,64,65]. Specialized Agility: Agility refers to the ability to rapidly change the speed or direction of the entire body in response to specific stimuli during movement [66]. The tertiary indicators of smash and net kill and the four-corner shuttle run assess the athlete's quick reaction and agility, which often determine the outcome of a match in critical moments [67]. The inclusion of specialized physical fitness ensures that the evaluation system is comprehensive and tailored to the specific demands of badminton, providing a more accurate assessment of the athletes' capabilities and performance potential in real match scenarios.

The evaluation results and test data of the ten athletes mentioned above in the aspects of body morphology, basic physical fitness, and specialized physical fitness were obtained, and a comprehensive assessment was conducted. The results indicate significant differences in the athletes' performance across various indicators. Thus, this paper further discusses the implications of these results for coaches in terms of training plan formulation and execution, as well as for athletes' training. From the coach's perspective, the physical fitness evaluation results highlight certain issues in the formulation and execution of training plans. Coaches need to develop more targeted training plans based on the individual differences of athletes. For athletes who perform poorly on certain physical fitness indicators, training should focus on improving these weaknesses. Coaches should also enhance the monitoring and evaluation of athletes' physical fitness changes, adjusting training content and intensity in a timely manner to ensure that athletes achieve optimal conditions in all aspects. From the athlete selection perspective, the selection process should not only focus on the technical level of the athletes but also comprehensively consider factors such as body morphology, basic physical fitness, and specialized physical fitness. This approach ensures the selection of truly promising athletes with development potential. For athletes whose physical fitness evaluation results are not ideal, coaches can improve their performance through targeted physical training. From the athletes' training perspective, the results of the physical fitness evaluation system provide clear directions for their training. Athletes should recognize their strengths and weaknesses based on the evaluation results and make improvements in their daily training. For example, athletes with weaker body morphology should increase the proportion of basic and specialized physical fitness training. Conversely, athletes with poor specialized physical fitness should focus more on specialized technical and related physical training. Athletes can comprehensively improve their physical fitness levels through systematic and scientific training arrangements, laying a solid foundation for achieving excellent performance in competitions.

This study has several limitations. First, the research subjects are limited to elite male singles badminton players from China, which may constrain the applicability of the findings to athletes of different genders, countries, and regions. Although Chinese athletes hold a prominent position in international badminton, the generalizability of these results to other cultural and training environments has not been established. Second, the consulted experts are all from China, and their professional opinions may have regional and cultural biases, lacking a comprehensive inclusion of professional perspectives from diverse global backgrounds. This might limit the applicability of the evaluation index system on a worldwide scale. Additionally, since the study focuses on the characteristics of high-level athletes in China, the constructed physical fitness evaluation index system may not fully capture the needs and characteristics of athletes at other levels or backgrounds. Therefore, future research should consider broader samples and more diverse expert opinions to enhance the generalizability and scientific validity of the findings.

5. Research Limitations and Future Research Directions

Although this study has achieved certain findings and results, it presents several unavoidable limitations. The subjects were elite male singles badminton players at the national level in China, selected using stringent criteria, thus limiting the number of athletes who could participate. Future research should increase the sample size to enhance the representativeness and scientific validity of the results. Moreover, the reliance on data from Chinese elite players may limit the generalizability of the results. This regional restriction could affect the applicability of the evaluation system in other cultural contexts. Future studies are planned to include athletes from various countries to enhance the global applicability of the system. Another limitation is the lack of longitudinal data, which restricts demonstrating the long-term effectiveness of the system and its practical value in improving athletic performance. Future research should focus on collecting and analyzing longitudinal data, which would help to assess the ongoing impact of the evaluation system on long-term training and performance, as well as determine if adjustments are needed in the evaluation metrics or methods. Moreover, although the Delphi method reduced individual biases through collective consensus, it could not completely eliminate the subjectivity of expert opinions, which might influence the study results. Additionally, the absence of a control group is a clear limitation. Ideally, future studies would include non-elite athletes or those using different evaluation metrics to provide a broader comparison and validate the effectiveness of the evaluation system. Finally, the study's focus on elite male singles players may limit the broader applicability of the findings. Including female athletes, doubles players, and athletes from other sports in future studies would enhance the universality and applicability of the evaluation system, helping to more fully understand and meet the diverse needs and characteristics of different athlete groups.

6. Conclusions

The established indicator system for evaluating the physical fitness of elite male singles badminton players comprises three primary indicators: "Body Morphology", "Basic Physical Fitness" and "Specialized Physical Fitness", alongside nine secondary and twenty-one tertiary indicators. The Analytic Hierarchy Process (AHP) results highlighted the predominance of specialized physical fitness among primary indicators. Secondary indicators such as specialized agility, strength, and endurance, along with tertiary indicators including four-corner shuttle runs, 400 m$\times$5 shuttle runs, smash and net kill, and vertical jump height, were weighted significantly. This system's application to elite badminton players validated its scientific accuracy and practical value, providing comprehensive fitness assessments that guide targeted training plans and optimize performance outcomes. The evaluation system and assigned weights proved to enhance the accuracy and objectivity of fitness assessments, demonstrating significant practical value in athlete training and development.

Author Contributions: H.Z. and B.Y. designed the study. H.G. and B.Y. performed the experiment and wrote the main manuscript text. Z.Y. and Z.H. analyzed the data. H.P. and G.L. guided the experiment and thesis writing. All authors have read and agreed to the published version of the manuscript.

Funding: The present study was funded by the Department of Science and Technology of Zhejiang Province, China (No. 2021C03128 and No. 2023C03197) and the Science and Technology Innovation Project of the General Administration of Sport of China (No. 23KJCX036).

Institutional Review Board Statement: The study was conducted in accordance with the Declaration of Helsinki, and approved by the Institutional Review Board of Zhejiang Normal University (protocol code ZSRT2024165).

Informed Consent Statement: Informed consent was obtained from all subjects involved in the study.

Data Availability Statement: Data will be made available on request.

Conflicts of Interest: The authors unequivocally assert that this research was undertaken while devoid of any commercial or financial affiliations that might be perceived as potential conflicts of interest.

References

1. Lei, L.C. Design and implementation of physical fitness training of China national badminton team in preparing for 2012 London olympic games. *J. Beijing Sport Univ.* **2016**, *39*, 86–91. [CrossRef]
2. Zhou, J.L.; Dai, J. An analysis on the characters of technique; tactic in "twenty-one point" system of badminton men's doubles. *J. Beijing Sport Univ.* **2010**, *33*, 111–113. [CrossRef]
3. Dai, J.B.; Guan, Y.; Lu, Z.Q.; Jiang, J. Influence of21-point system on development of men's singles technique and tactics of badminton. *J. Shanghai Univ. Sport* **2008**, *2*, 86–89. [CrossRef]
4. Sharif, M.A.S.; Ramezani, M.; Siavoshy, H. The effect of continuous aerobic and high-intensity interval training on some physical fitness factors in young football players. *Int. J. Sport Stud. Health* **2024**, *7*, 11–17.
5. Lam, W.K.; Wong, D.W.C.; Lee, W.C.C. Biomechanics of lower limb in badminton lunge: A systematic scoping review. *PeerJ* **2020**, *8*, e10300. [CrossRef] [PubMed]
6. Zhao, W.; Wang, C.; Bi, Y.; Chen, L. Effect of integrative neuromuscular training for injury prevention and sports performance of female badminton players. *Biomed Res. Int.* **2021**, *2021*, 5555853. [CrossRef] [PubMed]
7. Wang, D.; Wang, S.; Hou, J.; Yin, M. Construction of a sport-specific strength and conditioning evaluation index system for elite male wheelchair badminton athletes by the Delphi method. *J. Men's Health* **2023**, *19*, 82–91. [CrossRef]
8. Huang, H.; Lin, C.; Hu, C. Analysis of selection indicators of badminton players by the delphi method and analytic hierarchy process. *AIRCC's Int. J. Comput. Sci. Inf. Technol.* **2015**, *7*, 19–31. [CrossRef]
9. Jaworski, J.; Żak, M. The structure of morpho-functional conditions determining the level of sports performance of young badminton players. *J. Hum. Kinet.* **2015**, *47*, 215–223. [CrossRef]
10. Zhang, Y. Tianjin badminton team to prepare for the 2017 national games special physical training measurement index system. *Contemp. Sports Technol.* **2017**, *7*, 49–50. [CrossRef]
11. Ning, Q. Measurement index system construction for male teenagers' special physical training in badminton. *J. Chengdu Sport Univ.* **2015**, *41*, 91–96. [CrossRef]
12. Shizhujiang, Z. Research on the special physical training of vietnamese elite badminton players. *Contemp. Sports Technol.* **2021**, *11*, 54–56. [CrossRef]
13. Minas, H.; Jorm, A.F. Where there is no evidence: Use of expert consensus methods to fill the evidence gap in low-income countries and cultural minorities. *Int. J. Ment. Health Syst.* **2010**, *4*, 33. [CrossRef] [PubMed]
14. Rowe, G.; Wright, G.; Bolger, F. Delphi: A reevaluation of research and theory. *Technol. Forecast. Soc. Chang.* **1991**, *39*, 235–251. [CrossRef]
15. Goodman, C.M. The Delphi technique: A critique. *J. Adv. Nurs.* **1987**, *12*, 729–734. [CrossRef]
16. KEENEY, S.; Hasson, F.; Mckenna, H. Consulting the oracle: Ten lessons from using the Delphi technique in nursing research. *J. Adv. Nurs.* **2006**, *53*, 205–212. [CrossRef] [PubMed]
17. Mullen, P.M. Delphi: Myths and reality. *J. Health Organ. Manag.* **2003**, *17*, 37–52. [CrossRef] [PubMed]
18. Zhang, L. Physical strength training methods in badminton teaching and training. *Rev. Bras. Med. Esporte* **2023**, *29*, e2022_0253. [CrossRef]
19. Tomaszewski, P.; Kęska, A.; Tkaczyk, J.; Nowicki, D.; Sienkiewicz-Dianzenza, E. Somatic characteristics and motor fitness of elite and sub-elite polish male badminton players. *J. Sports Med. Phys. Fit.* **2018**, *58*, 1456–1464. [CrossRef]
20. Suppiah, P.K.; Joummy, A.J.; Samsir, M.S.; Mariappan, M.; Noordin, H.; Azmi, A.M.N. The effects of high intensity functional interval training on selected fitness components among young badminton players. In *International Conference on Movement, Health and Exercise*; Lecture Notes in Bioengineering; Springer: Singapore, 2020; pp. 42–53. [CrossRef]

21. Ooi, C.H.; Tan, A.; Ahmad, A.; Kwong, K.W.; Sompong, R.; Mohd Ghazali, K.A.; Liew, S.L.; Chai, W.J.; Thompson, M.W. Physiological characteristics of elite and sub-elite badminton players. *J. Sports Sci.* **2009**, *27*, 1591–1599. [CrossRef]
22. Andersen, L.L.; Larsson, B.; Overgaard, H.; Aagaard, P. Torque–velocity characteristics and contractile rate of force development in elite badminton players. *Eur. J. Sport Sci.* **2007**, *7*, 127–134. [CrossRef]
23. Wonisch, M.; Hofmann, P.; Schwaberger, G.; Von Duvillard, S.P.; Klein, W. Validation of a field test for the non-invasive determination of badminton specific aerobic performance. *Br. J. Sports Med.* **2003**, *37*, 115–118. [CrossRef] [PubMed]
24. Visnapuu, M.; Jürimäe, T. Relations of anthropometric parameters with scores on basic and specific motor tasks in young handball players. *Percept. Mot. Skills* **2009**, *108*, 670–676. [CrossRef] [PubMed]
25. Nikolaidis, P.T.; Chtourou, H.; Torres-Luque, G.; Rosemann, T.; Knechtle, B. The relationship of age and bmi with physical fitness in futsal players. *Sports* **2019**, *7*, 87. [CrossRef] [PubMed]
26. Ayuningtyas, N.T.; Susanto, H.; Suroto, S. Relationship between somatotype and physical Fitness: Study on badminton athletes of PB Djarum kudus. *J. Keolahragaan* **2021**, *9*, 128–136. [CrossRef]
27. Chin, M.; Wong, A.S.; So, R.C.; Siu, O.T.; Steininger, K.; Lo, D.T. Sport specific fitness testing of elite badminton players. *Br. J. Sports Med.* **1995**, *29*, 153–157. [CrossRef] [PubMed]
28. Ma, S.; Soh, K.G.; Japar, S.B.; Liu, C.; Luo, S.; Mai, Y.; Wang, X.; Zhai, M. Effect of core strength training on the badminton player's performance: A systematic review & meta-analysis. *PLoS ONE* **2024**, *19*, e305116. [CrossRef]
29. Cavedon, V.; Zancanaro, C.; Milanese, C. Anthropometry, body composition, and performance in sport-specific field test in female wheelchair basketball players. *Front. Physiol.* **2018**, *9*, 568. [CrossRef] [PubMed]
30. Hasson, F.; Keeney, S.; Mckenna, H. Research guidelines for the Delphi survey technique. *J. Adv. Nurs.* **2000**, *32*, 1008–1015. [CrossRef]
31. Xie, L.; Feng, M.; Cheng, J.; Huang, S. Developing a core competency training curriculum system for emergency trauma nurses in china: A modified delphi method study. *BMJ Open* **2023**, *13*, e66540. [CrossRef]
32. Wang, X.; Chen, T.; Zhang, Y.; Yang, H.H. Implications of the Delphi method in the evaluation of sustainability open education resource repositories. *Educ. Inf. Technol.* **2021**, *26*, 3825–3844. [CrossRef]
33. He, C.; Pei, C.; Ma, J. Developing an evaluation indicators of health literacy for cervical cancer among chinese women: A modified delphi method study. *BMC Cancer* **2023**, *23*, 863. [CrossRef]
34. Cancela, J.; Fico, G.; Waldmeyer, M.T.A. Using the analytic hierarchy process (AHP) to understand the most important factors to design and evaluate a telehealth system for parkinson's disease. *BMC Med. Inform. Decis. Mak.* **2015**, *15*, S7. [CrossRef]
35. Wang, K.; Wang, C.K.; Hu, C. Analytic hierarchy process with fuzzy scoring in evaluating multidisciplinary r&d projects in china. *IEEE Trans. Eng. Manag.* **2005**, *52*, 119–129. [CrossRef]
36. Vaidya, O.S.; Kumar, S. Analytic hierarchy process: An overview of applications. *Eur. J. Oper. Res.* **2006**, *169*, 1–29. [CrossRef]
37. Vidal, L.; Marle, F.; Bocquet, J. Using a Delphi process and the analytic hierarchy process (Ahp) to evaluate the complexity of projects. *Expert Syst. Appl.* **2011**, *38*, 5388–5405. [CrossRef]
38. Krejčí, J.; Stoklasa, J. Aggregation in the analytic hierarchy process: Why weighted geometric mean should be used instead of weighted arithmetic mean. *Expert Syst. Appl.* **2018**, *114*, 97–106. [CrossRef]
39. Pant, S.; Kumar, A.; Ram, M.; Klochkov, Y.; Sharma, H.K. Consistency indices in analytic hierarchy process: A review. *Mathematics* **2022**, *10*, 1206. [CrossRef]
40. Jünger, S.; Payne, S.; Brine, J.; Radbruch, L.; Brearley, S. Guidance on conducting and reporting Delphi studies (CREDES) in palliative care: Recommendations based on a methodological systematic review. *Palliat. Med.* **2017**, *31*, 684–706. [CrossRef]
41. Frost, J. Percentiles: Interpretations; calculations, Statistics by Jim (February 2023). 2019. Available online: https://statisticsbyjim.com/basics/percentiles (accessed on 10 May 2024).
42. Bornmann, L.; Leydesdorff, L.; Mutz, R. The use of percentiles and percentile rank classes in the analysis of bibliometric data: Opportunities and limits. *J. Informetr.* **2013**, *7*, 158–165. [CrossRef]
43. Sayyah, M.; Hosseini, M.E.; Dibahosseini, S.A. Comparing the physical fitness conditions of boys' school students in the city of kashan. *AI Tech Behav. Soc. Sci.* **2024**, *2*, 7–11. [CrossRef]
44. Duncan, M.J.; Woodfield, L.; Al-Nakeeb, Y. Anthropometric and physiological characteristics of junior elite volleyball players. *Br. J. Sports Med.* **2006**, *40*, 649–651. [CrossRef]
45. Gutnik, B.; Zuoza, A.; Zuozienė, I.; Alekrinskis, A.; Nash, D.; Scherbina, S. Body physique and dominant somatotype in elite and low-profile athletes with different specializations. *Medicina* **2015**, *51*, 247–252. [CrossRef]
46. de Faria Pastore, J.C.; de Azevedo Ferreira, C.A.; Da Costa, F.C.H.; João, P.V. Kinanthropometric profile of beach volleyball player of category under 19 and under 21 of European championship. *Chest* **2016**, *80*, 83–89. [CrossRef]
47. Moss, S.L.; Mcwhannell, N.; Michalsik, L.B.; Twist, C. Anthropometric and physical performance characteristics of top-elite, elite and non-elite youth female team handball players. *J. Sports. Sci.* **2015**, *33*, 1780–1789. [CrossRef]
48. Aini, T.R.N.; Rahfiludin, M.Z.; Kartini, A. Hubungan persen lemak tubuh dan kadar hemoglobin dengan kapasitas vo2 max atlet bulutangkis (Studi di UKM Bulutangkis Universitas Negeri Semarang dan Universitas Diponegoro). *Media Kesehat. Masy. Indones.* **2019**, *18*, 113–116. [CrossRef]
49. Brezdeniuk, O. Aerobic possibilities of 17–21 year students with different components of body mass. *Fiz. Akt. Zdorov'ia Sport* **2014**, *1*, 9–18.

50. Maher, E. Health-related quality of life of severely obese children and adolescents. *Child Care Health Dev.* **2004**, *30*, 94–95. [CrossRef]
51. Phomsoupha, M.; Laffaye, G. The science of badminton: Game characteristics, anthropometry, physiology, visual fitness and biomechanics. *Sports Med.* **2015**, *45*, 473–495. [CrossRef]
52. Knudson, D.V.; Knudson, D. *Fundamentals of Biomechanics*; Springer: New York, NY, USA, 2007. [CrossRef]
53. Tsoukos, A.; Drikos, S.; Brown, L.E.; Sotiropoulos, K.; Veligekas, P.; Bogdanis, G.C. Anthropometric and motor performance variables are decisive factors for the selection of junior national female volleyball players. *J. Hum. Kinet.* **2019**, *67*, 163–173. [CrossRef]
54. Liddle, D.; Murphy, M.; Bleakley, W. A comparison of the physiological demands of singles and doubles badminton: A heart rate and time/motion analysis. *J. Hum. Mov. Stud.* **1996**, *30*, 159–176.
55. Raman, D.; Nageswaran, A. Effect of game-specific strength training on selected physiological variables among badminton players. *SSB* **2013**, *1*, 57–563. [CrossRef]
56. Tiwari, L.M.; Rai, V.; Srinet, S. Relationship of selected motor fitness components with the performance of badminton player. *Asian J. Phys. Educ. Comput. Sci. Sports* **2011**, *5*, 88–91.
57. Liu, Z.; Wang, L. Impacts of lower limb training in badminton. *Rev. Bras. Med. Esporte* **2023**, *29*, e2022_0702. [CrossRef]
58. Churi, A.B.; Varadharajulu, G. Effect of core strengthening and conditioning programme on badminton beginners. *Indian J. Forensic Med. Toxicol.* **2020**, *14*, 156–161.
59. Manrique, D.C.; Gonzalez-Badillo, J.J. Analysis of the characteristics of competitive badminton. *Br. J. Sports Med.* **2003**, *37*, 62–66. [CrossRef]
60. Cabello, D.; Padial, P.; Lees, A.; Rivas, F. Temporal and physiological characteristics of elite women's and men's singles badminton. *Int. J. Appl. Sports Sci.* **2004**, *16*, 1–12.
61. Faude, O.; Meyer, T.; Rosenberger, F.; Fries, M.; Huber, G.; Kindermann, W. Physiological characteristics of badminton match play. *Eur. J. Appl. Physiol.* **2007**, *100*, 479–485. [CrossRef]
62. Singh, N.N.; Singh, R.R.; Singh, S.K. Study of trunk flexibility and body composition between football and badminton players. *J. Phys. Educ. Sport* **2011**, *11*, 18.
63. Wang, D.; Wang, L.; Liu, X.; Fan, H. A comparative study of lower limb strength during vertical jump of male college students in table tennis, badminton and tennis. *J. Men's Health* **2024**, *1*, 8. [CrossRef]
64. Li, J.; Li, R. Effect of double rocking jump rope training on lower limb muscle strength of badminton players. *Technol. Health Care* **2024**, 1–18. [CrossRef] [PubMed]
65. Tang, J. Effects of jump rope training on explosive power in badminton players. *Rev. Bras. Med. Esporte* **2023**, *29*, e2022_0627. [CrossRef]
66. Sheppard, J.M.; Young, W.B. Agility literature review: Classifications, training and testing. *J. Sports Sci.* **2006**, *24*, 919–932. [CrossRef] [PubMed]
67. Kusuma, D.W.Y.; Raharjo, H.P.; Taathadi, M.S. Introducing a new agility test in badminton. *Am. J. Sports Sci.* **2015**, *3*, 18–28. [CrossRef]

life

Article

Acute Effects of Breath-Hold Conditions on Aerobic Fitness in Elite Rugby Players

Wendi Wang [1,†], Dongzhe Wu [1,2,†], Hao Wang [1], Zhiqiang Zhang [3], Xuming Jiang [3], Shufeng Li [3], Yongjin Shi [3] and Xiaolin Gao [1,*]

1 Sports Rehabilitation Research Center, China Institute of Sport Science, Beijing 100061, China; wangwendi@ciss.cn (W.W.); wdz276640188@outlook.com (D.W.); wanghao@ciss.cn (H.W.)
2 School of Sport Science, Beijing Sport University, Beijing 100084, China
3 Department of Sports and Arts, China Agricultural University, Beijing 100083, China; johnny1974@sina.cn (Z.Z.); jxm7762@126.com (X.J.); lishufeng@cau.edu.cn (S.L.); shiyongjin@cau.edu.cn (Y.S.)
* Correspondence: gaoxiaolin@ciss.cn
† These authors contributed equally to this work.

Abstract: The effects of face immersion and concurrent exercise on the diving reflex evoked by breath-hold (BH) differ, yet little is known about the combined effects of different BH conditions on aerobic fitness in elite athletes. This study aimed to assess the acute effects of various BH conditions on 18 male elite rugby players (age: 23.5 ± 1.8 years; height: 183.3 ± 3.4 cm; body mass: 84.8 ± 8.5 kg) and identify the BH condition eliciting the greatest aerobic fitness activation. Participants underwent five warm-up conditions: baseline regular breathing, dynamic dry BH (DD), static dry BH (SD), wet dynamic BH (WD), and wet static BH (WS). Significant differences ($p < 0.05$) were found in red blood cells (RBCs), red blood cell volume (RGB), and hematocrit (HCT) pre- and post-warm-up. Peak oxygen uptake (VO_2peak) and relative oxygen uptake (VO_2/kgpeak) varied significantly across conditions, with BH groups showing notably higher values than the regular breathing group ($p < 0.05$). Interaction effects of facial immersion and movement conditions were significant for VO_2peak, VO_2/kgpeak, and the cardiopulmonary optimal point ($p < 0.05$). Specifically, VO_2peak and peak stroke volume (SVpeak) were significantly higher in the DD group compared to that in other conditions. Increases in VO_2peak were strongly correlated with changes in RBCs and HCT induced by DD warm-up ($r\Delta RBC = 0.84$, $r\Delta HCT = 0.77$, $p < 0.01$). In conclusion, DD BH warm-up appears to optimize subsequent aerobic performance in elite athletes.

Keywords: breath-hold; apnea; diving reflex; warm-up; aerobic fitness; cardiopulmonary optimal point; autonomic nervous system; peak oxygen uptake; stroke volume

Citation: Wang, W.; Wu, D.; Wang, H.; Zhang, Z.; Jiang, X.; Li, S.; Shi, Y.; Gao, X. Acute Effects of Breath-Hold Conditions on Aerobic Fitness in Elite Rugby Players. *Life* **2024**, *14*, 917. https://doi.org/10.3390/life14080917

Academic Editors: Ines Drenjančević and Marco Duca

Received: 14 June 2024
Revised: 9 July 2024
Accepted: 19 July 2024
Published: 23 July 2024

1. Introduction

Breath-hold (BH) is frequently utilized in sports such as diving and swimming to improve athletes' endurance. This practice has been associated with notable erythropoietic responses, leading to increased hemoglobin concentration and potentially enhancing athletic performance through the diving reflex (DR) [1,2]. DR involves bradycardia, decreased cardiac output, increased vascular resistance, and the redistribution of blood flow [3,4]. Its primary function is to reduce blood flow and oxygen supply to nonessential organs, redirecting resources to oxygen-dependent organs. Recent research has demonstrated that even a single maximal breath-hold can enhance athletic performance [5–7]. However, understanding the diverse factors influencing BH and their impact on athletic performance remains an ongoing area of investigation [8,9].

Initially, the DR was thought to be triggered exclusively by BH with facial immersion, stimulating the trigeminal nerve [10]. Emerging research indicates that a similar reflex can be elicited by BH even without facial immersion, leading to related physiological effects. Consequently, research on BH's physiological effects has predominantly focused on specific factors such as facial immersion [11], water temperature [12], and dive depth [13]. Nonetheless, caution is warranted regarding the potential adverse effects of cold-water-induced trigeminal nerve over-stimulation, which may excessively activate the parasympathetic nervous system and impede subsequent exercise performance.

Additionally, static BH differs from dynamic BH in several important ways. Static BH induces bradycardia and ensures blood supply to vital organs through α-adrenergic vasoconstriction [14], which constricts blood vessels, reducing oxygen consumption and cardiac load. In contrast, dynamic BH is followed by a higher heart rate (HR) [15], increased oxygen consumption, and blood redistribution to cardiac and skeletal muscles [16]. Moreover, BH, while reflexively enhancing muscle tone, is believed to augment muscle strength [17], facilitating more efficient activation of muscle groups and improving the body's energy expenditure rate. Thus, dynamic BH may produce superior physiological effects on DR [18]. However, current studies only compare the effects of regular lower limb exercise and face immersion individually, neglecting their combined impact or interactions. Given BH's significant impact on physiological parameters, it is crucial to investigate how these changes translate into athletic performance, especially during high-intensity exercise. The athletes' exercise performance serves as a critical dependent variable in understanding the practical applications of BH strategies in sports.

This study hypothesized that combining facial immersion (wet and dry) and movement conditions (dynamic and static) during breath-holding affects cardiopulmonary fitness in elite athletes. The primary objective was to explore how these different conditions influence athletes' exercise performance during subsequent high-intensity exercise. To achieve this, we selected cardiopulmonary exercise tests, hemodynamic parameters, and hematological parameters as key measures, providing a comprehensive evaluation of the physiological responses and their potential implications for performance. Additionally, the study aimed to assess the feasibility of incorporating BH into the warm-up process and investigate related physiological mechanisms.

2. Materials and Methods

2.1. Procedures

Sample size calculation was conducted using G*Power 3.1 software, confirming the sufficiency of the sample size for this study. Statistical power analysis indicated that, with a significance level (α) of 0.05, a medium effect size (Cohen's d = 0.65), and a statistical power of ≥0.7, a total of 17 participants were required for the survey.

The study utilized a self-controlled trial design, wherein each participant completed all five warm-up protocols in a randomized order across five separate trials, with a one-week interval between each trial. Participants' random trial sequence was determined by labeling the experiments 1–5 and using Microsoft Excel 2010. Before each trial, all subjects performed a uniform lower limb muscle stretch for 3 min and then entered a 10 min warm-up intervention phase. Fingerstick blood was collected before and after each warm-up intervention. Cardiopulmonary exercise testing was started immediately after the warm-up intervention. To account for possible circadian rhythm effects, all subjects were tested between 9:30 and 11:00 a.m. to ensure that both experiments on the same subject were conducted during the same time period. Participants were instructed not to consume caffeine or alcoholic beverages within the 48 h preceding the test day.

2.2. Participants

Eighteen male athletes (age: 23.5 ± 1.8 years; height = 183.3 ± 3.4 cm; BM: 84.8 ± 8.5 kg; BMI: 25.2 ± 2.0 kg/m^2) were voluntarily recruited to participate in this study. Prior to the commencement of the formal trial, participants were briefed on the trial's complete process and objectives. They provided informed consent by signing a consent form and underwent exercise risk screening. Additionally, participants were instructed to abstain from consuming caffeine or alcohol within the 48 h preceding the test day. The day before each experimental condition, participants were ensured a minimum of eight hours of sleep and were required to wear identical sneakers and clothing during the experiment. An experimental technician supervised the entirety of the exercise test. The Ethics Committee of the China Institute of Sports Science approved the experiment.

The inclusion criteria were as follows: ① being aged 18–25 years old, holding a national second-level or above athlete qualification certificate; ② being healthy and without recent movement disorders; ③ having no cardiovascular or cerebrovascular disease, and no family history of sudden death; ④ understanding the trial, signing the informed consent form, and committing to participate in and complete the entire trial process.

The exclusion criteria were as follows: ① a presence of cardiac, cerebral, or vascular diseases and a family history of sudden death; ② failure of routine ECG, blood pressure monitoring, and blood screening; ③ underlying medical problems or a history of ankle, knee, or back injuries; ④ any lower extremity reconstructive surgery or unresolved musculoskeletal disorders in the last two years; ⑤ previous BH dives or hypoxic training; ⑥ travel to a high-altitude area in the last six months.

2.3. Testing Procedures

When the subjects arrived at the laboratory, they initially sat quietly for 3 min. Following this period, they performed a 3 min lower limb muscle stretch designed to reduce injury risk during subsequent cycling. This was followed by pedaling on the power bike, during which various BH procedures were employed to assess the impact of different environmental factors on exercise performance.

The specific phases of the cardiopulmonary exercise test were as follows: (1) The quiet phase: 2 min of sitting still on the bike. (2) The first blood collection phase: peripheral blood collection from fingertips. (3) The warm-up phase: five different warm-up interventions were performed. (4) The second blood collection phase: peripheral blood collection from fingertips. (5) The exercise phase: the initial load of 40 W was maintained for 2 min, followed by a linear load increment of 25 W/min. The electronic tachometer controlled the pedaling speed at 60–70 rpm, with verbal encouragement provided. Participants meeting any one of these criteria were regarded as having reached the maximum load [19,20], which entailed the following: ① reaching 90% of the individual's maximum predicted HR (calculated as $210 - (\text{age} \times 0.65)$) or maintaining a stable HR for two minutes; ② reaching a rate of perceived exertion (RPE) ≥ 17; ③ showing a decline in oxygen uptake during [21] sustained exercise changes or a decrease in oxygen uptake during continuous exercise. (6) The recovery stage: pedaling at 20 W for 3 min followed by equipment unloading. (7) The end stage: subjects were allowed to leave the laboratory only after a 30 min rest with no observed abnormalities. After the 30 min rest period, subjects were permitted to leave the laboratory if no abnormalities were detected. The experimental flow is shown in Figure 1.

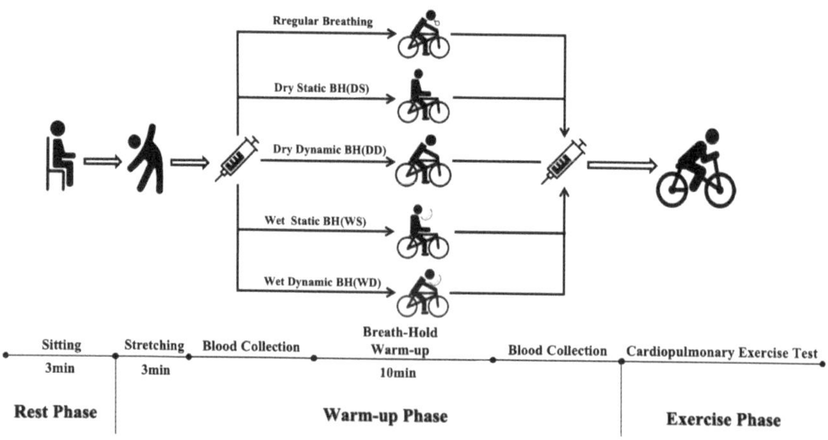

Figure 1. Procedures.

2.4. Testing Parameters

Cardiorespiratory exercise tests were conducted using a vertical power bike (Eegoselect100, Ergoline Academy, Würzburg, Germany) equipped with a telemetric exercise cardiorespiratory tester (Cortex MetaMax3B, Leipzig, Germany). The system measured gas exchange parameters including peak oxygen uptake (VO_2peak), peak oxygen uptake per kilogram of body weight (VO_2/kg peak), the anaerobic threshold (AT), and the anaerobic threshold per kilogram of body weight (AT/kg) [19]. Ventilatory efficiency, indicated by the lowest ventilatory equivalent ratio for oxygen (VE/VO_2), highlights cardiac or pulmonary limitations during exercise and formed a "U" curve during an incremental load exercise test. The lowest point, known as the cardiopulmonary optimal point (COP) [22], serves as a submaximal indicator to assess the cardiorespiratory efficiency of athletes [23,24].

We calculated deltas (Δ) to observe the differences between regular warm-up and breath-hold warm-up conditions. Each Δ was determined by subtracting the value measured under regular warm-up conditions from the value measured after the breath-hold warm-up intervention. Specifically, the calculations were as follows:

(1) ΔVO_2peak = VO_2peak (BH warm-up) − VO_2peak (Regular warm-up);
(2) ΔVO_2/kg peak = VO_2/kg peak (BH warm-up) − VO_2/kg peak (Regular warm-up);
(3) ΔCOP = COP (BH warm-up) − COP (Regular warm-up).

Additionally, a noninvasive cardiac output system (Cheetah NICOM, USA) was used to measure cardiovascular indicators at VO_2peak, including peak stroke volume (SVpeak), peak heart rate (HRpeak), and peak cardiac output (COpeak). Similarly, we calculated the deltas for additional key variables to assess differences in SVpeak, HRpeak, and COPEAK. These calculations were as follows:

(1) ΔSVpeak = SVpeak (BH warm-up) − SVpeak (Regular warm-up);
(2) ΔHRpeak = HRpeak (BH warm-up) − HRpeak (Regular warm-up);
(3) ΔCOpeak = COpeak (BH warm-up) − COpeak (Regular warm-up).

Hematological parameters were collected by professionals following the Chinese Consensus on the Operation of Capillary Blood Collection. Fingerstick blood samples were immediately transferred to the laboratory to measure various blood parameters, including red blood cells (RBCs), hemoglobin (HGB), and hematocrit (HCT). We used Δ to represent the changes in RBCs, HGB, and HCT before and after the warm-up intervention. Each Δ was calculated by subtracting the value measured before the warm-up from the value measured after the warm-up. Specifically, the calculations were as follows:

(1) ΔRBC = RBC (post-warm-up) − RBC (pre-warm-up);
(2) ΔHGB = HGB (post-warm-up) − HGB (pre-warm-up);
(3) ΔHCT = HCT (post-warm-up) − HCT (pre-warm-up).

2.5. Warm-Up Methods

Five warm-up methods were used for eighteen male athletes, one of which was a regular breathing warm-up, and the remaining four of which were BH warm-ups. The baseline regular warm-up was 10 min of zero-power pedaling while breathing normally, with all pedaling maintained at 60–70 r/min. The dynamic dry BH (DD) group performed a warm-up of 10 min of zero-power pedaling at the same time lasting for six times the maximum BH; static dry BH (SD) required the subject to sit quietly on a power bike for 10 min while performing the procedure for a duration of six- times the maximum BH; wet dynamic BH (WD) entailed 10 min of zero-power pedaling at the same time for a duration of six times the maximum BH when the participant's face was immersed in water; wet static BH (WS) required the subject to be on a power bike for 10 min while submerging their face in water for a duration of six times that of the BH.

All BH warm-ups involved six repetitions of maximal voluntary end-expiratory breath-holds, with a 30 s interval between each BH. BH commenced at the end of natural expiration, with participants instructed not to initiate the hold immediately after maximal inhalation. The criterion for each BH pause was that the impulse to resume breathing was more significant than the willingness to hold the breath before the subject could resume free breathing. Wet BH (including WD and WS) alternated between face immersion in cold water for maximal BH and free breathing with the face held above the water surface, with only the neck flexed during the face immersion BH to ensure that the entire face was immersed in water. Dry BH (DD & SD) refers to the maximal BH in the air with the neck flexed to hold the body in almost the same position. The water temperature was maintained between 10 and 11 °C.

2.6. Safety Monitoring

Throughout the entire testing procedure, heart rate was continuously monitored using a Polar heart rate monitor (Polar Electro, Kempele, Finland). Testing was discontinued if any three of the following conditions were met [19]: (1) the increment in oxygen uptake between two consecutive sessions was less than 2 mL/(kg·min) or it decreased; (2) the respiratory quotient was equal to or greater than 1.15; (3) the HR exceeded 180 beats per minute or did not increase within 2 min. During breath-hold phases, participants were closely observed for signs of hypoperfusion, such as changes in lip or fingertip color, cold sweats, or dizziness.

2.7. Statistical Analyses

Data were processed using SPSS 26.0 (IBM, Chicago, IL, USA) software. GraphPad Prism 9.0 was used to construct figures. Normality was examined with the Shapiro–Wilk test. Homoscedasticity was determined through Levene's test. Analysis was performed using a two-way repeated measures ANOVA [25] in the facial immersion condition (dry and wet) × 2-movement condition (dynamic and static) in four BH warm-up groups for gas metabolism parameters and hemodynamic parameters. When significant main or interaction effects were observed, pairwise comparisons were performed using the Bonferroni post hoc test. Effect size and partial eta squared ($\eta 2$) are presented where appropriate. Paired sample *t*-tests were conducted between the regular group and each of the four BH warm-up groups. A paired sample *t*-test was performed on RBCs, HCT, and HGB before and after five warm-ups to explore the hematological changes. Pearson correlation analyses were performed to determine the relationship between ΔVO_2peak and ΔRBC, ΔHCT, and ΔHGB. A general linear regression model was established when $r > 0.5$. The significance level was set at $p < 0.05$. $p = 0.000$ was reported as $p < 0.001$. Data are reported as means ± SD.

3. Results

All athletes completed the protocol without adverse reactions. All data passed the Shapiro–Wilk test and Levene's test.

3.1. Cardiopulmonary Exercise Test Parameters

Two-way repeated measures ANOVA revealed significant between-group differences in VO_2peak and VO_2/kgpeak ($p < 0.001$, Figure 2A,B). Static conditions exhibited a significantly lower VO_2peak values compared to Dynamic conditions ($F = 52.4$, $p < 0.001$, Static < Dynamic, Figure 2A). VO_2/kgpeak in all four BH groups was significantly higher than in the regular breathing group ($p < 0.05$, Figure 2B). COP after the WD warm-up was significantly higher than in the other BH groups ($p < 0.05$, Figure 2C). There was no significant difference in AT among the five groups (all $p > 0.05$, Figure 2D). Dry conditions exhibited significantly higher ΔVO_2peak values compared to wet conditions ($F = 39.9$, $p < 0.001$, Dry > Wet, Figure 2E). There was no significant intergroup difference in ΔCOP ($p = 0.178$, Figure 2E).

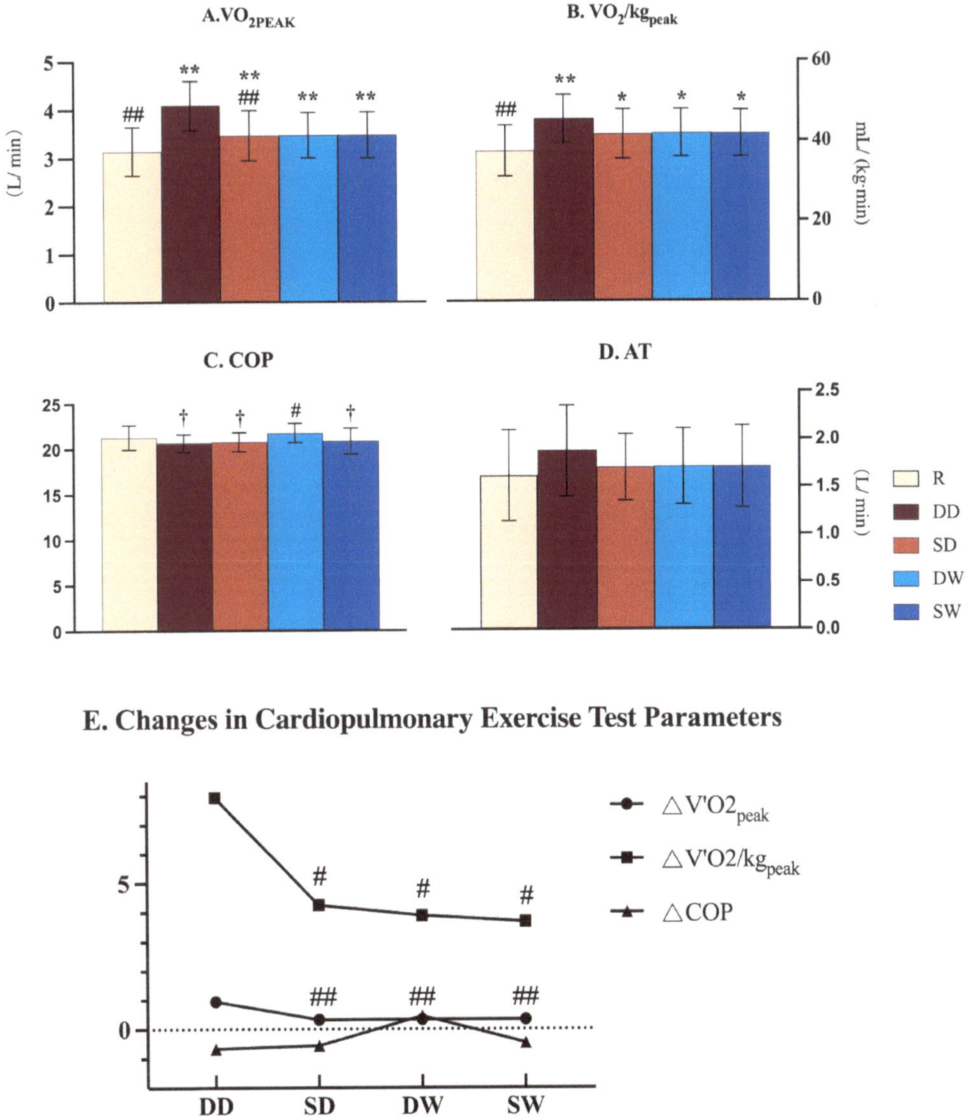

Figure 2. Gas metabolism parameters. R = regular breathing warm-up; DD = dynamic dry breath-hold; SD = static dry breath-hold; WD = dynamic wet breath-hold; WS = static wet breath-hold; VO_2peak = peak oxygen uptake; VO_2/kgpeak = relative peak oxygen uptake; COP = cardiopulmonary optimal point; AT = anaerobic threshold; △VO_2peak, △VO_2/kg peak, and △COP = BH warm-up(VO_2peak, VO_2/kg peak, COP)-Regular (VO_2peak, VO_2/kg peak, COP); significant difference vs. regular breathing group (* $p < 0.05$, ** $p < 0.01$); significant difference vs. DD group (# $p < 0.05$, ## $p < 0.01$); significant difference vs. DW group († $p < 0.05$).

There was a significant effect of the facial immersion condition on VO_2peak, VO_2/kgpeak, and COP ($p < 0.05$). There was also a significant effect of the movement condition on VO_2peak and VO_2/kgpeak ($p < 0.05$). A significant interaction between the facial immersion condition and movement condition was found for VO_2peak, VO_2/kgpeak, and COP ($p < 0.05$, Table 1).

Table 1. Warm-up condition effect. VO_2peak = peak oxygen uptake; VO_2/kgpeak = relative peak oxygen uptake; COP = cardiopulmonary optimal point; AT = anaerobic threshold; AT/kg = relative anaerobic threshold; COpeak = peak cardiac output; HRpeak = peak heart rate; SVpeak = peak stroke volume. "×" represents the interaction effects of the two conditions on the parameters.

	Facial Immersion Condition			Movement Condition			Facial Immersion Condition × Movement Condition		
	F(1,17)	Partial η^2	*p*	F(1,17)	Partial η^2	*p*	F(1,17)	Partial η^2	*p*
VO2peak (L/min)	39.942	0.755	<0.001	52.405	0.701	<0.001	95.239	0.849	<0.001
VO2/kg peak (mL/(kg·min)	13.011	0.434	0.002	17.984	0.514	0.001	44.197	0.722	<0.001
COP	5.765	0.253	0.028	2.501	0.128	0.132	5.223	0.235	0.035
AT (L/min)	1.695	0.091	0.21	1.102	0.061	0.308	1.064	0.059	0.317
AT/kg (mL/(kg·min))	5.401	0.241	0.033	0.889	0.05	0.359	1.637	0.088	0.218
SVpeak	17.98	0.514	0.001	0.75	0.042	0.398	7.176	0.297	0.016
HRpeak	1.911	0.101	0.185	1.986	0.105	0.177	9.157	0.35	0.008
COpeak	4.804	0.22	0.043	0.119	0.007	0.119	0.206	0.012	0.656

3.2. Hemodynamic Parameters

Analysis of variance with the groups as factors showed no significant differences in COpeak and HRpeak among the five groups ($p > 0.05$, Figure 3A,B). The SVpeak of the DD group was significantly higher than that at the baseline and that of the other three BH warm-up groups (Figure 3C). The DD group showed higher ΔSVpeak values compared to the other three BH warm-up groups ($p < 0.05$, Figure 3D).

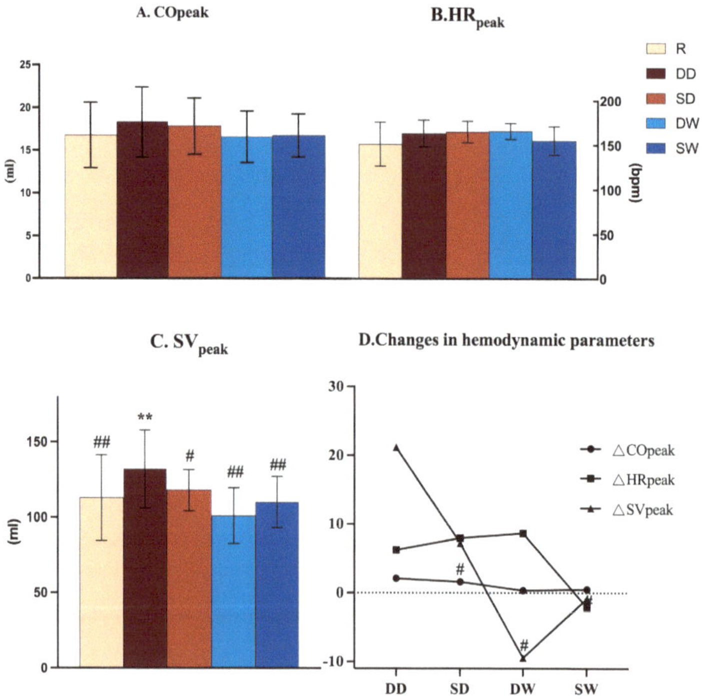

Figure 3. Hemodynamic parameters. R = regular breathing warm-up; DD = dynamic dry breath-hold; SD = static dry breath-hold; WD = dynamic wet breath-hold; WS = static wet breath-hold; COpeak = peak cardiac output; HRpeak = peak heart rate; SVpeak = peak stroke volume. ΔSVpeak = SVpeak (BH warm-up)−SVpeak (regular warm-up); ΔHRpeak = HRpeak (BH warm-up)−HRpeak (regular warm-up); ΔCOPEAK = COPEAK (BH warm-up)−COPEAK (regular warm-up). Significant difference vs. regular breathing group (** $p < 0.01$); significant difference vs. DD group (# $p < 0.05$, ## $p < 0.01$).

Statistical results revealed a significant interaction (Facial immersion × Movement) upon reaching SVpeak and HRpeak ($p < 0.05$). Dynamic HRpeak was higher than static HRpeak under wet conditions ($p = 0.025$).There was significant effect of the facial immersion condition on SVpeak and COpeak ($p < 0.05$).There was no significant effect of the movement condition on all hemodynamic parameters ($p > 0.05$) (Table 1).

3.3. Hematological Parameters

Before warm-up initiation, there were no significant differences observed among the five groups for RBCs, HGB, and HCT ($p > 0.05$). Paired-sample *t*-tests on RBCs, RGB, and HCT before and after the warm-up in four BH groups showed significant differences ($p < 0.05$, Figure 4A–C). ΔRBC and ΔHCT were significantly higher after the BH warm-up interventions compared to the corresponding indicators after the regular warm-up ($p = 0.021$, Figure 4D).

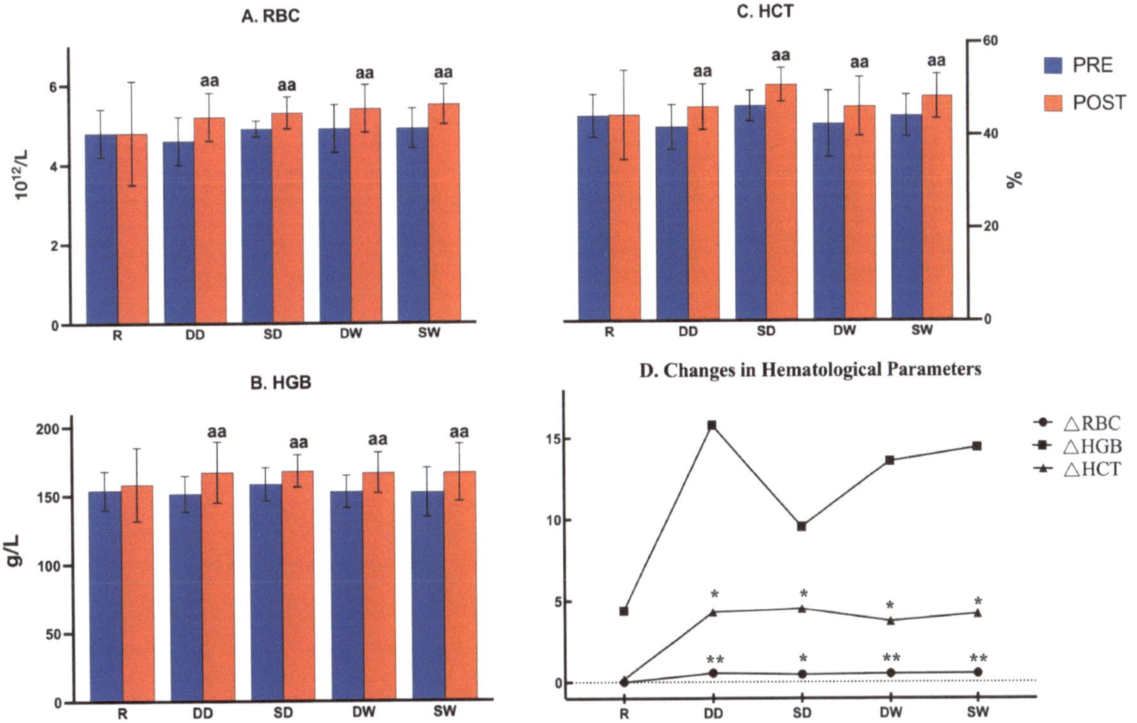

Figure 4. Hematological parameters. R = regular breathing warm-up; DD = dynamic dry breath-hold; SD = static dry breath-hold; WD = dynamic wet breath-hold; WS = static wet breath-hold; RBC = red blood cells; HGB = hemoglobin; HCT = hematocrit. ΔRBC = RBC (post-warm-up)−RBC; ΔHGB = HGB (post-warm-up)−HGB (pre-warm-up); ΔHCT = HCT (post-warm-up)−HCT (pre-warm-up). Significant difference vs. pre-breath-hold warm-up (aa $p < 0.01$); significant difference vs. regular breathing group (* $p < 0.05$, ** $p < 0.01$).

3.4. VO_2peak Correlation and Regression Analysis

There was a significant correlation between △VO_2peak and △RBC, △HCT in the DD group (rRBC = 0.84, rHCT = 0.77, $p < 0.01$). There was no significant correlation between △HGB and △VO_2peak.The linear regression model is detailed in Figure 5.

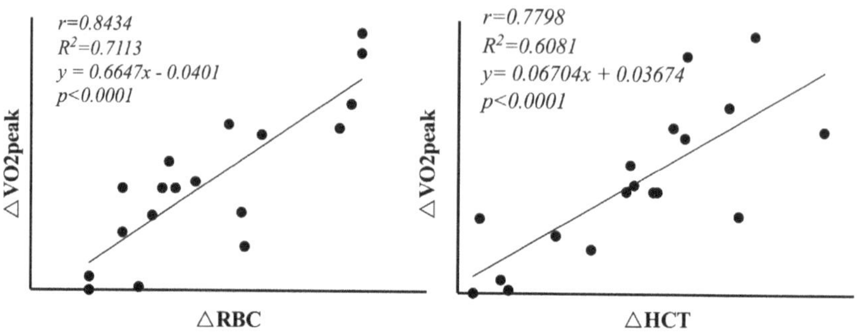

Figure 5. Linear regression model.

4. Discussion

To our knowledge, this is the first study to investigate the effects of four BH conditions on subsequent cardiopulmonary fitness in elite athletes. Our findings showed the the following: (1) there was a significant interaction between facial immersion conditions and movement conditions during BH warm-ups; (2) compared with other BH warm-ups, the dynamic dry (DD) warm-up significantly enhanced oxygen uptake and transportation, increased circulatory efficiency, and improved cardiovascular function, showing the most pronounced effect on cardiopulmonary fitness improvement. Therefore, these results strongly suggest that the DD warm-up can be used as an effective warm-up modality, offering better body activation effects.

4.1. Cardiorespiratory

The study findings indicate that VO_2/kgpeak following the BH warm-up surpassed that of the regular warm-up group. Previous research suggests that BH impacts various cardiorespiratory parameters, including cardiac output, venous oxygen reserve, and pulmonary oxygen reserve. Traditionally associated with decreased cardiac output and oxygen reserves, recent studies indicate potential variations in these effects [1,26]. Pre-exercise oxygen storage levels play a significant role in determining cardiopulmonary fitness. The development of warm-up methods targeting enhanced oxygen storage before exercise underscores the importance of optimizing this parameter. These findings underscore the potential for alternative warm-up strategies to improve VO_2peak through enhanced oxygen availability.

The study highlights the influence of environmental factors, such as movement conditions and facial immersion, on VO_2/kg peak. Specifically, it demonstrates that the movement conditions and facial immersion condition collectively explain 72.2% of the variability. Notably, dry BH (4.1 L/min) resulted in higher VO_2peak compared to wet BH(3.5 L/min) under similar movement conditions, suggesting a potential interaction between facial immersion and warm-up methods. Previous research suggests that the relationship between sympathovagal balance and cardiopulmonary fitness significantly influences athletic performance [27]. Pre-exercise parasympathetic dominance positively affects cardiorespiratory parameters and endurance exercise performance, such as running [28]. However, an imbalance between sympathetic and parasympathetic inputs to cardiac autonomic regulation may delay the withdrawal of parasympathetic nerves before exercise, subsequently affecting the timely activation of sympathetic nerves and reducing cardiorespiratory fitness [29]. This aligns with our study's outcomes, suggesting that DD BH may have a more robust interaction effect, thereby pre-activating sympathetic activity and favoring pre-exercise autonomic activation preparation.

Research indicates that higher CO_2 tolerance correlates with a lower COP, suggesting enhanced cardiorespiratory endurance [24]. The present study showed an interaction effect between facial immersion conditions and movement conditions for COP [30]. Under movement conditions, COP reduction after a dry BH was significantly higher than that after wet BH ($p = 0.001$). During incremental exercise, VO_2 increases linearly to a plateau, whereas ventilation shows a sharp inflection point of increase [22]. Higher PaCO2 stimulates peripheral chemoreceptors, inducing reflexive deep and fast respiration and increased blood circulation [31]. Dry BH reduces COP by increasing $PaCO_2$ earlier, increasing CO_2 tolerance, leading to a non-equivalent proportional change in VE and VO_2. Dry BH as a warm-up method may reduce COP and enhance respiratory oxygen consumption during exercise, improving overall cardiorespiratory efficiency.

A meta-analysis has shown that BH training facilitates improved anaerobic glycolysis [32]. Mechanisms may include increases in glycogen and phosphocreatine availability, enzyme activity, buffering capacity, or improved tolerance to exercise-induced acidosis [33]. Training adaptations to anaerobic glycolysis may include an increase in the maximal rate of lactate production and an improvement in muscle buffering capacity, leading to an increase in lactate clearance and minimizing pH disturbances in muscle [34]. The lack of direct measurement of blood lactate concentrations in this paper may have limited the exploration of improvements in anaerobic capacity, suggesting that future studies consider adding pre- and post-warm-up blood lactate measurements. Additionally, rugby players were selected as subjects for this study. The typical training emphasis of rugby players on high-intensity explosive activities such as sprinting, jumping, and tackling may shape their partial anaerobic threshold predominantly through their training regimen, potentially resulting in minimal impacts from a single BH warm-up. This could explain why the study did not detect significant differences following various warm-up modalities.

4.2. Hemodynamic Parameters

Different warm-up types induce changes in hemodynamic parameters upon reaching VO_2peak, primarily affecting stroke volume (SV) and HR. Following DD warm-up, SVpeak demonstrates a notable increase compared to that in other warm-up methods. Engaging in a DD BH warm-up before exercise enhances SV, thereby improving oxygen delivery capacity. Helgerud et al. suggested that enhanced cardiac contractility contributing to increased SV may explain the observed enhancement in cardiopulmonary fitness [35]. Increased SVpeak indicates improved blood pumping efficiency, crucial for rugby players requiring adequate oxygen and nutrient supply to muscles during intense exercise, thus sustaining athletic performance. Additionally, Woorons et al. noted that maximal dry BH may prolong heart filling time, augment left ventricular filling, and reduce right ventricular volume, thereby increasing left ventricular output and improving oxygen delivery to muscles [36].

Furthermore, this study's results revealed that facial immersion conditions significantly influenced SVpeak and COpeak, explaining 51.4% and 22% of their variability, respectively ($p < 0.05$). Submerging the face in water during dynamic warm-up decreased SVpeak, possibly due to increased parasympathetic activation from the DR, counteracting sympathetic activation during high-intensity exercise. Specifically, the DD group showed a significantly higher increase in SVpeak (ΔSVpeak) compared to the other three BH groups and the conventional warm-up group. Higher SVpeak indicates enhanced oxygen and nutrient delivery to muscles, critical for maintaining performance during intense competition and training.

There was a significant interaction effect (facial immersion × movement) on HRpeak ($p < 0.05$). BH and facial immersion increase parasympathetic tone, decreasing HR, while exercise onset stimulates sympathetic activation, leading to increased HR [29]. Typically, HR increases approximately 5–10 s after exercise onset due to the delay between the stimulus and physiological response [37,38]. In contrast, dynamic BH initiates HR changes at the BH onset (i.e., exercise start) due to DR activation.

Under wet conditions, the dynamic HRpeak was higher than the static HRpeak ($p = 0.025$). Butler and Woakes reported no significant difference between static and dynamic BH under dry conditions [39]. Sterba et al. conducted a study using 20 °C water cooling during static BH, resulting in a 12% HR decrease during and a 19% decrease after dynamic BH [40]. However, unlike previous studies focusing on recovery post-BH, ours assessed HRpeak during exercise immediately following BH. During the transition from warm-up to exercise, HR initially decreases due to reduced parasympathetic activation, followed by sympathetic tone increase. Previous research focused on factors affecting BH like water temperature, body area submerged, cooling method (immersion or spray) [41], BH mode (end-expiration or end-inspiration) [42], diving experience [43], and work-related activities, among others. Thus, future research can study the BH impact on cardiopulmonary fitness, controlling HR via the autonomic nervous system using indicators like HR variability.

High-lung-volume BH leads to a sudden intrapulmonary pressure rise, decreasing cardiac preload and increasing afterload, temporarily lowering CO, and reducing blood and oxygen supply to tissues. Continued ventilation restores sympathetic nerve dominance, increasing HR and contractility, restoring or slightly increasing CO to normal levels. COpeak defines cardiorespiratory fitness. Repeated end-expiratory BH warm-ups slightly increased COpeak compared to conventional warm-ups, consistent with prior research [20].

4.3. Hematologic Parameters

Previous research has demonstrated that prolonged breath-hold (BH) training induces physiological adaptations, including increased splanchnic volume, erythrocyte pressure volume, erythropoietin (EPO) production, hemoglobin (HGB) levels, and lung capacity [44–46]. These adaptations have been associated with improvements in aerobic fitness and maximal exercise capacity [47]. In our study, significant changes in red blood cell (RBC), HGB, and hematocrit (HCT) levels were observed before and after BH, indicating an acute response to hypoxic stimulation. This suggests that BH may enhance erythrocyte reserve and improve oxygen transport capacity, potentially contributing to improved aerobic fitness.

The DR induced by BH results in splenic contraction, causing a transient surge in hemoglobin concentration and erythrocyte pressure volume [48]. This response enhances oxygen carrying capacity and redirects blood flow from peripheral vasculature to critical organs such as the brain and heart. The Valsalva effect during BH may further trigger peripheral vasoconstriction, facilitating blood redistribution and optimizing oxygen delivery [49]. However, these effects are transient, typically lasting for 10 min after BH cessation. Multiple BH sessions, as employed in our study, may accumulate these effects, potentially leading to a more pronounced oxygen conservation effect post-warmup. Schagatay et al. concluded that this effect persists for at least 10 min after BH, suggesting that maximal exercise following BH warm-up can effectively utilize enhanced erythrocyte reserve and exhibit higher aerobic fitness [50].

Furthermore, the Valsalva effect induced during BH may trigger peripheral vasoconstriction, directing blood from the peripheral vasculature towards the central circulatory system, thereby increasing the concentration of RBCs within a given blood volume. Dynamic BH redistributes blood while inducing selective sympathetic-induced vasoconstriction in small arteries, particularly in peripheral and visceral capillary beds of nonessential organs and body extremities. This results in the preferentially oxygenated blood in critical organs such as the brain and heart. Consequently, nonessential organs shift metabolically from predominantly aerobic to predominantly anaerobic metabolism, further enhancing the oxygen conservation effect mediated by the diving response. However, due to a circulatory delay in oxygen storage within peripheral veins, this effect typically diminishes within 15 to 45 s following BH cessation. In our study, participants performed six maximal BH exercises at 30 s intervals, potentially accumulating delayed oxygen storage effects and leading to a more pronounced oxygen conservation effect post-warmup.

Another important finding of this study is that different conditions impact BH-induced splanchnic contraction, affecting erythrocyte circulation. Elia et al. demonstrated that dynamic dry BH induces more pronounced splanchnic contraction. In their study of eight non-apneic divers, they found no significant change in EPO concentration 30 min after the BH intervention compared to the baseline (6.62 $\pm$ 3.03 mlU/mL vs. 8.46 $\pm$ 2.21 mlU/mL; p = 0.109) [51]. Espersen et al. found an immediate decrease in the relative content of erythrocytes following static wet BH [52]. Conversely, Bouten et al. found that static dry BH led to a more significant elevation in HGB compared to dynamic dry BH. Furthermore, some studies have not observed changes in erythrocyte circulation-related parameters following a single BH [53].

In our study, SD BH elicited an increase in RBC from 4.9 to 5.3 $\times$ 10^{12}/L, while DD BH led to an increase from 4.6 to 5.2 $\times$ 10^{12}/L. Although both conditions resulted in higher RBC levels compared to the baseline, the difference was not statistically significant. During static BH without limb movement, core muscles contract more, increasing intra-abdominal pressure and exerting more pressure on the spleen. Consequently, this leads to a greater release of HGB and RBCs into the circulatory system. However, excessive erythrocyte release prior to exercise may elevate erythrocyte pressure production, increase blood viscosity, and slow blood flow, all detrimental to oxygen carrying capacity.

Notably, Bouten et al. concluded that acute BH did not improve performance in professional cycling over a 3 km distance. Similarly, Sperlich et al. supported that four BH sessions preceding a 4 km cycling time trial resulted in splanchnic constriction without notable changes in hemoglobin levels, erythrocyte pressure production, lateral femoral muscle oxygen saturation, or oxygen uptake [40]. In contrast, the robust correlation observed between changes in ΔRBC and ΔHCT were induced by dynamic BH warm-up and alterations in ΔVO_2/kg peak in our study. Research indicates that both animal and recent human studies are consistent in demonstrating a spleen contraction of 40–70% during maximal exercise. Regardless of whether dry or wet breath-holding is employed, the spleen contraction induced exceeds 25%. Qvist et al. found hematological changes occurring only with significant spleen contraction, defined as more than 25%, whether conducted underwater or in a laboratory setting with BH [54]. This study, limited by the mode of exercise, did not immediately measure spleen volume after BH. However, given the significant relationship between ΔRBC and ΔHCT under ΔVO_2/kg peak, we can reasonably infer that BH warm-up induces changes in erythrocyte circulation by eliciting spleen contraction.

4.4. Application in Training

Dynamic dry BH warm-up sessions should be strategically integrated into athletes' training routines, particularly before high-intensity workouts or competitions. Our study demonstrates that dynamic dry BH significantly enhances oxygen uptake, improves circulatory efficiency, and optimizes cardiovascular function, thereby enhancing immediate cardiopulmonary readiness. Coaches and trainers can tailor the duration and intensity of these sessions based on individual athlete fitness levels and specific sport demands, ensuring gradual adaptation and maximizing physiological benefits over time.

It is crucial to provide athletes with educational support regarding the physiological mechanisms underlying dynamic dry BH. This understanding not only enhances their compliance with the warm-up protocol but also fosters appreciation of its impact on performance. Regular monitoring of athletes' responses, including metrics such as heart rate variability, recovery rates, and perceived exertion, allows for adjustments to optimize warm-up effectiveness while minimizing potential risks associated with BH practices, such as the overstimulation of the parasympathetic nervous system [55].

Moreover, emphasizing safety protocols, especially concerning cold water exposure during warm-up procedures, ensures that the implementation of dynamic dry apnea warm-ups remains safe and effective. By following these guidelines, coaches and athletes can leverage the benefits identified in our study to optimize athletic performance through a scientifically supported warm-up strategy that enhances cardiopulmonary fitness and overall readiness.

4.5. Limitations

Our study focused on elite rugby players, known for their confrontational sport demands and distinct physiological characteristics, such as intermittent high-intensity activities, frequent collisions, and diverse movement patterns. These factors can significantly influence how athletes respond to warm-up strategies like dynamic dry BH, impacting its effectiveness in enhancing cardiopulmonary fitness outcomes. While our findings offer valuable insights into this specific athlete group, their direct applicability to other athletes, particularly those with different physiological adaptations and training focuses, may be limited. To enhance the generalizability of our results, future studies should expand participant inclusion across various sports and disciplines. This approach would provide a broader understanding of how dynamic dry BH warm-up strategies affect different physiological profiles and athletic demands.

5. Conclusions

BH warm-up serves as an effective preparatory measure, significantly enhancing oxygen uptake and delivery, improving circulatory efficiency, and enhancing cardiovascular function. Variations in BH conditions, such as the presence or absence of facial immersion in cold water and dynamic activity, result in different physiological effects. Among these conditions, dynamic dry BH shows the most promising potential for improving immediate cardiopulmonary fitness. Our findings underscore the value of dynamic dry BH warm-up as a beneficial preparatory strategy, positively impacting athletes' cardiopulmonary fitness. Future research should explore the detailed physiological mechanisms and optimal protocols of dynamic dry BH warm-up, comparing its effectiveness with traditional warm-up methods. Additionally, assessing its long-term impact on fitness, performance, and psychological factors will provide a comprehensive understanding of its benefits for athletes.

Author Contributions: Conceptualization, W.W. and D.W.; methodology, W.W. and D.W.; software, W.W. and D.W.; validation, H.W.; formal analysis, Z.Z., X.J. and S.L.; investigation, X.G. and Y.S.; resources, Z.Z., X.J. and S.L.; data curation, H.W.; writing—original draft preparation, W.W.; writing—review and editing, D.W.; visualization, W.W.; supervision, X.G. and Y.S.; project administration, D.W.; funding acquisition, D.W. All authors have read and agreed to the published version of the manuscript.

Funding: Project 22–46, 23–36 was supported by the Fundamental Research Funds for the China Institute of Sport Science.

Institutional Review Board Statement: The animal study protocol was approved by the Institutional Ethics Committee of China Institute of Sport Science.

Informed Consent Statement: Informed consent was obtained from all subjects involved in the study. Written informed consent was obtained from the patients to publish this paper.

Data Availability Statement: The data used in this study are available from the corresponding author upon reasonable request.

Acknowledgments: We are grateful to all the participants whose involvement made this study possible. Your dedication and cooperation were essential to our research.

Conflicts of Interest: The authors declare no conflicts of interest.

References

1. Andersson, J.P.; Linér, M.H.; Fredsted, A.; Schagatay, E.K. Cardiovascular and respiratory responses to apneas with and without face immersion in exercising humans. *J. Appl. Physiol.* **2004**, *96*, 1005–1010. [CrossRef] [PubMed]
2. Schagatay, E. Predicting performance in competitive apnea diving, part II: Dynamic apnoea. *Diving Hyperb. Med.* **2010**, *40*, 11. [PubMed]
3. Flamm, S.D.; Taki, J.; Moore, R.; Lewis, S.F.; Keech, F.; Maltais, F.; Ahmad, M.; Callahan, R.; Dragotakes, S.; Alpert, N.; et al. Redistribution of regional and organ blood volume and effects on cardiac function in relation to upright exercise intensity in healthy human subjects. *Circulation* **1990**, *81*, 1550–1559. [CrossRef] [PubMed]
4. Guyenet, P.G. Regulation of breathing and autonomic outflows by chemoreceptors. *Compr. Physiol.* **2014**, *4*, 1511. [CrossRef] [PubMed]
5. Bouten, J.; De Bock, S.; Bourgois, G.; de Jager, S.; Dumortier, J.; Boone, J.; Bourgois, J.G. Heart Rate and Muscle Oxygenation Kinetics During Dynamic Constant Load Intermittent Breath-Holds. *Front. Physiol.* **2021**, *12*, 712629. [CrossRef]

6. Cherouveim, E.D.; Botonis, P.G.; Koskolou, M.D.; Geladas, N.D. Effect of gender on maximal breath-hold time. *Eur. J. Appl. Physiol.* **2013**, *113*, 1321–1330. [CrossRef] [PubMed]
7. Elia, A.; Barlow, M.J.; Deighton, K.; Wilson, O.J.; O'Hara, J.P. Erythropoietic responses to a series of repeated maximal dynamic and static apnoeas in elite and non-breath-hold divers. *Eur. J. Appl. Physiol.* **2019**, *119*, 2557–2565. [CrossRef] [PubMed]
8. Castro, C.L.B.D.; Franca, J.F.; Bottino, A.; Myers, J.; Araújo, C.G.S.D. Cardiorespiratory optimal point in professional soccer players: A novel submaximal variable during exercise. *Int. J. Cardiovasc. Sci.* **2018**, *31*, 323–332. [CrossRef]
9. Parkes, M.J. Breath-holding and its breakpoint. *Exp. Physiol.* **2006**, *91*, 1–15. [CrossRef]
10. Lemaitre, F.; Chowdhury, T.; Schaller, B. The trigeminocardiac reflex—A comparison with the diving reflex in humans. *Arch Med. Sci.* **2015**, *11*, 419–426. [CrossRef]
11. Kawakami, Y.O.S.H.I.K.A.Z.U.; Natelson, B.H.; DuBois, A.R. Cardiovascular effects of face immersion and factors affecting diving reflex in man. *J. Appl. Physiol.* **1967**, *23*, 964–970. [CrossRef] [PubMed]
12. Furedy, J.J.; Morrison, J.W.; Heslegrave, R.J.; Arabian, J.M. Effects of water temperature on some noninvasively measured components of the human dive reflex: An experimental response-topography analysis. *Psychophysiology* **1983**, *20*, 569–570. [CrossRef]
13. Ekaterina, P.; Tatyana, Z.; Lavrentij, S.; Tatyana, B. Features of hemodynamics of pulmonary circulation during the diving reflex. *Biol. Commun.* **2020**, *65*, 244–251. [CrossRef]
14. Bakovic, D.; Valic, Z.; Eterovic, D.; Vukovic, I.; Obad, A.; Marinovic-Terzic, I.; Dujic, Z. Spleen volume and blood flow response to repeated breath-hold apneas. *J. Appl. Physiol.* **2003**, *95*, 1460–1466. [CrossRef]
15. Hanly, P.J.; George, C.F.; Millar, T.W.; Kryger, M.H. Heart rate response to breath-hold, valsalva and Mueller maneuvers in obstructive sleep apnea. *Chest* **1989**, *95*, 735–739. [CrossRef]
16. Linér, M.H. Cardiovascular and pulmonary responses to breath-hold diving in humans. *Acta Physiol. Scand. Suppl.* **1994**, *620*, 1–32. [PubMed]
17. Finnoff, J.T.; Smith, J.; Low, P.A.; Dahm, D.L.; Harrington, S.P. Acute hemodynamic effects of abdominal exercise with and without breath holding. *Arch. Phys. Med. Rehabil.* **2003**, *84*, 1017–1022. [CrossRef] [PubMed]
18. Fitz-Clarke, J.R. Breath-hold diving. *Compr. Physiol.* **2011**, *8*, 585–630. [CrossRef]
19. Wang, Z.Z. *ACSM Guide to Exercise Testing and Exercise Prescription*; Beijing Sports University Press: Beijing, China, 2015; (In Chinese). [CrossRef]
20. Woorons, X.; Lemaitre, F.; Claessen, G.; Woorons, C.; Vandewalle, H. Exercise with end-expiratory breath holding induces large increase in stroke volume. *Int. J. Sports Med.* **2021**, *42*, 56–65. [CrossRef]
21. Gelbart, M.; Ziv-Baran, T.; Williams, C.A.; Yarom, Y.; Dubnov-Raz, G. Prediction of maximal heart rate in children and adolescents. *Clin. J. Sport Med.* **2017**, *27*, 139–144. [CrossRef]
22. Ramos, P.S.; Ricardo, D.R.; Araújo, C.G.S.D. Cardiorespiratory optimal point: A submaximal variable of the cardiopulmonary exercise testing. *Arq. Bras. De Cardiol.* **2012**, *99*, 988–996. [CrossRef] [PubMed]
23. Ramos, P.S.; Araújo, C.G.S. Cardiorespiratory optimal point during exercise testing as a predictor of all-cause mortality. *Rev. Port. De Cardiol.* **2017**, *36*, 261–269. [CrossRef]
24. Ramos, P.S.; Sardinha, A.; Nardi, A.E.; Araújo, C.G.S.D. Cardiorespiratory optimal point: A submaximal exercise variable to assess panic disorder patients. *PLoS ONE* **2014**, *9*, e104932. [CrossRef]
25. Andersson, J.P.; Linér, M.H.; Runow, E.; Schagatay, E.K. Diving response and arterial oxygen saturation during apnea and exercise in breath-hold divers. *J. Appl. Physiol.* **2002**, *93*, 882–886. [CrossRef] [PubMed]
26. Iellamo, F. Neural mechanisms of cardiovascular regulation during exercise. *Auton. Neurosci.* **2001**, *90*, 66–75. [CrossRef]
27. Hautala, A.J.; Kiviniemi, A.M.; Tulppo, M.P. Individual responses to aerobic exercise: The role of the autonomic nervous system. *Neurosci. Biobehav. Rev.* **2009**, *33*, 107–115. [CrossRef] [PubMed]
28. Wijayasiri, U.K.; Wimalasekere, S.W.; Ishikawa, Y.; Waidyasekara, H.; Sivayogan, S. A Physiological Approach to Cardio-endurance Training: Indicators of Optimal Parasympathetic Input on Cardiovascular Regulation are Better Predictors of Running Performance of Distance Runners. *Eur. J. Sport Sci.* **2023**, *2*, 1–9. [CrossRef]
29. Iellamo, F.; Legramante, J.M.; Pigozzi, F.; Spataro, A.; Norbiato, G.; Lucini, D.; Pagani, M. Conversion from vagal to sympathetic predominance with strenuous training in high-performance world class athletes. *Circulation* **2002**, *105*, 2719–2724. [CrossRef] [PubMed]
30. Oyarzo-Aravena, A.; Arce-Alvarez, A.; Salazar-Ardiles, C.; Ramirez-Campillo, R.; Alvarez, C.; Toledo, C.; Izquierdo, M.; Andrade, D.C. Cardiorespiratory optimal point as a submaximal evaluation tool in endurance athletes: An exploratory study. *Front. Physiol.* **2023**, *14*, 1087829. [CrossRef]
31. Brice, A.G.; Forster, H.V.; Pan, L.G.; Funahashi, A.; Lowry, T.F.; Murphy, C.L.; Hoffman, M.D. Ventilatory and PaCO2 responses to voluntary and electrically induced leg exercise. *J. Appl. Physiol.* **1988**, *64*, 218–225. [CrossRef]
32. de Asís-Fernández, F.; Sereno, D.; Turner, A.P.; González-Mohíno, F.; González-Ravé, J.M. Effects of apnoea training on aerobic and anaerobic performance: A systematic review and meta-analysis. *Front. Physiol.* **2022**, *13*, 964144. [CrossRef] [PubMed]
33. Bishop, D.; Girard, O.; Mendez-Villanueva, A. Repeated-sprint ability-part II: Recommendations for training. *Sports Med.* **2011**, *41*, 741–756. [CrossRef]
34. Sharp, R.L.; Costill, D.L.; Fink, W.J.; King, D.S. Effects of eight weeks of bicycle ergometer sprint training on human muscle buffer capacity. *Int. J. Sports Med.* **1986**, *7*, 13–17. [CrossRef] [PubMed]

35. Helgerud, J.; Høydal, K.; Wang, E.; Karlsen, T.; Berg, P.; Bjerkaas, M.; Simonsen, T.; Helgesen, C.; Hjorth, N.; Bac, R.; et al. Aerobic high-intensity intervals improve V˙ O2max more than moderate training. *Med. Sci. Sports Exerc.* **2007**, *39*, 665–671. [CrossRef] [PubMed]
36. Woorons, X.; Bourdillon, N.; Vandewalle, H.; Lamberto, C.; Mollard, P.; Richalet, J.P.; Pichon, A. Exercise with hypoventilation induces lower muscle oxygenation and higher blood lactate concentration: Role of hypoxia and hypercapnia. *Eur. J. Appl. Physiol.* **2010**, *110*, 367–377. [CrossRef] [PubMed]
37. Eriksen, M.O.R.T.E.N.; Waaler, B.A.; Walløe, L.; Wesche, J. Dynamics and dimensions of cardiac output changes in humans at the onset and at the end of moderate rhythmic exercise. *J. Physiol.* **1990**, *426*, 423–437. [CrossRef]
38. Michael, S.; Graham, K.S.; Davis, G.M. Cardiac autonomic responses during exercise and post-exercise recovery using heart rate variability and systolic time intervals—A review. *Front. Physiol.* **2017**, *8*, 259883. [CrossRef]
39. Butler, P.J.; Woakes, A.J. Heart rate in humans during underwater swimming with and without breath-hold. *Respir. Physiol.* **1987**, *69*, 387–399. [CrossRef]
40. Sterba, J.A.; Lundgren, C.E. Breath-hold duration in man and the diving response induced by face immersion. *Undersea Biomed. Res.* **1988**, *15*, 361–375. [PubMed]
41. Stevens, C.J.; Kittel, A.; Sculley, D.V.; Callister, R.; Taylor, L.; Dascombe, B.J. Running performance in the heat is improved by similar magnitude with pre-exercise cold-water immersion and mid-exercise facial water spray. *J. Sports Sci.* **2017**, *35*, 798–805. [CrossRef]
42. Trincat, L.; Woorons, X.; Millet, G.P. Repeated-sprint training in hypoxia induced by voluntary hypoventilation in swimming. *Int. J. Sports Physiol. Perform.* **2017**, *12*, 329–335. [CrossRef] [PubMed]
43. Schagatay, E.; Lunde, A.; Nilsson, S.; Palm, O.; Lodin-Sundström, A. Spleen contraction elevates hemoglobin concentration at high altitude during rest and exercise. *Eur. J. Appl. Physiol.* **2020**, *120*, 2693–2704. [CrossRef] [PubMed]
44. Godek, D.; Freeman, A.M. *Physiology, Diving Reflex*; StatPearls Publishing: Treasure Island, FL, USA, 2019. [PubMed]
45. Panneton, W.M.; Gan, Q. The mammalian diving response: Inroads to its neural control. *Front. Neurosci.* **2020**, *14*, 541224. [CrossRef] [PubMed]
46. Panneton, W.M.; Anch, A.M.; Panneton, W.M.; Gan, Q. Parasympathetic preganglionic cardiac motoneurons labeled after voluntary diving. *Front. Physiol.* **2014**, *5*, 8. [CrossRef]
47. Laub, M.I.C.H.A.E.L.; Hvid-Jacobsen, K.E.L.D.; Hovind, P.E.T.E.R.; Kanstrup, I.L.; Christensen, N.J.; Nielsen, S.L. Spleen emptying and venous hematocrit in humans during exercise. *J. Appl. Physiol.* **1993**, *74*, 1024–1026. [CrossRef] [PubMed]
48. Sperlich, B.; Zinner, C.; Pfister, R.; Holmberg, H.C.; Michels, G. Repeated apnea-induced contraction of the spleen in cyclists does not enhance performance in a subsequent time-trial. *Eur. J. Appl. Physiol.* **2015**, *115*, 205–212. [CrossRef] [PubMed]
49. Engan, H.K.; Lodin-Sundström, A.; Schagatay, F.; Schagatay, E. The effect of climbing Mount Everest on spleen contraction and increase in hemoglobin concentration during breath holding and exercise. *High Alt. Med. Biol.* **2014**, *15*, 52–57. [CrossRef] [PubMed]
50. Richardson, M.; Bruijn, R.D.; Holmberg, H.C.; Björklund, G.; Haughey, H.; Schagatay, E. Increase of hemoglobin concentration after maximal apneas in divers, skiers, and untrained humans. *Can. J. Appl. Physiol.* **2005**, *30*, 276–281. [CrossRef]
51. Elia, A.; Barlow, M.J.; Wilson, O.J.; O'Hara, J.P. Six weeks of dynamic apnoeic training stimulates erythropoiesis but does not increase splenic volume. *Eur. J. Appl. Physiol.* **2021**, *121*, 827–838. [CrossRef]
52. Espersen, K.; Frandsen, H.; Lorentzen, T.; Kanstrup, I.L.; Christensen, N.J. The human spleen as an erythrocyte reservoir in diving-related interventions. *J. Appl. Physiol.* **2002**, *92*, 2071–2079. [CrossRef]
53. Bosco, G.; Di Tano, G.; Zanon, V.; Fanò, G. Breath-hold diving: A point of view. *Sport Sci. Health* **2007**, *2*, 47–54. [CrossRef]
54. Qvist, J.; Hurford, W.E.; Radermacher, P.; Guyton, G.P.; Falke, K.J.; Park, Y.S.; Ahn, D.W.; Hong, S.K.; Stanek, K.S.; Zapol, W.M. Arterial blood gas tensions during breath-hold diving in the Korean Ama. *J. Appl. Physiol.* **1993**, *75*, 285–293. [CrossRef] [PubMed]
55. Zimmermann, P.; Wüstenfeld, J.; Zimmermann, L.; Schöffl, V.; Schöffl, I. Physiological aspects of world elite competitive German winter sport athletes. *Int. J. Environ. Res. Public Health* **2022**, *19*, 5620. [CrossRef] [PubMed]

life

MDPI

Systematic Review

Effects of Inspiratory Muscle Training in People with Chronic Obstructive Pulmonary Disease: A Systematic Review and Meta-Analysis

Bing Han [1,2,†], Zhuying Chen [2,†], Bing Ruan [3], Yongjie Chen [2], Yuanyuan Lv [1,4], Cui Li [5] and Laikang Yu [1,2,*]

1 Key Laboratory of Physical Fitness and Exercise, Ministry of Education, Beijing Sport University, Beijing 100084, China; 13520753687@163.com (B.H.); sunflowerlyy@bsu.edu.cn (Y.L.)
2 Department of Strength and Conditioning Assessment and Monitoring, Beijing Sport University, Beijing 100084, China; zhuying20232120126@126.com (Z.C.); youngJ1012@163.com (Y.C.)
3 School of Sport Medicine and Rehabilitation, Beijing Sport University, Beijing 100084, China; ruanbing@bsu.edu.cn
4 China Institute of Sport and Health Science, Beijing Sport University, Beijing 100084, China
5 School of Physical Education (Main Campus), Zhengzhou University, Zhengzhou 450001, China; lc@zzu.edu.cn
* Correspondence: yulaikang@126.com
† These authors contributed equally to this work.

Abstract: This study aimed to investigate the effects of inspiratory muscle training (IMT) on inspiratory muscle strength, dyspnea, and quality of life (QOL) in COPD patients. A comprehensive search was undertaken on the Web of Science, Scopus, Embase, Cochrane, and PubMed databases, encompassing data published up to 31 March 2024. A meta-analysis was subsequently conducted to quantify the standardized mean difference (SMD) and 95% confidence interval (CI) for the effects of IMT in COPD patients. Sixteen studies met the inclusion criteria. IMT significantly improved inspiratory muscle strength (SMD, 0.86, $p < 0.00001$), dyspnea (SMD = −0.50, $p < 0.00001$), and QOL (SMD = 0.48, $p = 0.0006$). Subgroup analysis showed that <60% maximal inspiratory muscle pressure (PImax) IMT (inspiratory muscle strength, SMD = 1.22, $p = 0.005$; dyspnea, SMD = −0.92, $p < 0.0001$), IMT conducted for ≤20 min (inspiratory muscle strength, SMD = 0.97, $p = 0.008$; dyspnea, SMD = −0.63, $p = 0.007$; QOL, SMD = 1.66, $p = 0.007$), and IMT conducted >3 times per week (inspiratory muscle strength, SMD = 1.06, $p < 0.00001$; dyspnea, SMD = −0.54, $p < 0.00001$; QOL, SMD = 0.48, $p = 0.0009$) had greater effects. This meta-analysis provides clinicians with evidence supporting the recommendation that COPD patients engage in IMT at <60% PImax for more than 3 times per week, with each session lasting no more than 20 min, to improve inspiratory muscle strength, dyspnea, and QOL.

Keywords: inspiratory muscle training; chronic obstructive pulmonary disease; inspiratory muscle strength; dyspnea; quality of life

Citation: Han, B.; Chen, Z.; Ruan, B.; Chen, Y.; Lv, Y.; Li, C.; Yu, L. Effects of Inspiratory Muscle Training in People with Chronic Obstructive Pulmonary Disease: A Systematic Review and Meta-Analysis. *Life* **2024**, *14*, 1470. https://doi.org/10.3390/life14111470

Academic Editor: Chin Kook Rhee

Received: 13 September 2024
Revised: 9 October 2024
Accepted: 4 November 2024
Published: 12 November 2024

1. Introduction

Chronic obstructive pulmonary disease (COPD) is a preventable and treatable disease characterized by persistent, often progressive airflow obstruction due to abnormalities in the airways (bronchitis, bronchiolitis) and/or alveoli (emphysema) which leads to chronic respiratory symptoms such as dyspnea, cough, sputum, and/or aggravation [1]. COPD symptoms significantly impact a person's activity, health status, and quality of life (QOL), particularly as dyspnea is a major contributor to COPD-related anxiety and disability. These symptoms can also affect family life and the individual's ability to perform daily activities, like household chores and climbing stairs. Apart from pulmonary symptoms, COPD may also manifest with systemic symptoms, including fatigue, weight loss, and sleep disturbances, as well as psychiatric symptoms like depression and anxiety, which severely compromise QOL [2]. Studies have shown that varying degrees of COPD are associated with reduced inspiratory muscle function [3]. Patients with COPD experience a decline in the functional strength of inspiratory muscles, particular those with moderate to severe COPD, where muscle strength typically falls between 40% and 60% of the predicted normal value, with varying degrees of decline. This deterioration can stem from a multitude of factors, including lung hyperinflation, muscle mass loss, frequent exacerbations of respiratory infection symptoms (once or more times a year), increased resting energy expenditure, corticosteroid use, individual muscle strength weakness, and aging [3,4].

As COPD progresses, the obstruction of airflow is accompanied by chest hyperinflation, which mechanically disadvantages the inspiratory muscles in terms of the length-tension relationship. The diaphragm becomes shortened and flattened, reducing its resting length and maximum tension-generating capacity [5]. This diminishes the diaphragm's ability to exert force, weakening the inspiratory muscles' function, evident in a decrease in maximal inspiratory muscle pressure (PImax). The extent to which hyperinflation contributes to the decline in inspiratory muscle strength remains unclear, but even minor increases in hyperinflation can significantly reduce PImax [3]. Consequently, reduced inspiratory muscle strength exacerbates dyspnea and puts individuals at risk of respiratory muscle fatigue. In addition, intensified breathing difficulties can limit physical activity, significantly affecting QOL.

At present, non-drug therapy constitutes an essential component of the comprehensive treatment of COPD. Inspiratory muscle training (IMT) has been extensively employed in both healthy individuals and those with illnesses, particularly the elderly, and its effects have been proven to positively influence not only inspiratory muscle strength but also dyspnea, exercise capacity, QOL, and various other health parameters [6]. A recent meta-analysis showed significant improvements in balance among both healthy participants and participants with diseases following IMT treatment, though the effect on functional activity remained inconclusive [6]. In addition, Bissett et al. [7] demonstrated that IMT is a safe and viable option for ventilator-dependent intensive care unit (ICU) patients, potentially enhancing weaning and improving inspiratory muscle strength and QOL. Furthermore, Ferraro et al. [8] conducted an 8-week unsupervised family-based IMT program for healthy elderly individuals, resulting in significant improvements in inspiratory muscle strength. Moreover, Figueiredo et al. [9] demonstrated that IMT can improve the strength and endurance of the respiratory muscles, thereby alleviating feelings of respiratory fatigue and dyspnea, significantly improving respiratory muscle function.

Numerous studies have demonstrated that IMT exerts a positive influence on COPD patients, but the specific advantages and ultimate conclusions remain inconclusive. Lötters F et al. [10] evaluated the effectiveness of IMT in COPD patients and concluded that IMT serves as a vital adjunct to pulmonary rehabilitation programs for COPD patients. Nevertheless, the assessment of IMT's impact on inspiratory muscle strength and QOL was found to be imprecise. Additionally, Crowe et al. [11] conducted a meta-analysis on the effects of IMT and alternative interventions in adult COPD patients, reinforcing the notion that IMT enhances improve inspiratory muscle strength and endurance. Furthermore, in the meta-analyses by O'Brien et al. [12] and Beaumont et al. [13], IMT or a combination of IMT with exercise/pulmonary rehabilitation (PR) was compared to other rehabilitation interventions in adult COPD patients. Their findings indicated that combining IMT with exercise may significantly improve inspiratory muscle strength and exercise tolerance outcomes in COPD patients. The use of threshold devices for IMT improved inspiratory muscle strength, exercise capacity, and QOL while mitigating dyspnea. Nevertheless, when compared to PR alone, IMT did not confer additional benefits in alleviating breathing difficulties during PR. Moreover, a recent meta-analysis evaluating the effectiveness of various breathing exercises in COPD patients revealed that IMT was solely beneficial for dyspnea relief, without notable improvements in inspiratory muscle strength or QOL [14]. Notably, in the aforementioned study, the control groups received interventions other than IMT, impeding an accurate and holistic assessment of IMT's effects on COPD patients.

However, in the control groups of the studies we included, apart from usual care and sham IMT, all other IMT interventions were of low intensity and did not involve any additional interventions. Notably, we conducted subgroup analyses based on session time, frequency, and intensity, which can provide an optimal exercise prescription for COPD patients.

Therefore, we conducted a comprehensive systematic review and meta-analysis of randomized controlled trials (RCTs) to investigate the effects of IMT on inspiratory muscle strength, dyspnea, and QOL in COPD patients.

2. Materials and Methods

2.1. Design

This systematic review was conducted following the Preferred Reporting Items for Systematic Evaluation and Meta-Analyses (PRISMA, 2020) guidelines [15]. The protocol is registered with PROSPERO (CRD42024554445).

2.2. Search Strategy

We searched the Web of Science, Scopus, Embase, Cochrane, and PubMed databases from inception to 31 March 2024, for RCTs using the following keywords and MESH terms: inspiratory muscle training and chronic obstructive pulmonary disease (Supplementary Materials). The screening and selection procedure was autonomously carried out by two authors (BH and ZC). In instances of disagreement, a third author (LY) was engaged in the deliberation, promoting a collaborative effort until a unanimous decision was achieved.

2.3. Eligibility Criteria

The inclusion criteria were as follows: (1) studies adopting a RCT design; (2) involving participants diagnosed with COPD; (3) containing both an intervention group and a control group; and (4) using inspiratory muscle strength, dyspnea, or QOL as the outcome measures.

The excluded criteria were as follows: (1) publications not in English; (2) conference articles; (3) review articles; (4) studies including outcome indicators that could not be transformed into mean and standard deviation (SD); and (5) research lacking a control group.

2.4. Data Extraction

Utilizing a standardized data extraction form, two authors (BH and ZC) independently extracted the data. In cases where discrepancies arose in extracted data, they collaboratively conducted a second round of extraction to validate the accuracy of the information. The following data and information were extracted: characteristics of the included studies (the primary author's surname, publication year), sample size, participant characteristics (age, sex, COPD prevalence), exercise protocols (type, intensity, intervention duration, frequency, session duration), and the mean values with SD of the key outcomes.

2.5. Methodological Quality Assessment

To evaluate the methodological quality of the included studies, we employed the Cochrane risk of bias tool (RoB2) [16,17]. The RoB2 includes seven key domains: random sequence generation (selection bias), allocation concealment (selection bias), blinding of participants and personnel (performance bias), blinding of outcome assessment (detection bias), incomplete outcome data (attrition bias), selective reporting (reporting bias), and other biases [18]. Each study was appraised along these dimensions, categorizing the risk as "low", "uncertain", or "high". Two authors (BH and ZC) independently assessed the methodological quality of the included studies, and any discrepancies were resolved through the involvement of a third author (LY).

2.6. Statistical Analysis

For the purpose of meta-analysis, the mean values accompanied by their SD were extracted from each included study. In instances where the research data were presented as standard error (SEM), a conversion was made to SD [19,20]. To synthesize the data, a random-effects model was used in this study, yielding standardized mean differences (SMDs) and 95% confidence intervals (CIs). The extent of heterogeneity among the included studies was evaluated using I^2, with interpretations as follows: $I^2 < 25\%$ indicating no heterogeneity, 25% to 50% indicating low heterogeneity, 50% to 75% indicating moderate heterogeneity, and $I^2 > 75\%$ indicating high heterogeneity [21,22]. Forest plots were generated using RevMan 5.4 software and funnel plots were generated using Stata 17 software, with statistical significance set at $p < 0.05$ for all results.

3. Results

3.1. Study Selection

As shown in Figure 1, the collaborative efforts of two authors yielded a total of 1618 retrieval records. Following the elimination of duplicate studies, the pool was narrowed down to 971 studies. Subsequently, a rigorous screening process involving the examination of titles and abstracts led to the exclusion of an additional 898 studies that did not meet the inclusion criteria. After a meticulous full-text assessment, 57 studies were disqualified based on several criteria: (1) the intervention not encompassing IMT (n = 35); (2) an absence of COPD patient involvement (n = 12); (3) the full text not being accessible for review (n = 8); (4) the outcomes studied being deemed irrelevant to the research objectives (n = 2). Finally, 16 studies [23–38] met the inclusion criteria and were selected for further analysis.

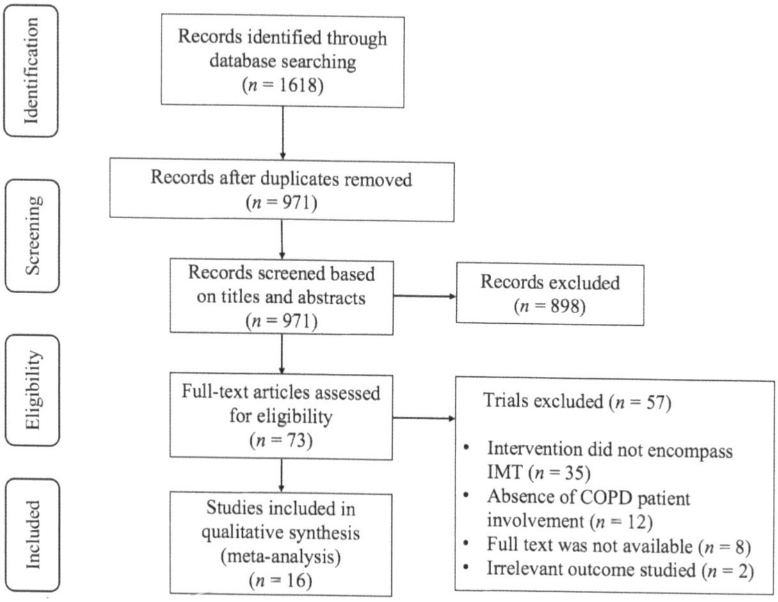

Figure 1. PRISMA flowchart of study selection.

3.2. Characteristics of the Included Studies

Table S1 summarizes the included studies' composition, encompassing 700 participants in the intervention group and 644 participants in the control group, all diagnosed with COPD at the GOLD II–IV stages. The studies accommodated multiple intervention groups within a single study, while the control groups were subjected to either a placebo intervention or no additional intervention beyond their daily routines, medications, and usual care. The intervention duration varied significantly across the studies, ranging from as brief as 3 weeks to as extensive as 15 months, with session duration lasting from 3 min to 60 min. Notably, 10 studies provided data on inspiratory muscle strength [23–26,28,29,32,36–38], primarily measured using PImax. Additionally, 11 studies provided data on dyspnea [24,25,27,28,30–33,35,36,38], primarily assessed using the baseline dyspnea index/transition dyspnea index (BDI/TDI), modified Medical Research Council (mMRC) scale, Borg scale, and San Diego Shortness of Breath Questionnaire (SOBQ). Furthermore, seven studies provided data on QOL [26,27,31–34,38], primarily evaluated using the Chronic Respiratory Disease Questionnaire (CRDQ), St. George's Respiratory Questionnaire (SGRQ), Short Form 36-Item Health Survey (SF-36), Chronic Airway Test (CAT), and Clinical COPD Questionnaire (CCQ).

3.3. Meta-Analysis

3.3.1. Effects of IMT on Inspiratory Muscle Strength in COPD Patients

As shown in Figure 2, our results showed that IMT significantly enhanced inspiratory muscle strength in COPD patients (SMD = 0.86; 95% CI: 0.52 to 1.19, $p < 0.00001$, $I^2 = 71\%$).

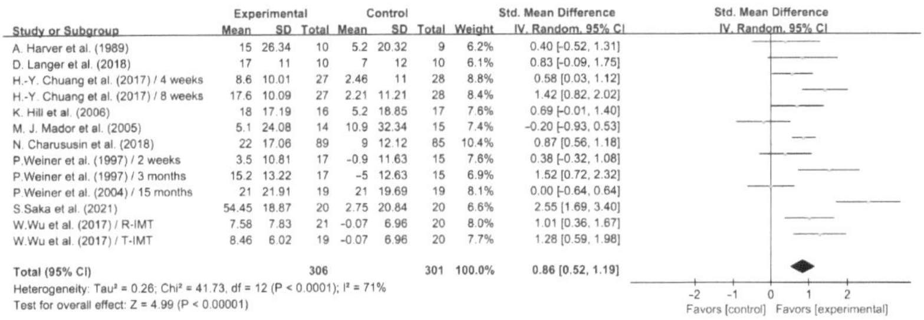

Figure 2. Meta-analysis results on the effects of IMT on inspiratory muscle strength in COPD patients [23–26,28,29,32,36–38].

Subgroup analysis showed that IMT conducted for ≤20 min (SMD = 0.97; 95% CI: 0.26 to 1.69, $p = 0.008$, $I^2 = 80\%$), between 20 and 30 min (SMD = 0.90; 95% CI: 0.36 to 1.43, $p = 0.001$, $I^2 = 57\%$), and ≥30 min (SMD = 0.68; 95% CI: 0.13 to 1.23, $p = 0.01$, $I^2 = 72\%$, Figure 3) significantly enhanced inspiratory muscle strength in COPD patients, with IMT conducted for ≤20 min having a greater effect.

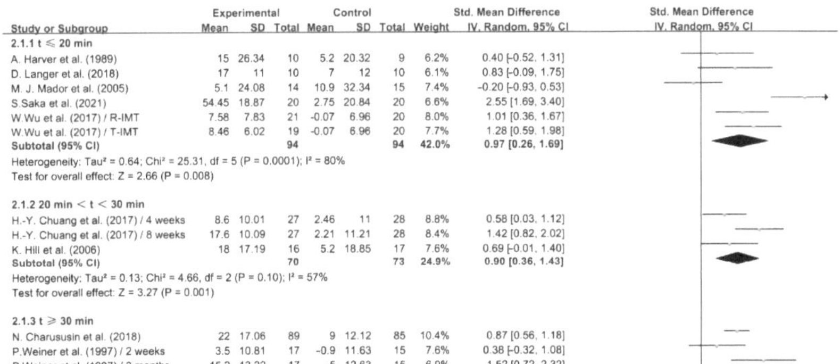

Figure 3. Meta-analysis results on the effects of session duration on inspiratory muscle strength in COPD patients [23–26,28,29,32,36–38].

In addition, when analyzing the subgroups by frequency, IMT conducted >3 times per week significantly enhanced inspiratory muscle strength in COPD patients (SMD = 1.06;

95% CI: 0.72 to 1.40, $p < 0.00001$, $I^2 = 63\%$), while IMT conducted ≤3 times per week had no significant effect on enhancing inspiratory muscle strength in COPD patients (SMD = 0.16; 95% CI: −0.35 to 0.68, $p = 0.53$, $I^2 = 40\%$, Figure 4).

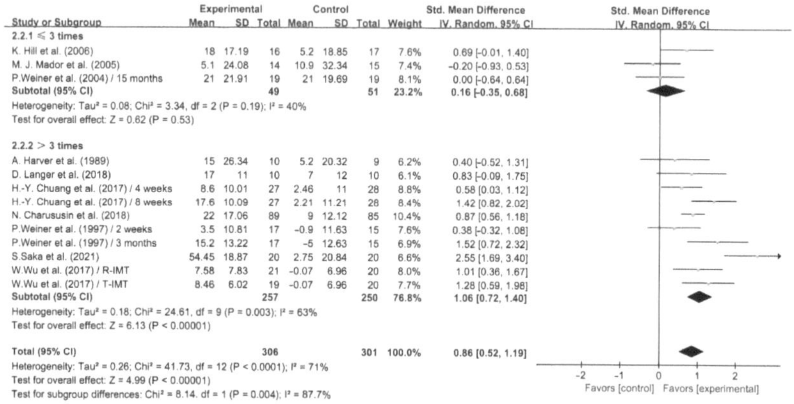

Figure 4. Meta-analysis results on the effects of frequency of intervention on inspiratory muscle strength in COPD patients [23–26,28,29,32,36–38].

Furthermore, when analyzing the subgroup by intensity, <60% PImax IMT (SMD = 1.22; 95% CI: 0.36 to 2.08, $p = 0.005$, $I^2 = 83\%$) and ≥60% PImax IMT (SMD = 0.81; 95% CI: 0.50 to 1.12, $p < 0.00001$, $I^2 = 46\%$, Figure 5) significantly enhanced inspiratory muscle strength in COPD patients, with <60% PImax IMT having a greater effect.

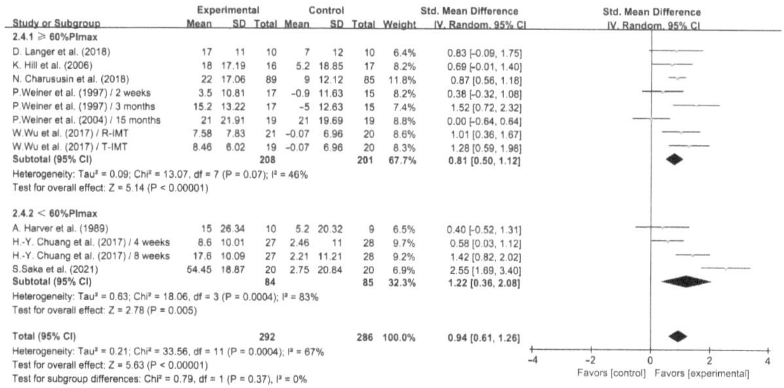

Figure 5. Meta-analysis results on the effects of intensity of intervention on inspiratory muscle strength in COPD patients [23–26,28,32,36–38].

3.3.2. Effects of IMT on Dyspnea in COPD Patients

As shown in Figure 6, IMT significantly improved dyspnea in COPD patients (SMD = −0.50; 95% CI: −0.71 to −0.29, $p < 0.00001$, $I^2 = 66\%$).

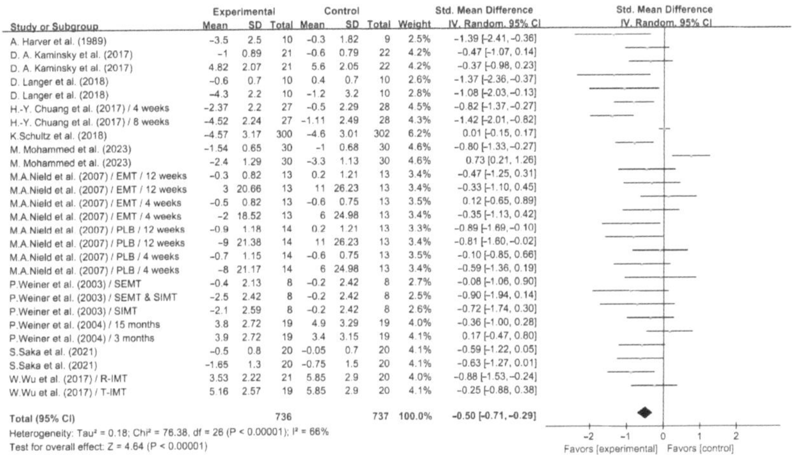

Figure 6. Meta-analysis results on the effects of IMT on dyspnea in COPD patients [24,25,27,28,30–33,35,36,38].

Subgroup analysis showed that IMT conducted for ≤20 min (SMD = −0.63; 95% CI: −1.09 to −0.17, $p = 0.007$, $I^2 = 75\%$), between 20 and 30 min (SMD = −0.50; 95% CI: −0.84 to −0.16, $p = 0.004$, $I^2 = 72\%$), and ≥30 min (SMD = −0.33; 95% CI: −0.60 to −0.05, $p = 0.02$, $I^2 = 0\%$, Figure 7) significantly alleviated dyspnea in COPD patients, with IMT conducted for ≤20 min having a greater effect.

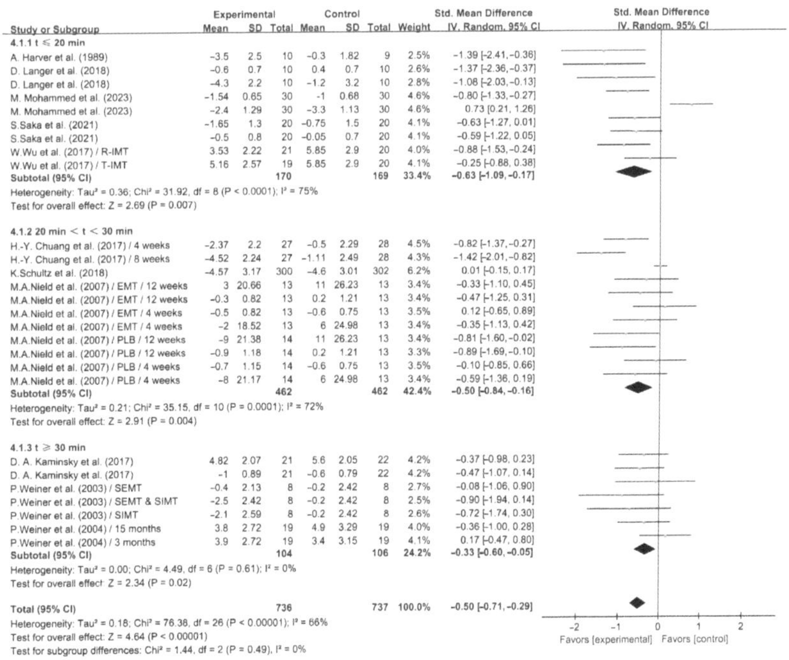

Figure 7. Meta-analysis results on the effects of session duration on dyspnea in COPD patients [24,25,27,28,30–33,35,36,38].

In addition, when analyzing the subgroups by frequency, IMT conducted >3 times per week significantly alleviated dyspnea in COPD patients (SMD = −0.54; 95% CI: −0.76 to

−0.31, $p < 0.00001$, $I^2 = 68\%$), while IMT conducted ≤3 times per week had no significant effect on alleviating dyspnea in COPD patients (SMD = −0.09; 95% CI: −0.61 to 0.42, $p = 0.72$, $I^2 = 22\%$, Figure 8).

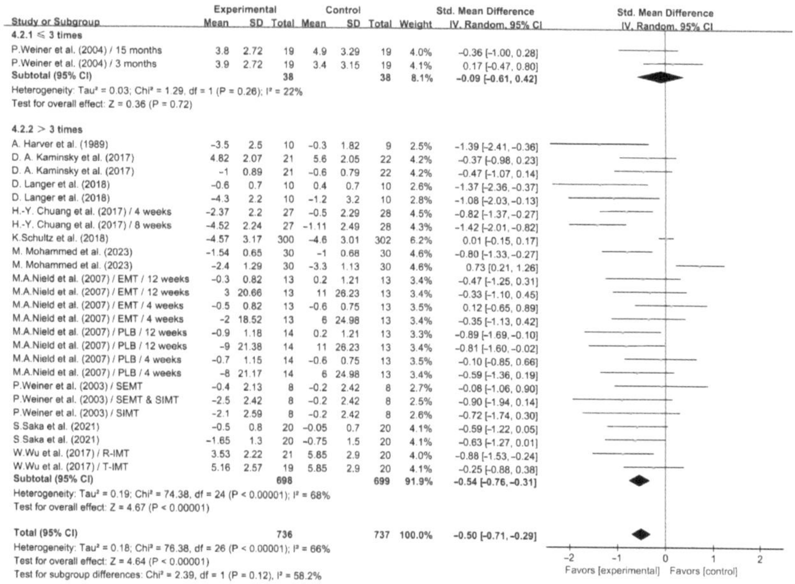

Figure 8. Meta-analysis results on the effects of frequency of intervention on dyspnea in COPD patients [24,25,27,28,30–33,35,36,38].

Furthermore, when analyzing the subgroups by intensity, <60% PImax IMT (SMD = −0.92; 95% CI: −1.27 to −0.58, $p < 0.0001$, $I^2 = 27\%$) and ≥60% PImax IMT (SMD = −0.38; 95% CI: −0.71 to −0.05, $p = 0.03$, $I^2 = 72\%$, Figure 9) significantly alleviated dyspnea in COPD patients, with <60% PImax IMT having a greater effect.

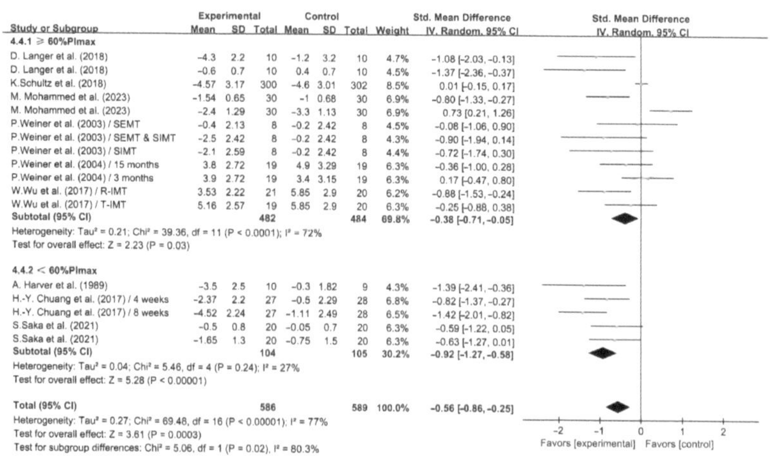

Figure 9. Meta-analysis results on the effects of intensity of intervention on dyspnea in COPD patients [24,25,28,30,32,33,35,36,38].

3.3.3. Effects of IMT on QOL in COPD Patients

As shown in Figure 10, IMT significantly improved QOL in COPD patients (SMD = 0.48; 95% CI: 0.21 to 0.75, $p = 0.0006$, $I^2 = 85\%$).

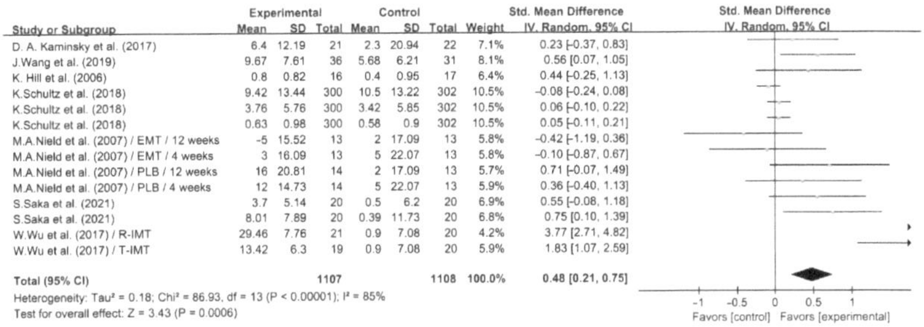

Figure 10. Meta-analysis results on the effects of IMT on QOL in COPD patients [26,27,31–34,38].

Subgroup analysis showed that IMT conducted for ≤20 min significantly enhanced QOL in COPD patients (SMD = 1.66; 95% CI: 0.46 to 2.86, $p = 0.007$, $I^2 = 90\%$); IMT conducted between 20 and 30 min had no significant effect on enhancing QOL in COPD patients (SMD = 0.03; 95% CI: −0.08 to 0.14, $p = 0.58$, $I^2 = 17\%$); and IMT conducted ≥30 min (SMD = 0.43; 95% CI: 0.05 to 0.81, $p = 0.03$, $I^2 = 0\%$, Figure 11) significantly enhanced QOL in COPD patients, with IMT conducted for ≤20 min having a greater effect.

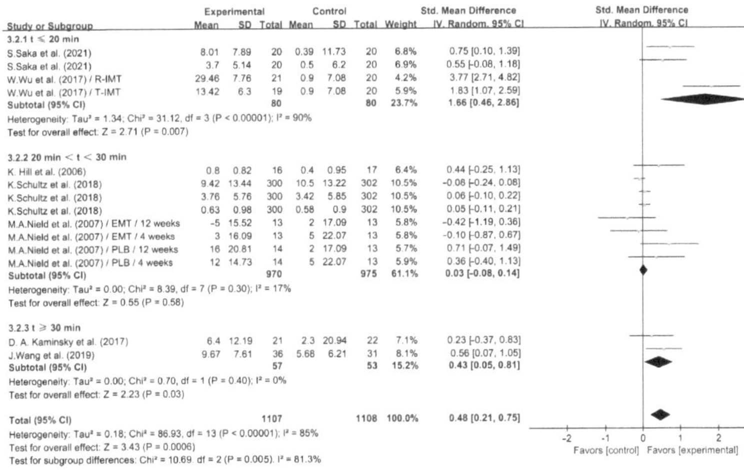

Figure 11. Meta-analysis results on the effects of session duration on QOL in COPD patients [26,27,31–34,38].

In addition, when analyzing the subgroups by frequency, IMT conducted >3 times per week significantly enhanced QOL in COPD (SMD = 0.48; 95% CI: 0.20 to 0.77, $p = 0.0009$, $I^2 = 86\%$), while IMT conducted ≤3 times per week had no significant effect on enhancing QOL in COPD patients (SMD = 0.44; 95% CI: −0.25 to 1.13, $p = 0.21$, Figure 12).

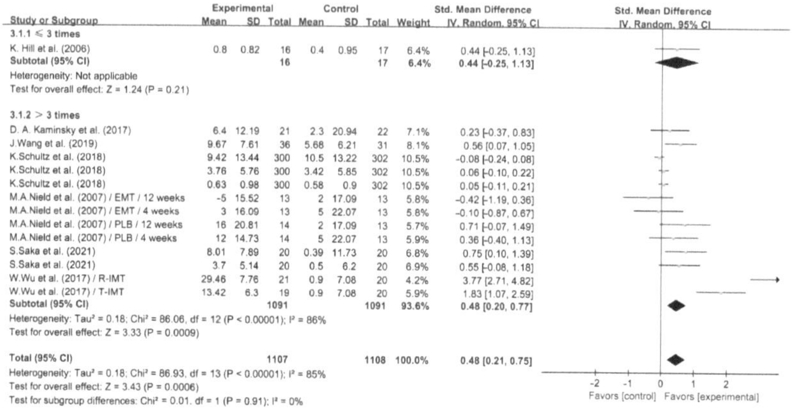

Figure 12. Meta-analysis results on the effects of frequency of intervention on QOL in COPD patients [26,27,31–34,38].

Furthermore, when analyzing the subgroups by intensity, <60% PImax IMT (SMD = 0.65; 95% CI: 0.20 to 1.10, $p = 0.005$, $I^2 = 0\%$) and ≥60% PImax IMT (SMD = 0.65; 95% CI: 0.23 to 1.07, $p = 0.002$, $I^2 = 93\%$, Figure 13) significantly enhanced QOL in COPD. However, the effect of IMT at <60% PImax and ≥60% PImax on enhancing the QOL of COPD patients was equivalent.

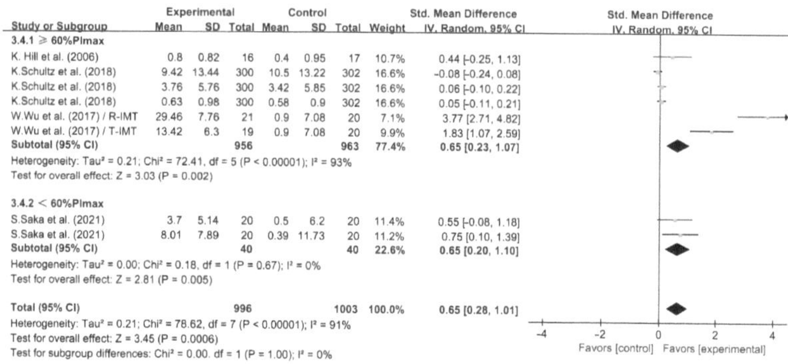

Figure 13. Meta-analysis results on the effects of intensity of intervention on QOL in COPD patients [26,32,33,38].

3.4. Risk of Bias

The RoB2 tool was utilized to evaluate the methodological quality of the included literature, emphasizing six domains: random bias, performance bias, detection bias, reporting bias, and other bias. As depicted in Figure S1, the overall quality of the studies was stratified into three levels: low, moderate, and high. Specifically, seven studies demonstrated a low risk of bias, four studies showed a moderate risk, and five studies were categorized as having a high risk of bias.

3.5. Sensitivity Analyses

Sensitivity analysis indicated that the positive effects of exercise on inspiratory muscle strength (Figure S2), dyspnea (Figure S3), and QOL (Figure S4) in COPD patients remained stable and consistent in both direction and magnitude, regardless of the exclusion of any individual study.

3.6. Publication Bias

To evaluate the possibility of publication bias, funnel plots were generated (Figures S3–S5). The results of Egger's test suggested that the small-sample-size studies did not significantly impact the overall results, particularly for inspiratory muscle strength ($p = 0.834$). However, exceptions were observed for dyspnea ($p = 0.001$) and QOL ($p = 0.013$, Table S2), suggesting potential bias in these areas. In response, Dsuval and Tweedie's trim and fill procedure was applied, and the subsequent analysis confirmed the absence of evidence for publication bias in relation to dyspnea and QOL.

4. Discussion

The purpose of this study was to explore the effects of IMT on inspiratory muscle strength, dyspnea, and QOL in COPD patients. A total of 16 studies were included, with the results conclusively demonstrating that IMT significantly improved inspiratory muscle strength, dyspnea, and QOL in COPD patients.

4.1. Effects of IMT on Inspiratory Muscle Strength in COPD Patients

We found that IMT can significantly improve inspiratory muscle strength in COPD patients. This finding is consistent with previous studies [39] showing that 12 weeks of home-based IMT significantly improved PImax. Figueiredo et al. [9] showed that inspiratory muscle strength can be notably improved through IMT alone or in combination with other interventions. In addition, Wanke et al. [40] showed that a rehabilitation program incorporating both IMT and cardiorespiratory exercise training (CET) could improve inspiratory muscle function, whereas CET alone did not produce the same effect. Furthermore, Tounsi et al. [41] demonstrated that a combination of IMT and educational exercises, performed once daily for 4–5 min over 7 days per week, led to a more pronounced improvement in inspiratory muscle strength compared to educational exercises alone. Moreover, Shahin et al. [42] revealed that, in patients with severe COPD, IMT conducted six times a week can improve inspiratory muscle strength, alleviate dyspnea, and improve prognosis regarding inspiratory function (IF).

The primary abnormalities in respiratory muscle function in COPD patients are believed to stem from mechanical defects caused by overinflation, which shorten the diaphragm fibers and force it to work on an ineffective portion of its length–tension curve [43]. Specifically, the loss of lung elastic recoil and the development of expiratory flow limitation both contribute to progressive air trapping, increased end-expiratory lung volume, and reduced inspiratory capacity [44]. A systematic review has shown that IMT is more beneficial for inspiratory muscle strength in COPD patients when the training intensity is reduced [9]. Consequently, IMT in COPD patients can improve inspiratory muscle strength by improving the contractility and relaxation of inspiratory muscle. Recent research suggests that, when combined with pulmonary rehabilitation (PR), IMT may not necessarily improve dyspnea, respiratory muscle strength, or QOL, though larger effects in participants with weak respiratory muscles and longer training durations remain to be confirmed [45]. This may be intimately tied to the intensity, frequency, and duration of respiratory muscle training.

Based on the subgroup analysis of inspiratory muscle strength, when the intervention duration of IMT is $\leq$20 min per session, the improvement in inspiratory muscle strength in COPD patients is optimal. This finding concurs with a previous study [46], which demonstrated a significant enhancement in PImax in COPD patients who underwent 15 min of home-based IMT per session. This may stem from the fact that excessively brief intervention times may lack efficacy, whereas excessively long intervention times can induce respiratory muscle fatigue, subsequently diminishing inspiratory muscle improvement.

In addition, our subgroup analysis revealed that when IMT was conducted >3 times per week, the improvement in inspiratory muscle strength was most pronounced, which was consistent with previous studies. Weiner et al. [36] demonstrated that IMT leads to an improvement in inspiratory muscle strength; specifically, PImax significantly increased in COPD patients who underwent IMT six times per week. Battaglia et al. [36], on the other hand, reported on the combined use of a novel expiratory device alongside a previously evaluated inspiratory device, which enhanced inspiratory muscle strength in COPD patients receiving IMT seven times per week IMT. Furthermore, the ACSM recommends that COPD patients engage in 8 to 10 sessions per week of activities aimed at maintaining or enhancing muscle strength and endurance to maximize strength gains. Moreover, our findings indicated that IMT with an intensity of <60% PImax was more efficacious in improving inspiratory muscle strength in COPD patients, consistent with previous studies. Saka et al. [32] demonstrated that in the IMT group, intensity was initially set at 30% PImax and subsequently adjusted based on weekly PImax values. Following IMT, a statistically significant improvement in PImax was observed. Xu et al. [32], on the other hand, reported that IMT with an intensity of 45% PImax alone is an effective method for improving inspiratory muscle strength. In conclusion, IMT sessions lasting $\leq$ 20 min, conducted > 3 times per week, and at an intensity of <60% PImax are more beneficial for enhancing inspiratory muscle strength in COPD patients.

4.2. Effects of IMT on Dyspnea in COPD Patients

Our results demonstrated that IMT significantly alleviated dyspnea in COPD patients, which is consistent with previous studies. For example, Beaumont et al. [47] showed that threshold IMT can significantly improve dyspnea in COPD patients. In addition, Battaglia et al. [48] reported that combining home exercise using expiratory and inspiratory devices can significantly improve dyspnea in patients with mild to severe COPD. Furthermore, Shahin et al. [42] presented results indicating that IMT not only improves the perception of dyspnea but also improves IF outcomes in patients with severe COPD. Previous studies have underscored the pivotal role of respiratory muscle weakness in the development of dyspnea. Therefore, dyspnea is frequently used as a prognostic indicator in COPD studies [49]. Dyspnea in COPD patients arises from an increased respiratory rate due to pulmonary hyperinflation, leading to rapid and shallow breathing. This breathing pattern decreases the gas exchange rate and an exacerbates respiratory muscle fatigue. During exercise, COPD patients experience increased lactic acidosis and are unable to meet their ventilation demands. IMT fortifies muscle strength and endurance by exercising the inspiratory muscles, mainly the diaphragm, and augments patients' ventilation capacity, thereby mitigating dyspnea [50]. It alleviates dyspnea in COPD patients by enhancing inspiratory muscle strength. Nevertheless, some researchers remain skeptical regarding the dyspnea-improving effects of IMT. Figueiredo et al. [9] showed that isolated IMT did not alleviate dyspnea, and the presence of inspiratory muscle weakness did not alter this outcome. In addition, Koch et al. [51] reported that a small number of subjects exhibited a placebo effect following independent IMT, and the effectiveness of independent IMT did not significantly differ from that of the control group. These studies suggest that the effectiveness of IMT in improving dyspnea in COPD patients is influenced by factors such as the patient's disease status, intervention type, intensity, and frequency, and a unified conclusion has yet to be reached.

Based on the subgroup analysis of dyspnea, IMT conducted for $\leq$20 min yielded the most pronounced improvement in dyspnea in COPD patients, which is consistent with previous studies. Battaglia et al. [32] reported that COPD patients progressed with IMT for 15 min per session. The results showed that COPD patients' dyspnea significantly alleviated. Buran et al. [52] demonstrated that after 15 min of IMT combined with manual therapy (MT), COPD patients' dyspnea significantly decreased. In addition, IMT conducted >3 times per week was more effective in alleviating dyspnea in COPD patients, which aligns with previous studies. Weiner et al. [36] showed that IMT resulted in significant improvement in dyspnea when COPD patients received IMT six times per week. Battaglia et al. [48] reported that after COPD patients performed IMT seven times per week, dyspnea values were significantly improved. Furthermore, IMT at an intensity of <60% PImax was more effective in alleviating dyspnea in COPD patients, which is in agreement with the findings of Saka et al. [32], who showed that a 30% PImax IMT load can facilitate the alleviation of dyspnea.

4.3. Effects of IMT on QOL in COPD Patients

Decreased QOL serves as a predictor for mortality and re-hospitalization in COPD patients. Research has established that both inspiratory and expiratory muscle training can improve the QOL in COPD patients. Notably, the improvement in QOL observed in IMT groups was significantly greater than that in control groups, likely due to the more substantial reduction in dyspnea achieved through IMT. Addressing dyspnea through respiratory muscle training contributes to improving QOL [53]. COPD is characterized by its progressive and insidious nature, with lung function often declining by 50% before symptoms manifest. A sudden acute exacerbation of symptoms significantly impact patients' QOL and treatment costs. As a standalone treatment, IMT can mitigate dyspnea and improve exercise tolerance, thereby improving the QOL of COPD patients by reducing dyspnea and slowing disease progression [50].

In this study, we have discovered that IMT improves the QOL in COPD patients by alleviating dyspnea, which is consistent with previous studies. Abedi et al. [50] has shown that a combination of IMT and aerobic exercise has a marked effect on improving QOL. In addition, Sánchez et al. [54] demonstrated that 6 days of weekly, targeted IMT significantly contributed to improving QOL. Furthermore, Ozsoy et al. [49] revealed that both 4 weeks of basic IMT and 4 weeks of functional IMT (aiming to enhance all muscle functions for greater benefits) notably improved the QOL in elderly COPD patients.

Concurrently, some researchers have expressed doubts about the extent of IMT's benefits on QOL in COPD patients. While IMT can significantly improve respiratory muscle function, its impact on lung function, clinical outcomes, and QOL remains unsubstantiated by robust scientific evidence. Polkey et al. [55] suggested that the decrease in diaphragmic pressure observed in COPD patients is primarily attributed to hyperinflation, which they believe is unlikely to be ameliorated by IMT. They further posited that the link between the decline in inspiratory pressure caused by hyperinflation and adverse outcomes might be an epiphenomenon. O'Brien et al. [12] found that IMT significantly improved inspiratory muscle strength and endurance, yet the benefits in terms of sports performance and QOL were not evident.

According to the subgroup analysis of QOL, when the intervention duration of IMT is $\leq$20 min per session, the improvement in QOL in COPD patients is optimal, which aligns with previous studies. Beaumont et al. [47] reported that IMT for 15 min per session combined with a pulmonary rehabilitation program (PRP) is an effective way to improve the QOL of COPD patients. Buran et al. [52] revealed that after 15 min of IMT combined with MT, a comparison between COPD patients' QOL and the control group showed significant different changes in the total and all subscale scores of QOL. In addition, IMT conducted > 3 times per week was more effective in improving QOL, which is consistent with previous studies. Xu et al. [53] showed that IMT at a frequency of seven times per week can improve symptoms of depression, anxiety, insomnia, and QOL in COPD patients. Sánchez et al. [54] demonstrated that targeted IMT six times per week improves QOL in COPD patients. However, an intervention frequency of $\leq$3 times per week did not affect the improvement of QOL. Since only one study with $\leq$3 times per week was included in this subgroup, further studies are needed to verify this result in the future. Furthermore, our subgroup analysis indicated that both intervention intensities of <60% PImax and $\geq$60% PImax were effective in improving QOL, which is in line with previous studies. Xu et al. [53] showed that 45% PImax IMT significantly improved QOL compared to sham training. Beaumont et al. [47] reported that in severe and very severe COPD patients, 60% PImax IMT performed during a PRP is associated with an improvement in QOL.

4.4. Strengths and Limitations of This Study

There are some potential limitations to this meta-analysis. Firstly, all the included studies are RCTs of IMT interventions, which cannot be fully blinded, potentially introducing a degree of bias in the quality evaluation due to subjective factors. Secondly, while the included studies did not explicitly identify intervention-related adverse events, it is uncertain whether the investigators comprehensively documented all possible adverse events. Finally, the subgroup analysis did not encompass the monitoring of intervention measures and rest periods during exercise. Consequently, there may be a necessity for future research endeavors featuring larger sample populations and enhanced methodological rigor to augment and validate our current findings.

5. Conclusions

IMT is an effective approach to improving inspiratory muscle strength, alleviating dyspnea, and enhancing QOL in COPD patients. Conducting IMT for $\leq$20 min and >3 times per week is the optimal way to improve inspiratory muscle strength, dyspnea, and QOL, whereas IMT performed at an intensity of <60% PImax is more effective in improving inspiratory muscle strength and dyspnea in COPD patients. Since the effect of IMT at <60% PImax and $\geq$60% PImax on enhancing the QOL of COPD patients is equivalent, this meta-analysis provides clinicians with evidence to recommend that COPD patients engage in IMT at <60% PImax for more than three times per week, with each session lasting no more than 20 min, to improving inspiratory muscle strength, dyspnea, and QOL.

Supplementary Materials: The following supporting information can be downloaded at https://www.mdpi.com/article/10.3390/life14111470/s1, Figure S1: Results of Cochrane risk-of-bias tool; Figure S2: Sensitivity analysis results on inspiratory muscle strength; Figure S3: Sensitivity analysis results on dyspnea; Figure S4: Sensitivity analysis results on QOL; Figure S5: Funnel plot of inspiratory muscle strength; Figure S6: Funnel plot of inspiratory dyspnea; Figure S7: Funnel plot of inspiratory QOL; Table S1: Characteristics of the studies included in this meta-analysis; Table S2: Results of Egger's test.

Author Contributions: Conceptualization, B.H. and L.Y.; methodology, B.H.; software, B.H.; validation, B.R., Y.C., Y.L. and C.L.; formal analysis, B.H. and Z.C.; investigation, B.H. and Z.C.; resources, L.Y.; data curation, B.R., Y.C., Y.L. and C.L.; writing—original draft preparation, B.H.; writing—review and editing, B.H. and L.Y.; visualization, B.H.; supervision, L.Y.; project administration, L.Y.; funding acquisition, L.Y. All authors have read and agreed to the published version of the manuscript.

Funding: This research was funded by the National Key R&D Program of China, grant number 2022YFC3600201, Humanities and Social Science Fund of Ministry of Education of China, grant number 24YJC890065, Chinese Universities Scientific Fund, grant number 2024JNPD004, and Project of Education and Teaching Reform at Beijing Sport University, grant number BSUJG2022KCJSB05.

Institutional Review Board Statement: Not applicable.

Informed Consent Statement: Not applicable.

Data Availability Statement: All data generated or analyzed during this study are included in the article/Supplementary Material.

Conflicts of Interest: The authors declare no conflicts of interest.

References

1. Agustí, A.; Celli, B.R.; Criner, G.J.; Halpin, D.; Anzueto, A.; Barnes, P.; Bourbeau, J.; Han, M.K.; Martinez, F.J.; Montes de Oca, M.; et al. Global Initiative for Chronic Obstructive Lung Disease 2023 Report: GOLD Executive Summary. *Eur. Respir. J.* **2023**, *61*, 2300239. [CrossRef] [PubMed]
2. Vogelmeier, C.F.; Román-Rodríguez, M.; Singh, D.; Han, M.K.; Rodríguez-Roisin, R.; Ferguson, G.T. Goals of COPD treatment: Focus on symptoms and exacerbations. *Respir. Med.* **2020**, *166*, 105938. [CrossRef] [PubMed]
3. Larson, J.L.; Covey, M.K.; Corbridge, S. Inspiratory muscle strength in chronic obstructive pulmonary disease. *AACN Clin. Issues* **2002**, *13*, 320–332. [CrossRef] [PubMed]

4. Polkey, M.I.; Kyroussis, D.; Hamnegard, C.H.; Mills, G.H.; Green, M.; Moxham, J. Diaphragm strength in chronic obstructive pulmonary disease. *Am. J. Respir. Crit. Care Med.* **1996**, *154*, 8912741. [CrossRef]
5. Kim, M.J.; Druz, W.S.; Danon, J.; Machnach, W.; Sharp, J.T.; Leduc, D.; Cappello, M.; Gevenois, P.A.; Wilson, T.A.; De Troyer, A.; et al. Mechanics of the canine diaphragm. *J. Appl. Physiol.* **1976**, *41*, 369–382. [CrossRef]
6. Sheraz, S.; Ferraro, F.V.; Siddiqui, F.A.; Tariq, H.; Anthony Faghy, M.; Malik, A.N. The effects of inspiratory muscle training on balance and functional mobility: A systematic review. *Postgrad. Med.* **2023**, *135*, 690–700. [CrossRef]
7. Bissett, B.; Leditschke, I.A.; Green, M.; Marzano, V.; Collins, S.; Van Haren, F. Inspiratory muscle training for intensive care patients: A multidisciplinary practical guide for clinicians. *Aust. Crit. Care* **2019**, *32*, 249–255. [CrossRef]
8. Ferraro, F.V.; Gavin, J.P.; Wainwright, T.; McConnell, A. The effects of 8 weeks of inspiratory muscle training on the balance of healthy older adults: A randomized, double-blind, placebo-controlled study. *Physiol. Rep.* **2019**, *7*, 14076. [CrossRef]
9. Figueiredo, R.I.N.; Azambuja, A.M.; Cureau, F.V.; Sbruzzi, G. Inspiratory Muscle Training in COPD. *Respir. Care* **2020**, *65*, 1189–1201. [CrossRef]
10. Lotters, F.; van Tol, B.; Kwakkel, G.; Gosselink, R. Effects of controlled inspiratory muscle training in patients with COPD: A meta-analysis. *Eur. Respir. J.* **2002**, *20*, 570–576. [CrossRef]
11. Crowe, J.; Reid, W.D.; Geddes, E.L.; O'Brien, K.; Brooks, D. Inspiratory Muscle Training Compared with Other Rehabilitation Interventions in Adults with Chronic Obstructive Pulmonary Disease: A Systematic Literature Review and Meta-Analysis. *COPD J. Chronic Obstr. Pulm. Dis.* **2005**, *2*, 319–329. [CrossRef] [PubMed]
12. O'Brien, K.; Geddes, E.L.; Reid, W.D.; Brooks, D.; Crowe, J. Inspiratory muscle training compared with other rehabilitation interventions in chronic obstructive pulmonary disease: A systematic review update. *J. Cardiopulm. Rehabil. Prev.* **2008**, *28*, 128–141. [CrossRef] [PubMed]
13. Beaumont, M.; Forget, P.; Couturaud, F.; Reychler, G. Effects of inspiratory muscle training in COPD patients: A systematic review and meta-analysis. *Clin. Respir. J.* **2018**, *12*, 2178–2188. [CrossRef]
14. Cai, Y.; Ren, X.; Wang, J.; Ma, B.; Chen, O. Effects of Breathing Exercises in Patients with Chronic Obstructive Pulmonary Disease: A Network Meta-analysis. *Arch. Phys. Med. Rehabil.* **2024**, *105*, 558–570. [CrossRef]
15. Moher, D.; Shamseer, L.; Clarke, M.; Ghersi, D.; Liberati, A.; Petticrew, M.; Shekelle, P.; Stewart, L.A.; Prisma-P Group. Preferred reporting items for systematic review and meta-analysis protocols (PRISMA-P) 2015 statement. *Syst. Rev.* **2015**, *4*, 1. [CrossRef]
16. Du, L.; Xi, H.; Zhang, S.; Zhou, Y.; Tao, X.; Lv, Y.; Hou, X.; Yu, L. Effects of exercise in people with multiple sclerosis: A systematic review and meta-analysis. *Front. Public. Health* **2024**, *12*, 1387658. [CrossRef]
17. Qiu, B.; Zhou, Y.; Tao, X.; Hou, X.; Du, L.; Lv, Y.; Yu, L. The effect of exercise on flow-mediated dilation in people with type 2 diabetes mellitus: A systematic review and meta-analysis of randomized controlled trials. *Front. Endocrinol.* **2024**, *15*, 1347399. [CrossRef]
18. Li, G.; Tao, X.; Lei, B.; Hou, X.; Yang, X.; Wang, L.; Zhang, S.; Lv, Y.; Wang, T.; Yu, L. Effects of exercise on post-stroke cognitive function: A systematic review and meta-analysis of randomized controlled trials. *Top. Stroke Rehabil.* **2024**, *31*, 645–666. [CrossRef]
19. Wu, W.; Chen, Z.; Zhou, H.; Wang, L.; Li, X.; Lv, Y.; Sun, T.; Yu, L. Effects of Acute Ingestion of Caffeine Capsules on Muscle Strength and Muscle Endurance: A Systematic Review and Meta-Analysis. *Nutrients* **2024**, *16*, 1146. [CrossRef]
20. Grgic, J.; Del Coso, J. Ergogenic Effects of Acute Caffeine Intake on Muscular Endurance and Muscular Strength in Women: A Meta-Analysis. *Int. J. Environ. Res. Public Health* **2021**, *18*, 5773. [CrossRef]
21. Li, G.; You, Q.; Hou, X.; Zhang, S.; Du, L.; Lv, L.; Yu, L. The effect of exercise on cognitive function in people with multiple sclerosis: A systematic review and meta-analysis of randomized controlled trials. *J. Neurol.* **2023**, *270*, 2908–2923. [CrossRef] [PubMed]
22. Chen, Y.; Su, Q.; Yang, J.; Li, G.; Zhang, S.; Lv, Y.; Yu, L. Effects of rest interval and training intensity on jumping performance: A systematic review and meta-analysis investigating post-activation performance enhancement. *Front. Physiol.* **2023**, *14*, 1202789. [CrossRef] [PubMed]
23. Charususin, N.; Gosselink, R.; Decramer, M.; Demeyer, H.; McConnell, A.; Saey, D.; Maltais, F.; Derom, E.; Vermeersch, S.; Heijdra, Y.F.; et al. Randomised controlled trial of adjunctive inspiratory muscle training for patients with COPD. *Thorax* **2018**, *73*, 942–950. [CrossRef]
24. Chang, H.Y.; Fang, Y.Y.; Guo, S.E. The effects of threshold inspiratory muscle training in patients with chronic obstructive pulmonary disease: Arandomised experimental study. *J. Clin. Nurs.* **2017**, *26*, 4830–4838. [CrossRef]
25. Harver, A.; Mahler, D.A.; Daubenspeck, J.A. Targeted inspiratory muscle training improves respiratory muscle function and reduces dyspnea in patients with chronic obstructive pulmonary disease. *Ann. Intern. Med.* **1989**, *111*, 117–124. [CrossRef]
26. Hill, K.; Jenkins, S.C.; Philippe, D.L.; Cecins, N.; Shepherd, K.L.; Green, D.J.; Hillman, D.R.; Eastwood, P.R. High-intensity inspiratory muscle training in COPD. *Eur. Respir. J.* **2006**, *27*, 1119–1128. [CrossRef]
27. Kaminsky, D.A.; Guntupalli, K.K.; Lippmann, J.; Burns, S.M.; Brock, M.A.; Skelly, J.; DeSarno, M.; Pecott-Grimm, H.; Mohsin, A.; LaRock-McMahon, C.; et al. Effect of Yoga Breathing (Pranayama) on Exercise Tolerance in Patients with Chronic Obstructive Pulmonary Disease: A Randomized, Controlled Trial. *J. Altern. Complement. Med.* **2017**, *23*, 696–704. [CrossRef]
28. Langer, D.; Ciavaglia, C.; Faisal, A.; Webb, K.A.; Neder, J.A.; Gosselink, R.; Dacha, S.; Topalovic, M.; Ivanova, A.; O'donnell, D.E. Inspiratory muscle training reduces diaphragm activation and dyspnea during exercise in COPD. *J. Appl. Physiol.* **2018**, *125*, 381–392. [CrossRef]

29. Mador, M.J.; Deniz, O.; Aggarwal, A.; Shaffer, M.; Kufel, T.J.; Spengler, C.M. Effect of respiratory muscle endurance training in patients with COPD undergoing pulmonary rehabilitation. *Chest* **2005**, *128*, 1216–1224. [CrossRef]
30. Mohammed, M.; Mehani, S.; Aziz, A.A.; Mohamed, M.F.; El Nahas, N. Efficacy of threshold inspiratory muscle trainer versus diaphragmatic plus pursed lip breathing in occupational COPD. *Beni-Suef Univ. J. Basic Appl. Sci.* **2023**, *12*, 73. [CrossRef]
31. Nield, M.A.; Soo Hoo, G.W.; Roper, J.M.; Santiago, S. Efficacy of pursed-lips breathing: A breathing pattern retraining strategy for dyspnea reduction. *J. Cardiopulm. Rehabil. Prev.* **2007**, *27*, 237–244. [CrossRef] [PubMed]
32. Saka, S.; Gurses, H.N.; Bayram, M. Effect of inspiratory muscle training on dyspnea-related kinesiophobia in chronic obstructive pulmonary disease: A randomized controlled trial. *Complement. Ther. Clin. Pract.* **2021**, *44*, 101418. [CrossRef]
33. Schultz, K.; Jelusic, D.; Wittmann, M.; Krämer, B.; Huber, V.; Fuchs, S.; Lehbert, N.; Wingart, S.; Stojanovic, D.; Göhl, O.; et al. Inspiratory muscle training does not improve clinical outcomes in 3-week COPD rehabilitation: Results from a randomised controlled trial. *Eur. Respir. J.* **2018**, *51*, 1702000. [CrossRef]
34. Wang, J.; Guo, S.; Zeng, M.; Yu, P.; Mo, W. Observation of the curative effect of device-guided rehabilitation on respiratory function in stable patients with chronic obstructive pulmonary disease. *Medicine* **2019**, *98*, e14034. [CrossRef]
35. Weiner, P.; Magadle, R.; Beckerman, M.; Weiner, M.; Berar-Yanay, N. Comparison of specific expiratory, inspiratory, and combined muscle training programs in COPD. *Chest* **2003**, *124*, 1357–1364. [CrossRef]
36. Weiner, P.; Magadle, R.; Beckerman, M.; Weiner, M.; Berar-Yanay, N. Maintenance of inspiratory muscle training in COPD patients: One year follow-up. *Eur. Respir. J.* **2004**, *23*, 61–65. [CrossRef] [PubMed]
37. Weiner, P.; Man, A.; Weiner, M.; Rabner, M.; Waizman, J.; Magadle, R.; Zamir, D.; Greiff, Y. The effect of incentive spirometry and inspiratory muscle training on pulmonary function after lung resection. *J. Thorac. Cardiovasc. Surg.* **1997**, *113*, 552–557. [CrossRef] [PubMed]
38. Wu, W.; Guan, L.; Zhang, X.; Li, X.; Yang, Y.; Guo, B.; Ou, Y.; Lin, L.; Zhou, L.; Chen, R. Effects of two types of equal-intensity inspiratory muscle training in stable patients with chronic obstructive pulmonary disease: A randomised controlled trial. *Respir. Med.* **2017**, *132*, 84–91. [CrossRef]
39. Gosselink, R.; De Vos, J.; Van Den Heuvel, S.P.; Segers, J.; Decramer, M.; Kwakkel, G. Impact of inspiratory muscle training in patients with COPD: What is the evidence? *Eur. Respir. J.* **2011**, *37*, 416–425. [CrossRef]
40. Wanke, T.; Formanek, D.; Lahrmann, H.; Brath, H.; Wild, M.; Wagner, C.; Zwick, H. Effects of combined inspiratory muscle and cycle ergometer training on exercise performance in patients with COPD. *Eur. Respir. J.* **1994**, *7*, 2205–2211. [CrossRef]
41. Tounsi, B.; Acheche, A.; Lelard, T.; Tabka, Z.; Trabelsi, Y.; Ahmaidi, S. Effects of specific inspiratory muscle training combined with whole-body endurance training program on balance in COPD patients: Randomized controlled trial. *PLoS ONE* **2021**, *16*, e0257595. [CrossRef] [PubMed]
42. Shahin, B.; Germain, M.; Kazem, A.; Anna, G. Benefits of short inspiratory muscle training on exercise capacity, dyspnea, and inspiratory fraction in COPD patients. *Int. J. Chronic Obstr. Pulm. Dis.* **2008**, *3*, 423–427. [CrossRef] [PubMed]
43. Diaz, O.; Villafranca, C.; Ghezzo, H.; Borzone, G.; Leiva, A.; Milic-Emil, J.; Lisboa, C. Role of inspiratory capacity on exercise tolerance in COPD patients with and without tidal expiratory flow limitation at rest. *Eur. Respir. J.* **2000**, *16*, 269–275. [CrossRef]
44. O'Donnell, D.E.; D'Arsigny, C.; Webb, K.A. Effects of hyperoxia on ventilatory limitation during exercise in advanced chronic obstructive pulmonary disease. *Am. J. Respir. Crit. Care Med.* **2001**, *163*, 892–898. [CrossRef]
45. Ammous, O.; Feki, W.; Lotfi, T.; Khamis, A.M.; Gosselink, R.; Rebai, A.; Kammoun, S. Inspiratory muscle training, with or without concomitant pulmonary rehabilitation, for COPD: State of the evidence from a Cochrane review. *Eur. Respir. J.* **2023**, *1*, CD013778.
46. Chang, W.; Lin, H.C.; Liu, H.E.; Han, C.Y.; Chang, P.J. The Effectiveness of Home-Based Inspiratory Muscle Training on Small Airway Function and Disease-Associated Symptoms in Patients with Chronic Obstructive Pulmonary Disease. *Healthcare* **2023**, *11*, 2310. [CrossRef]
47. Beaumont, M.; Mialon, P.; Le Ber, C.; Le Mevel, P.; Péran, L.; Meurisse, O.; Morelot-Panzini, C.; Dion, A.; Couturaud, F. Effects of inspiratory muscle training on dyspnoea in severe COPD patients during pulmonary rehabilitation: Controlled randomised trial. *Eur. Respir. J.* **2018**, *51*, 1701107. [CrossRef]
48. Battaglia, E.; Fulgenzi, A.; Ferrero, M.E. Rationale of the combined use of inspiratory and expiratory devices in improving maximal inspiratory pressure and maximal expiratory pressure of patients with chronic obstructive pulmonary disease. *Arch. Phys. Med. Rehabil.* **2009**, *90*, 913–918. [CrossRef]
49. Ozsoy, I.; Kahraman, B.O.; Ozsoy, G.; Ilcin, N.; Tekin, N.; Savci, S. Effects of an Integrated Exercise Program Including "Functional" Inspiratory Muscle Training in Geriatric Individuals with and without Chronic Obstructive Pulmonary Disease. *Ann. Geriatr. Med. Res.* **2021**, *25*, 45–54. [CrossRef]
50. Abedi Yekta, A.H.; Poursaeid Esfahani, M.; Salehi, S.; Hassabi, M.; Khosravi, S.; Kharabian, S. Assessment of the Effects of Inspiratory Muscle Training (IMT) and Aerobic Training on the Quality of Life of Patients with Chronic Obstructive Pulmonary Disease. *Tanaffos* **2019**, *18*, 223–229.
51. Koch, R.; Augusto, T.R.d.L.; Ramos, A.G.; Müller, P.d.T. Inspiratory Muscle Training Potentiates the Beneficial Effects of Proportional Assisted Ventilation on Exertional Dyspnea and Exercise Tolerance in COPD: A Proof-of-Concept Randomized and Controlled Trial. *Copd-J. Chronic Obstr. Pulm. Dis.* **2020**, *17*, 384–391. [CrossRef] [PubMed]
52. Buran Cirak, Y.; Yilmaz Yelvar, G.D.; Durustkan Elbasi, N. Effectiveness of 12-week inspiratory muscle training with manual therapy in patients with COPD: A randomized controlled study. *Clin. Respir. J.* **2022**, *16*, 317–328. [CrossRef] [PubMed]

53. Xu, W.; Li, R.; Guan, L.; Wang, K.; Hu, Y.; Xu, L.; Zhou, L.; Chen, R.; Chen, X. Combination of inspiratory and expiratory muscle training in same respiratory cycle versus different cycles in COPD patients: A randomized trial. *Respir. Res.* **2018**, *19*, 225. [CrossRef]
54. Riera, H.S.; Rubio, T.M.; Ruiz, F.O.; Ramos, P.C.; Otero, D.D.C.; Hernandez, T.E.; Gomez, J.C. Inspiratory muscle training in patients with COPD: Effect on dyspnea, exercise performance, and quality of life. *Chest* **2001**, *120*, 748–756. [CrossRef]
55. Polkey, M.I.; Moxham, J.; Green, M. The case against inspiratory muscle training in COPD. *Eur. Respir. J.* **2011**, *37*, 236–237. [CrossRef]

life

MDPI

Systematic Review

Effects of Exercise on Cancer-Related Fatigue in Breast Cancer Patients: A Systematic Review and Meta-Analysis of Randomized Controlled Trials

Runyu Zhou [1,2,†], Zhuying Chen [2,†], Shiyan Zhang [2], Yushu Wang [3], Chiyang Zhang [3], Yuanyuan Lv [1,3,*] and Laikang Yu [1,2,*]

1 Beijing Key Laboratory of Sports Performance and Skill Assessment, Beijing Sport University, Beijing 100084, China; 18953366022@163.com
2 Department of Strength and Conditioning Assessment and Monitoring, Beijing Sport University, Beijing 100084, China; zhuying20232120126@126.com (Z.C.); 17634977464@163.com (S.Z.)
3 China Institute of Sport and Health Science, Beijing Sport University, Beijing 100084, China; m18756902144@167.com (Y.W.); 13215580747@163.com (C.Z.)
* Correspondence: sunflowerlyy@bsu.edu.cn (Y.L.); yulaikang@126.com (L.Y.)
† These authors have contributed equally to this work.

Abstract: The primary objective of this study was to assess the influence of exercise interventions on cancer-related fatigue (CRF), specifically in breast cancer patients, with the ultimate goal of establishing an optimal exercise prescription for breast cancer patients. A comprehensive search was undertaken across multiple databases, including Embase, PubMed, Cochrane Library, Web of Science, and Scopus, covering data published up to 1 September 2023. A meta-analysis was conducted to calculate the standardized mean difference (SMD) along with its corresponding 95% confidence interval (CI), thereby quantifying the effectiveness of exercise in alleviating CRF in the breast cancer patient population. Twenty-six studies met the inclusion criteria. Aerobic exercise (SMD, −0.17, $p = 0.02$), resistance exercise (SMD, −0.37, $p = 0.0009$), and combined exercise (SMD, −0.53, $p < 0.0001$) significantly improved CRF in breast cancer patients. In addition, exercise intervention conducted ≥3 times per week (SMD, −0.47, $p = 0.0001$) for >60 min per session (SMD, −0.63, $p < 0.0001$) and ≥180 min per week (SMD, −0.79, $p < 0.0001$) had greater effects on improving CRF in breast cancer patients, especially middle-aged patients (SMD, −0.42, $p < 0.0001$). Exercise is an effective approach to improving CRF in breast cancer patients. When devising an exercise program, the primary consideration should be the incorporation of combined exercise as the principal intervention. This entails ensuring that participants engage in the program at least three times weekly, with each session lasting for more than 60 min. The ultimate aim is to achieve a total weekly exercise duration of 180 min by progressively increasing the frequency of exercise sessions.

Keywords: exercise; cancer-related fatigue; breast cancer; systematic review; meta-analysis

Citation: Zhou, R.; Chen, Z.; Zhang, S.; Wang, Y.; Zhang, C.; Lv, Y.; Yu, L. Effects of Exercise on Cancer-Related Fatigue in Breast Cancer Patients: A Systematic Review and Meta-Analysis of Randomized Controlled Trials. *Life* **2024**, *14*, 1011. https://doi.org/10.3390/life14081011

Academic Editor: Nicola Smania

Received: 15 June 2024
Revised: 15 July 2024
Accepted: 12 August 2024
Published: 14 August 2024

1. Introduction

Cancer stands as one of the most pervasive health conditions globally, encompassing an astonishing array of over 200 identified types that have been linked to cause more than 60 dysfunctions [1]. Alarmingly, both the global incidence and mortality rates of cancer have been escalating steadily over the past few decades, portending a grim future where it is projected to emerge as the primary cause of mortality and the foremost impediment to extending human life expectancy in the 21st century [2,3].

Notably, breast cancer occupies a particularly ominous position, ranking as the most frequently occurring cancer among women and the leading contributor to cancer-related fatalities [3]. According to the American Cancer Society, the incidence of breast cancer has continued to increase, with an annual increase of 0.6–1% from 2015 to 2019 [4,5]. In recent years, with the development and application of effective anti-tumor therapies, the mortality rate and risk of postoperative recurrence in breast cancer patients have been significantly reduced [6–8]. However, as survival rates increase, more patients are facing a range of quality-of-life issues related to breast cancer treatment, such as cancer-related fatigue (CRF), premature menopause, cognitive dysfunction, depression, and anxiety. Previous studies have shown that the overall quality of life of young women with breast cancer is significantly reduced [9–11].

CRF is a prevalent symptom encountered in breast cancer patients, with a staggering 60% of them enduring moderate to severe fatigue even a year post-diagnosis [12,13], which can significantly affect patients' quality of life [14]. As per the National Comprehensive Cancer Network, CRF is characterized as a distressing, relentless, and subjective experience of exhaustion or a combination of physical, emotional, and/or cognitive fatigue, which stems from either the cancer itself or from its treatment. Notably, this fatigue does not abate with rest or sleep and significantly hampers an individual's ability to function normally [15,16]. There is a suggested correlation between CRF and various physiological factors, including pro-inflammatory cytokines, hypothalamic–pituitary–adrenal axis dysregulation, circadian rhythm desynchronization, and skeletal muscle atrophy, but the precise mechanisms causing CRF remain incompletely understood [16–19]. CRF affects the quality of life of cancer patients and their ability to reintegrate into normal daily life [20].

It was once believed that cancer patients should avoid physical activity and prioritize rest to facilitate cancer treatment and recovery. However, excessive physical inactivity may lead to a deterioration in fitness and physical functioning, thereby promoting the development of CRF [21,22]. Currently, studies show that various exercise modalities can reduce CRF in breast cancer patients. Notably, aerobic exercise combined with relaxation training has been proven effective in substantially alleviating CRF in breast cancer patients [23]. Courneya et al. [24] emphasized the efficacy of aerobic exercise in postmenopausal breast cancer patients. In addition, Milne et al. [25] further validated the positive impact of exercise on post-treatment fatigue and physical function. However, Pagola et al. [26] found no significant reduction in fatigue after 16 weeks of combined exercise intervention. Similarly, Ergun et al. [27] found no notable differences in fatigue scores between pre- and post-intervention groups, regardless of exercise supervision or type. Furthermore, Furmaniak et al. [21] revealed that exercise during adjuvant therapy for breast cancer did not yield a clear improvement in fatigue. This discrepancy underscores the uncertainty surrounding the optimal exercise regimen (type, frequency, and duration) for effectively mitigating CRF in breast cancer patients.

Therefore, the present meta-analysis, which builds upon rigorous randomized controlled trials (RCTs), aims to assess the effects of exercise on CRF and establish a definitive exercise prescription tailored to the needs of breast cancer patients.

2. Materials and Methods

2.1. Design

This study adhered to the rigorous guidelines outlined in the Preferred Reporting Items for Systematic Evaluation and Meta-Analysis (PRISMA, 2020) [28], ensuring the highest standards of methodology and reporting. The protocol has been officially registered with PROSPERO under the identification number CRD42023457710.

2.2. Search Strategy

To gather an exhaustive collection of relevant RCTs, a comprehensive literature search was conducted across 5 prestigious databases: Embase, PubMed, Cochrane Library, Web of Science, and Scopus. The search was limited to studies published up until 1 September 2023 and utilized a combination of the following keywords and Medical Subject Headings (MESH) terms: exercise, cancer, and fatigue. To supplement the search, the reference lists of the identified studies were manually screened for any additional potentially eligible articles. The screening and selection process was independently undertaken by two researchers

(R.Z. and Z.C.). In cases where a disagreement arose, a third reviewer (L.Y.) was involved in the discussion, fostering a collaborative approach until a consensus was reached.

2.3. Eligibility Criteria

The inclusion criteria for the study were as follows: (1) RCT design; (2) participants were breast cancer patients; (3) there were both an intervention group and a control group; and (4) outcomes were assessed using a specific fatigue scale.

The exclusion criteria were as follows: (1) non-English publications; (2) review articles; (3) conference articles; (4) outcome indicators that could not be converted into mean and standard deviation (SD); and (5) studies without a control group.

2.4. Data Extraction

The process of data extraction was conducted independently by two authors (R.Z. and Z.C.), with a focus on the following key elements: (1) the primary author's surname and the year in which the study was published; (2) sample size, age, and tumor stage; (3) the type of intervention, intervention duration, frequency, and session duration; (4) the outcome metrics that captured the variation in CRF.

2.5. Methodological Quality Assessment

The assessment of the risk of bias was carried out independently by two authors (R.Z. and Z.C.), and any discrepancies were resolved through discussion. The assessment was conducted using the Cochrane Randomized Trials Risk of Bias Tool (RoB-2) [29], which scrutinizes six domains: randomization sequence generation, allocation concealment, blinding, incomplete outcome data, choice of outcome report, and other biases. Each domain was assigned a risk level of "low", "high", or "unclear" [30].

2.6. Statistical Analysis

Since fatigue was assessed using different questionnaires, the data analysis employed a random-effects model to derive a standardized mean difference (SMD) alongside a 95% confidence interval (CI). Heterogeneity was assessed using the I^2 statistic, with values of 0%, 25%, 50%, and 75% interpreted as indicating no, low, moderate, and high heterogeneity, respectively [31]. In case of high heterogeneity ($I^2 > 50\%$), additional analytical steps were undertaken, including subgroup analysis, meta-regression, and sensitivity analysis, to provide deeper insights into the results. The publication bias of the included studies was visualized by funnel plots.

During the subgroup analyses, we endeavored to classify the included studies according to various intervention characteristics: the type of exercise (aerobic, resistance, combined exercise), frequency (less than 3 times weekly, 3 or more times weekly), session duration (up to 60 min per session, over 60 min per session), weekly time (less than 180 min weekly, 180 min or more weekly), and participants' age (middle-aged, $45 \leq$ age < 60; elderly, age ≥ 60). We utilized RevMan.5 software to create forest plots, while for meta-regression, sensitivity analysis, and the generation of funnel plots, Stata 17 software was employed. Outcomes were considered statistically significant if the p-value was less than 0.05.

3. Results

3.1. Studies Selection

As depicted in Figure 1, a comprehensive search across five databases yielded a total of 8987 pertinent studies. After eliminating duplicates, 3600 studies were assessed by reading titles and abstracts, resulting in the exclusion of 3500. Following a thorough evaluation of the full text, 74 studies were excluded due to the following reasons: (1) the studies did not involve breast cancer patients ($n = 55$); (2) the intervention did not involve exercise ($n = 9$); (3) the full text was not available ($n = 6$); (4) the investigated outcomes were irrelevant ($n = 3$); and (5) study protocol ($n = 1$). Finally, 26 studies [32–57] met the inclusion criteria.

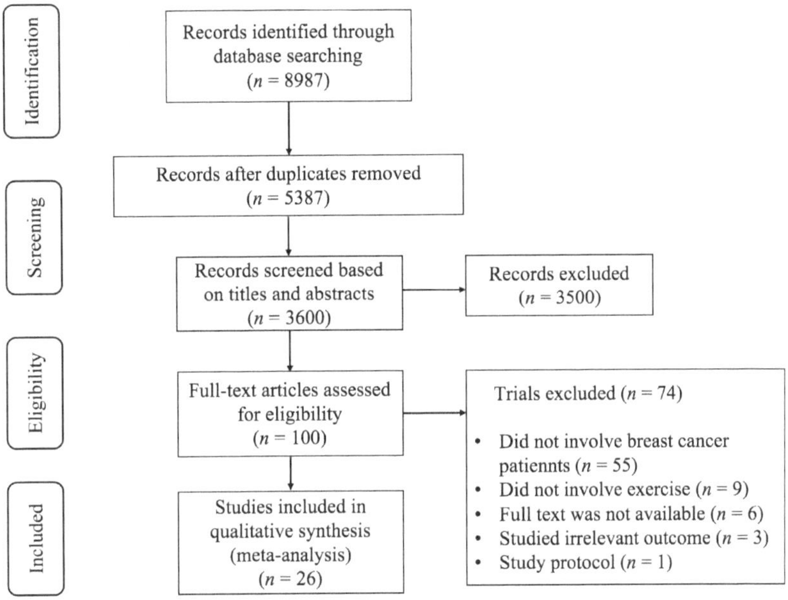

Figure 1. PRISMA flowchart of study selection.

3.2. Characteristics of the Included Studies

The key features of the interventions and participants are summarized in Table S1. The pooled studies encompassed 1258 patients in the intervention groups and 1049 patients in the control groups. Sample sizes ranged from 20 to 223 individuals across various studies. Eight studies utilized samples of over 100 breast cancer patients [37,40,42,43,51,54,56,57]. The age of the patients varied widely, with a mean range spanning from 45 to 66.6 years, and their breast cancer stages encompassed the entire spectrum, from stage 0 to stage 4. CRF was tested using the Functional Assessment of Chronic Illness Therapy-Fatigue (FACIT-F, five studies) [32,38,41,49,55]; the European Organization for Research and Treatment of Cancer Core Quality of Life Questionnaire-C30 (EORTC QLQ-C30, four studies) [33,36,42,47]; the Piper Fatigue Scale/the Revised Piper Fatigue Scale (PFS, five studies) [35,43,44,46,54]; the Multidimensional Fatigue Inventory (MFI/MFI-20, four studies) [40,42,56,57]; the Brief Fatigue Inventory (BFI, two studies) [39,52]; the Fatigue Quality List (FQL, two studies) [56,57]; the Profile of Mood States (POMS, one study) [34]; the Functional Assessment of Cancer Therapy-Anemia scale (FACT-An, one study) [37]; the Fatigue Severity Scale (FSS, one study) [45]; a 10 cm linear analog scale (one study) [48]; the Pittsburgh Fatigability Scale (PFS, one study) [50]; the Functional Assessment of Cancer Therapy-Endocrine Symptoms (FACT-ES, one study) [50]; the Fatigue Symptom Inventory (FSI, one study) [51]; and the Fatigue Assessment Questionnaire (FAQ, one study) [53]. Seven studies involved aerobic exercise [32,37,43,48,49,51,57], seven studies involved resistance exercise [36,37,41,44,46,52,53], and thirteen studies combined aerobic and resistance exercise [33–35,38–40,42,45,47,54–57]. The weekly intervention frequency varied, with some occurring as frequently as 5 times per week and as infrequently as once a week, averaging out to 3.2 times per week. The session duration ranged from a minimum of 15 min to a maximum of 90 min, with an average duration of 51.5 min per session. Lastly, the weekly time ranged from 40 min to 390 min.

3.3. Meta-Analysis

Exercise was found to have a significant effect on improving CRF in breast cancer patients (SMD, −0.42; 95% CI, −0.55 to −0.28, $p < 0.0001$, $I^2 = 70\%$, Figure 2). To delve deeper into the variability among the studies and identify potential factors that could be modified to optimize exercise effects, further analyses were conducted, including meta-regression, subgroup, and sensitivity analyses.

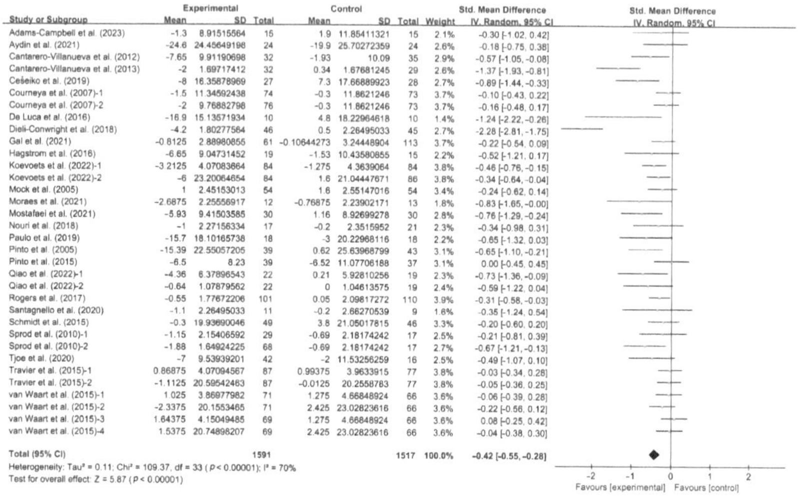

Figure 2. Meta-analysis results of the effects of exercise on CRF in breast cancer patients [32–57]. The size of the shaded squares was proportional to the percentage weight of each study. Diamonds indicated the effect size of each study summarized as SMD.

3.4. Meta-Regression Analysis

Meta-regression analysis was applied to investigate the relationship between CRF improvement, various intervention attributes (intervention duration, session duration, frequency, and weekly time), and participant age. However, no statistically significant associations were observed between CRF enhancement and any of these factors, including intervention duration ($p = 0.929$), frequency ($p = 0.387$), session duration ($p = 0.364$), weekly time ($p = 0.362$), or age ($p = 0.651$), as depicted in Figure S1.

3.5. Subgroup Analysis

Stratifying the analysis by types of intervention, aerobic exercise (SMD, −0.17; 95% CI, −0.33 to −0.02, $p = 0.02$, $I^2 = 24\%$), resistance exercise (SMD, −0.37; 95% CI, −0.59 to −0.15, $p = 0.0009$, $I^2 = 14\%$), and combined exercise (SMD, −0.53; 95% CI, −0.77 to −0.29, $p < 0.0001$, $I^2 = 81\%$, Figure 3 and Table 1) significantly improved CRF in breast cancer patients, with combined exercise being the most effective intervention.

In addition, when analyzing the subgroups by frequency, interventions conducted for <3 times per week (SMD, −0.28; 95% CI, −0.44 to −0.11, $p = 0.0009$, $I^2 = 39\%$) and ≥3 times per week (SMD, −0.47; 95% CI, −0.71 to −0.23, $p = 0.0001$, $I^2 = 80\%$, Figure 4 and Table 1) significantly improved CRF in breast cancer patients, with interventions conducted for ≥3 times per week having a greater effect.

Furthermore, when analyzing the subgroups by session duration, interventions conducted for ≤60 min per session (SMD, −0.28; 95% CI, −0.40 to −0.15, $p < 0.0001$, $I^2 = 53\%$) and >60 min per session (SMD, −0.63; 95% CI, −0.94 to −0.32, $p < 0.0001$, $I^2 = 0\%$, Figure 5 and Table 1) significantly improved CRF in breast cancer patients, with interventions conducted for >60 min per session having a greater effect.

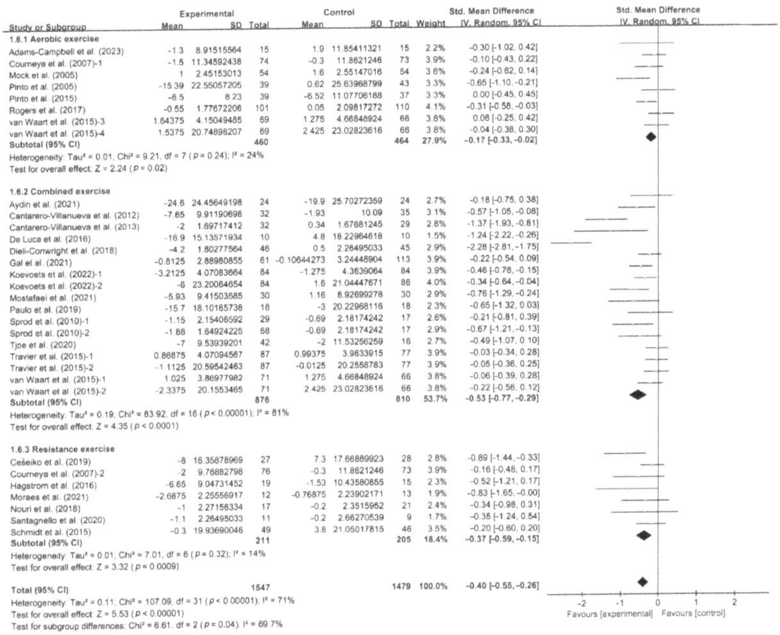

Figure 3. Meta-analysis results of the effects of types of intervention on CRF in breast cancer patients [32–49,51–57]. The size of the shaded squares was proportional to the percentage weight of each study. Diamonds indicated the effect size of each study summarized as SMD.

Table 1. Results of moderator analysis.

Moderator	SMD (95% CI)	I^2	*p* Value
Overall	−0.42 (−0.55, −0.28)	70%	<0.0001
Types of intervention			
Aerobic exercise	−0.17 (−0.33, −0.02)	24%	0.02
Resistance exercise	−0.37 (−0.59, −0.15)	14%	0.0009
Combined exercise	−0.53 (−0.77, −0.29)	81%	<0.0001
Frequency			
<3 times per week	−0.28 (−0.44, −0.11)	39%	0.0009
≥3 times per week	−0.47 (−0.71, −0.23)	80%	0.0001
Session duration			
≤60 min per session	−0.28 (−0.40, −0.15)	53%	<0.0001
>60 min per session	−0.63 (−0.94, −0.32)	0%	<0.0001
Weekly time			
<180 min per week	−0.24 (−0.35, −0.13)	35%	<0.0001
≥180 min per week	−0.79 (−1.18, −0.40)	83%	<0.0001
Age			
45 ≤ Age < 60	−0.42 (−0.57, −0.27)	72%	<0.0001
Age ≥60	−0.37 (−0.75, 0.01)	0%	0.05

Abbreviations: 95% CI, 95% confidence interval.

Moreover, when analyzing the subgroups by weekly time, interventions conducted for <180 min per week (SMD, −0.24; 95% CI, −0.35 to −0.13, $p < 0.0001$, $I^2 = 35\%$) and ≥180 min per week (SMD, −0.79; 95% CI, −1.18 to −0.40, $p < 0.0001$, $I^2 = 83\%$, Figure 6 and Table 1) significantly improved CRF in breast cancer patients, with interventions conducted for ≥180 min per week having a greater effect.

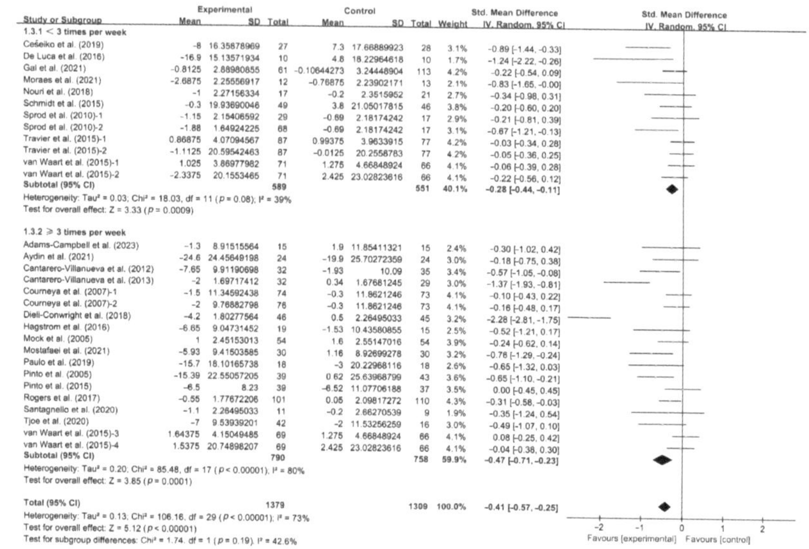

Figure 4. Meta-analysis results of the effects of frequency of intervention on CRF in breast cancer patients [32–41,43–49,51–57]. The size of the shaded squares was proportional to the percentage weight of each study. Diamonds indicated the effect size of each study summarized as SMD.

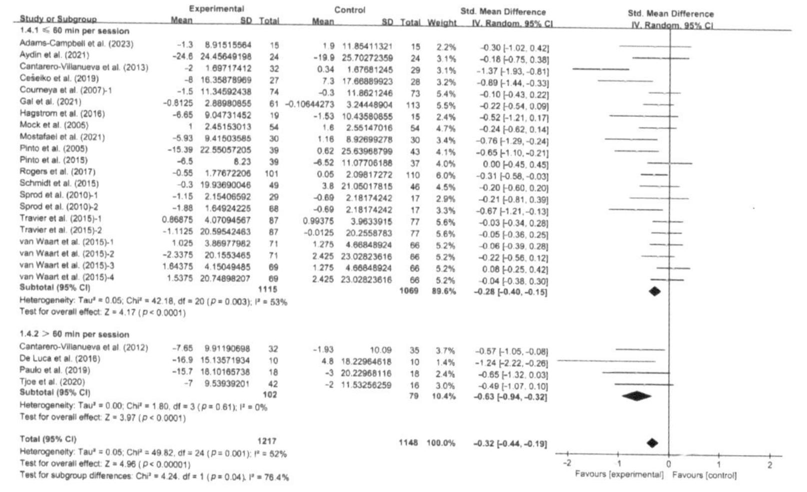

Figure 5. Meta-analysis results of the effects of duration of intervention per session on CRF in breast cancer patients [32–38,40,41,43,45,47–49,51,53–57]. The size of the shaded squares was proportional to the percentage weight of each study. Diamonds indicated the effect size of each study summarized as SMD.

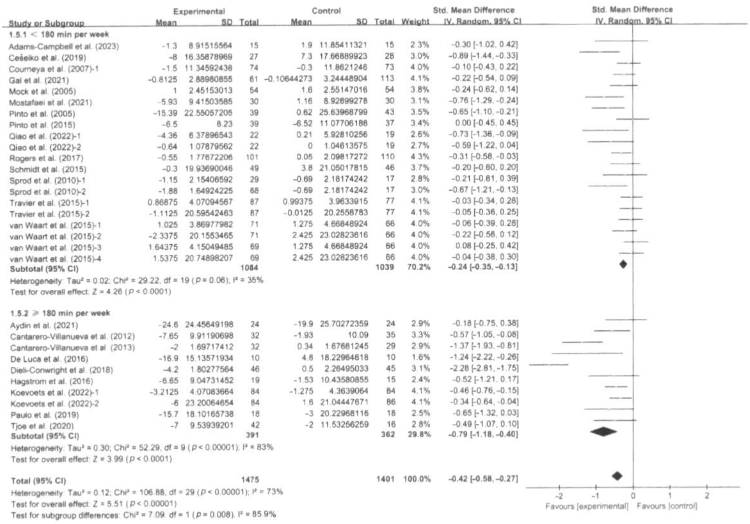

Figure 6. Meta-analysis results of the effects of duration of intervention per week on CRF in breast cancer patients [32–43,45,47–51,53–57]. The size of the shaded squares was proportional to the percentage weight of each study. Diamonds indicated the effect size of each study summarized as SMD.

Finally, when analyzing the subgroups by participant age, exercise significantly improved CRF in middle-aged breast cancer patients (SMD, −0.42; 95% CI, −0.57 to −0.27, $p < 0.0001$, $I^2 = 72\%$), while exercise had no significant effect on improving CRF in elderly breast cancer patients (SMD, −0.37; 95% CI, −0.75 to 0.01, $p = 0.05$, $I^2 = 0\%$, Figure 7 and Table 1).

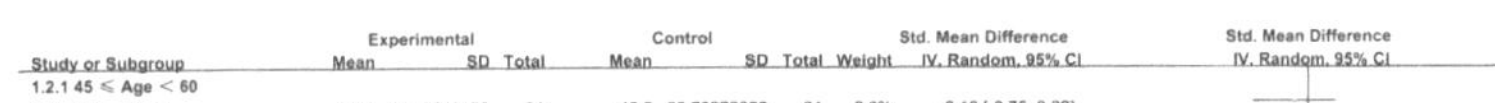

Figure 7. Meta-analysis results of the effects of exercise on CRF in middle-aged and elderly breast cancer patients [32–57]. The size of the shaded squares was proportional to the percentage weight of each study. Diamonds indicated the effect size of each study summarized as SMD.

3.6. Risk of Bias

The RoB-2 tool was utilized to evaluate the risk of bias in the included studies, considering factors such as selection, performance, detection, attrition, reporting, and other biases. As illustrated in Figure S2, the overall quality of the studies was categorized into three levels: low, moderate, and high. Two studies posed a low risk of bias, twenty-one studies presented a moderate risk, and three studies had a high risk of bias.

3.7. Sensitivity Analyses

Sensitivity analyses indicated that the positive impact of exercise on CRF in breast cancer patients remained stable and consistent in both direction and magnitude, regardless of the exclusion of any individual study (Figure S3).

3.8. Publication Bias

To further assess the potential for publication bias, a funnel plot analysis was conducted (Figure S4). The observed asymmetry indicates the presence of publication bias.

4. Discussion

4.1. Main Findings

The present study aimed to assess the effects of exercise on CRF and establish a definitive exercise prescription tailored to the needs of breast cancer patients. A total of 26 studies were included, with the results conclusively demonstrating that exercise significantly improved CRF in breast cancer patients. Further subgroup analyses revealed that combined exercise, undertaken at a frequency of at least three times weekly, each session lasting over 60 min, and accumulating a total of 180 min or more per week, proved to be the most efficacious in improving CRF, particularly in middle-aged breast cancer patients.

4.2. Effects of Exercise on CRF in Breast Cancer Patients

This study suggested that exercise holds the potential to improve CRF in breast cancer patients, which is consistent with previous studies [58,59]. There are numerous explanations for the potential mechanisms of how exercise improves CRF in breast cancer patients, and the following are some possible mechanisms that could account for the effects of exercise.

Firstly, research has consistently demonstrated that exercise boosts anti-inflammatory cytokines while reducing pro-inflammatory adipokines [60]. While only a few studies have investigated changes in inflammatory mediators in breast cancer patients following exercise interventions, all of these studies consistently reported a decrease in inflammatory factors like interleukin-6 (IL-6), C-reactive protein (CRP), and tumor necrosis factor-α (TNF-α) post-exercise [61,62]. Since inflammation is a potential cause of CRF, the anti-inflammatory effect of exercise likely contributes to the reduction in fatigue. Additionally, the increase in the number of lymphocytes stimulated by exercise may explain the positive effect of exercise on CRF from an immunological perspective [63].

Secondly, resistance exercise is likely to mitigate muscle function decline, such as muscle atrophy caused by cancer [64]. Studies have demonstrated that resistance exercise improves cytokine responses [65] and enhances generalized muscle strength in cancer patients [66]. However, aerobic exercise enhances energy metabolism processes, where carbohydrates and fats are thoroughly oxidized into water and carbon dioxide within the mitochondria, yielding adenosine triphosphate (ATP) as a stored energy source in cells, thereby enhancing cardiorespiratory fitness among patients [67]. These improvements may enable breast cancer patients to perform daily activities with more ease and at the same intensity as before, thereby reducing the perception of fatigue.

Finally, exercise can also have positive effects on mental health. For instance, achieving daily activity goals can boost self-confidence and self-efficacy, indirectly reducing fatigue. Previous studies have shown that self-efficacy is a mediating factor in reducing fatigue in breast cancer patients [68,69]. Additionally, exercise improves sleep, mood, and cognition, all of which indirectly affects fatigue [70].

However, our results were inconsistent with some previous studies. For instance, Cramp et al. [71] suggested that aerobic exercise significantly improved CRF, while other forms of exercise did not exhibit such an effect. Concurrently, there exists uncertainty in the literature regarding the definitive benefits of exercise on CRF in adult populations [72]. The inconsistency may be attributed to the fact that these studies did not guarantee that all included patients suffered from breast cancer, but that merely a majority did, and the varying characteristics of the interventions focused on in different studies may also have contributed to the inconsistent results.

4.3. Subgroup Analysis

Our subgroup analysis showed that aerobic exercise, resistance exercise, and combined exercise significantly improved CRF in breast cancer patients, with combined exercise emerging as the most effective approach, mirroring prior research findings. Steindorf et al. [73] found that resistance exercise significantly reduced CRF after a twelve-week intervention in breast cancer patients. In addition, Yang et al. [74] suggested that moderate-intensity aerobic exercise also achieved improvements, while Mijwel et al. [75] proposed that the combination of resistance and high-intensity interval training (HIIT) was superior to conventional controls in reducing CRF. Furthermore, numerous studies have demonstrated the unique benefits of combined exercise. Milne et al. [25] corroborates our results, revealing that a blend of aerobic and resistance exercises significantly improved health outcomes, including diminished fatigue, within a brief timeframe. However, both aerobic and resistance exercise, as separate modalities, also possess therapeutic effects, we can still justify the use of aerobic and resistance exercise individually in actual treatment, considering the patient's physical capabilities. For patients without an exercise foundation, we can prioritize aerobic exercise to develop their cardiorespiratory function before gradually incorporating resistance training and ultimately forming a combined exercise regimen. However, there is still no definitive conclusion on how to design the ratio of aerobic and resistance in the combined exercise mode.

In regard to intervention frequency, both less than three and at least three sessions per week significantly improved CRF in breast cancer patients, aligning with prior studies [76,77]. Notably, a frequency of at least three sessions weekly had a more pronounced effect on CRF, potentially due to its role in fostering a regular exercise routine [78]. However, we did not dismiss the potential benefits of interventions conducted less than 3 times per week, which may still be considered in practical applications, taking into account factors such as session duration.

Our subgroup analysis indicated that interventions conducted for up to 60 min per session and over 60 min per session significantly improved CRF in breast cancer patients, which aligns with previous studies. Meneses-Echávez et al. [79] showed that a supervised exercise intervention of 40 min per session significantly reduced CRF. In addition, a meta-analysis conducted by Sweegers et al. [80] on moderators of exercise in cancer patients (two-thirds of whom were breast cancer patients) also showed a significant effect of exercise intervention within 60 min. Furthermore, our results showed that interventions conducted for over 60 min per session had a greater effect on improving CRF, which is consistent with previous studies. For instance, Zhou et al. [81] showed that engaging in exercise for more than 60 min per session significantly alleviate CRF in breast cancer patients. In addition, Danhauer et al. [82] also concluded that even 75 min of low-intensity exercise can improve fatigue symptoms in breast cancer patients. However, it is crucial to acknowledge that excessive exercise durations may not yield additional health benefits and could potentially have adverse effects. Li et al. [83] found that exercise interventions exceeding 60 min did not significantly impact cognitive function in multiple sclerosis patients, suggesting a threshold beyond which further exercise may not be beneficial. Moreover, insufficient exercise durations fail to elicit improvements, while excessive exercise can induce fatigue and compromise brain plasticity. Given that our intervention targeted cancer patients, it is worth noting that exercise tolerance in this population is reduced compared to the healthy population, mainly due to the negative impact of CRF on exercise tolerance [84]. Therefore, we caution against blindly increasing session durations and advocate for potentially more effective improvements through increased frequency.

Nevertheless, our study revealed that merely focusing on frequency and session duration was insufficient to mitigate the influence of other confounding variables. Consequently, we devised a method to calculate the weekly exercise time by combining both these factors. Our findings indicated that interventions totaling at least 180 min per week had a more pronounced effect on improving CRF in breast cancer patients. Therefore, the combination of a frequency of at least three interventions per week and a duration of at least 180 min of intervention per week achieves a better effect, suggesting that the recommended exercise pattern for breast cancer patients should be to appropriately reduce the duration of each exercise intervention to avoid side effects such as muscle injury or functional impairment caused by decreased exercise tolerance, while ensuring an adequate volume of exercise by increasing the frequency of exercise per week to achieve a better therapeutic effect.

In the current study, participants comprised middle-aged and older individuals, and our subgroup analysis showed that exercise significantly improved CRF in middle-aged breast cancer patients. The lack of a significant effect in older patients may be attributed to their reduced exercise tolerance [85], as well as decreased motivation and adherence to exercise compared to middle-aged patients [86]. However, only three trials in this meta-analysis involved participants with a mean age of 60 years or older, necessitating further clinical trials to validate the therapeutic effects of exercise on CRF in older adults.

Exercise interventions during or after other treatments were excluded from this study, which could potentially impact treatment effectiveness based on previous studies. Juvet et al. [77] found that exercise initiated after radiotherapy or chemotherapy was more effective in reducing CRF in breast cancer patients compared to exercise during treatment. Additionally, Hilfiker et al. [59] showed that relaxation exercises were effective during cancer-related treatment, but their effectiveness significantly diminished post-treatment. It has also been proposed that exercise during treatment may have a more favorable and faster effect on mobility compared to post-treatment [87], though further evidence is needed to substantiate this claim.

4.4. Strengths and Limitations of This Study

Our strength lies in the thorough analysis of the intervention-related factors, including type of exercise, frequency, session duration, and weekly time. However, this study had certain limitations. The blinding quality of the included studies could not be assured, potentially affecting the robustness of the evidence. In addition, exercise intensity, the supervision of exercise interventions, and rest intervals during exercise were not statistically analyzed. Future studies with larger sample sizes and higher quality may be needed to complement our findings. Finally, there is a high degree of heterogeneity in this study, necessitating careful and appropriate handling of the results.

5. Conclusions

Exercise is an effective approach to improving CRF in breast cancer patients. When devising an exercise program, the primary consideration should be the incorporation of combined exercise as the principal intervention. This entails ensuring that participants engage in the program at least three times weekly, with each session lasting for more than 60 min. The ultimate aim is to achieve a total weekly exercise duration of 180 min by progressively increasing the frequency of exercise sessions.

Supplementary Materials: The following supporting information can be downloaded at: https://www.mdpi.com/article/10.3390/life14081011/s1, Figure S1: Meta-regression analysis results; Figure S2: Results of Cochrane risk of bias tool; Figure S3: Sensitivity analyses results [32–57]; Figure S4: Funnel plot; Table S1: Characteristics of the studies included in this meta-analysis.

Author Contributions: Conceptualization, R.Z. and L.Y.; methodology, R.Z., Z.C., Y.L. and L.Y.; software, R.Z. and Z.C.; validation, S.Z. and Y.W.; formal analysis, R.Z. and Z.C.; investigation, R.Z., Z.C., S.Z., Y.W., C.Z., Y.L. and L.Y.; resources, L.Y.; data curation, C.Z. and Y.L.; writing—original draft preparation, R.Z.; writing—review and editing, R.Z., Z.C., S.Z., Y.W., C.Z., Y.L. and L.Y.; visualization, R.Z., Z.C. and L.Y.; supervision, L.Y.; project administration, L.Y.; funding acquisition, Y.L. and L.Y. All authors have read and agreed to the published version of the manuscript.

Funding: This research was funded by the National Key R&D Program of China, grant number 2022YFC3600201, and by the Chinese Universities Scientific Fund, grant number 2022QN015, 2024JNPD004.

Institutional Review Board Statement: Not applicable.

Informed Consent Statement: Not applicable.

Data Availability Statement: All data generated or analyzed during this study are included in the article/Supplementary Materials.

Conflicts of Interest: The authors declare no conflicts of interest.

References

1. Song, Q.; Merajver, S.D.; Li, J.Z. Cancer classification in the genomic era: Five contemporary problems. *Hum. Genom.* **2015**, *9*, 27. [CrossRef] [PubMed]
2. GBD 2017 Causes of Death Collaborators. Global, regional, and national age-sex-specific mortality for 282 causes of death in 195 countries and territories, 1980–2017: A systematic analysis for the Global Burden of Disease Study 2017. *Lancet* **2018**, *392*, 1736–1788. [CrossRef]
3. Bray, F.; Ferlay, J.; Soerjomataram, I.; Siegel, R.L.; Torre, L.A.; Jemal, A. Global cancer statistics 2018: GLOBOCAN estimates of incidence and mortality worldwide for 36 cancers in 185 countries. *CA Cancer J. Clin.* **2018**, *68*, 394–424. [CrossRef]
4. Siegel, R.L.; Giaquinto, A.N.; Jemal, A. Cancer statistics, 2024. *CA Cancer J. Clin.* **2024**, *74*, 12–49. [CrossRef]
5. Siegel, R.L.; Miller, K.D.; Fuchs, H.E.; Jemal, A. Cancer statistics, 2022. *CA Cancer J. Clin.* **2022**, *72*, 7–33. [CrossRef] [PubMed]
6. Early Breast Cancer Trialists' Collaborative Group (EBCTCG); McGale, P.; Taylor, C.; Correa, C.; Cutter, D.; Duane, F.; Ewertz, M.; Gray, R.; Mannu, G.; Peto, R.; et al. Effect of radiotherapy after mastectomy and axillary surgery on 10-year recurrence and 20-year breast cancer mortality: Meta-analysis of individual patient data for 8135 women in 22 randomised trials. *Lancet* **2014**, *383*, 2127–2135. [CrossRef] [PubMed]
7. De Laurentiis, M.; Cancello, G.; D'Agostino, D.; Giuliano, M.; Giordano, A.; Montagna, E.; Lauria, R.; Forestieri, V.; Esposito, A.; Silvestro, L.; et al. Taxane-based combinations as adjuvant chemotherapy of early breast cancer: A meta-analysis of randomized trials. *J. Clin. Oncol.* **2008**, *26*, 44–53. [CrossRef]
8. Early Breast Cancer Trialists' Collaborative Group (EBCTCG); Peto, R.; Davies, C.; Godwin, J.; Gray, R.; Pan, H.C.; Clarke, M.; Cutter, D.; Darby, S.; McGale, P.; et al. Comparisons between different polychemotherapy regimens for early breast cancer: Meta-analyses of long-term outcome among 100,000 women in 123 randomised trials. *Lancet* **2012**, *379*, 432–444. [CrossRef]
9. Avis, N.E.; Crawford, S.; Manuel, J. Quality of life among younger women with breast cancer. *J. Clin. Oncol.* **2005**, *23*, 3322–3330. [CrossRef]
10. Bower, J.E.; Ganz, P.A.; Desmond, K.A.; Rowland, J.H.; Meyerowitz, B.E.; Belin, T.R. Fatigue in breast cancer survivors: Occurrence, correlates, and impact on quality of life. *J. Clin. Oncol.* **2000**, *18*, 743–753. [CrossRef]
11. Ganz, P.A. Monitoring the physical health of cancer survivors: A survivorship-focused medical history. *J. Clin. Oncol.* **2006**, *24*, 5105–5111. [CrossRef]
12. Nowe, E.; Friedrich, M.; Leuteritz, K.; Sender, A.; Stöbel-Richter, Y.; Schulte, T.; Hinz, A.; Geue, K. Cancer-Related Fatigue and Associated Factors in Young Adult Cancer Patients. *J. Adolesc. Young Adult Oncol.* **2019**, *8*, 297–303. [CrossRef]
13. Thong, M.S.Y.; van Noorden, C.J.F.; Steindorf, K.; Arndt, V. Correction to: Cancer-Related Fatigue: Causes and Current Treatment Options. *Curr. Treat. Options Oncol.* **2022**, *23*, 450–451. [CrossRef] [PubMed]
14. Silver, J.K.; Baima, J.; Mayer, R.S. Impairment-driven cancer rehabilitation: An essential component of quality care and survivorship. *CA Cancer J. Clin.* **2013**, *63*, 295–317. [CrossRef]
15. Ye, Z.-h.; Du, F.-r.; Wu, Y.-p.; Yang, X.; Zi, Y. Interpretation of NCCN clinical practice guidelines in oncology: Cancer-related fatigue. *J. Int. Transl. Med.* **2016**, *4*, 1–8. [CrossRef]
16. Wang, Z.; Qi, F.; Cui, Y.; Zhao, L.; Sun, X.; Tang, W.; Cai, P. An update on Chinese herbal medicines as adjuvant treatment of anticancer therapeutics. *Biosci. Trends* **2018**, *12*, 220–239. [CrossRef]
17. Jager, A.; Sleijfer, S.; van der Rijt, C.C. The pathogenesis of cancer related fatigue: Could increased activity of pro-inflammatory cytokines be the common denominator? *Eur. J. Cancer* **2008**, *44*, 175–181. [CrossRef] [PubMed]
18. Berger, A.M.; Wielgus, K.; Hertzog, M.; Fischer, P.; Farr, L. Patterns of circadian activity rhythms and their relationships with fatigue and anxiety/depression in women treated with breast cancer adjuvant chemotherapy. *Support. Care Cancer* **2010**, *18*, 105–114. [CrossRef]

19. Kilgour, R.D.; Vigano, A.; Trutschnigg, B.; Hornby, L.; Lucar, E.; Bacon, S.L.; Morais, J.A. Cancer-related fatigue: The impact of skeletal muscle mass and strength in patients with advanced cancer. *J. Cachexia. Sarcopenia Muscle* **2010**, *1*, 177–185. [CrossRef]
20. Garabeli Cavalli Kluthcovsky, A.C.; Urbanetz, A.A.; de Carvalho, D.S.; Pereira Maluf, E.M.; Schlickmann Sylvestre, G.C.; Bonatto Hatschbach, S.B. Fatigue after treatment in breast cancer survivors: Prevalence, determinants and impact on health-related quality of life. *Support. Care Cancer* **2012**, *20*, 1901–1909. [CrossRef] [PubMed]
21. Furmaniak, A.C.; Menig, M.; Markes, M.H. Exercise for women receiving adjuvant therapy for breast cancer. *Cochrane Database Syst. Rev.* **2016**, *9*, Cd005001. [CrossRef] [PubMed]
22. Lucía, A.; Earnest, C.; Pérez, M. Cancer-related fatigue: Can exercise physiology assist oncologists? *Lancet Oncol.* **2003**, *4*, 616–625. [CrossRef] [PubMed]
23. Cohen, J.; Rogers, W.A.; Petruzzello, S.; Trinh, L.; Mullen, S.P. Acute effects of aerobic exercise and relaxation training on fatigue in breast cancer survivors: A feasibility trial. *Psychooncology* **2021**, *30*, 252–259. [CrossRef]
24. Courneya, K.S.; Mackey, J.R.; Bell, G.J.; Jones, L.W.; Field, C.J.; Fairey, A.S. Randomized controlled trial of exercise training in postmenopausal breast cancer survivors: Cardiopulmonary and quality of life outcomes. *J. Clin. Oncol.* **2003**, *21*, 1660–1668. [CrossRef] [PubMed]
25. Milne, H.M.; Wallman, K.E.; Gordon, S.; Courneya, K.S. Effects of a combined aerobic and resistance exercise program in breast cancer survivors: A randomized controlled trial. *Breast Cancer Res. Treat.* **2008**, *108*, 279–288. [CrossRef]
26. Pagola, I.; Morales, J.S.; Alejo, L.B.; Barcelo, O.; Montil, M.; Oliván, J.; Álvarez-Bustos, A.; Cantos, B.; Maximiano, C.; Hidalgo, F.; et al. Concurrent Exercise Interventions in Breast Cancer Survivors with Cancer-related Fatigue. *Int. J. Sports Med.* **2020**, *41*, 790–797. [CrossRef]
27. Ergun, M.; Eyigor, S.; Karaca, B.; Kisim, A.; Uslu, R. Effects of exercise on angiogenesis and apoptosis-related molecules, quality of life, fatigue and depression in breast cancer patients. *Eur. J. Cancer Care* **2013**, *22*, 626–637. [CrossRef] [PubMed]
28. Page, M.J.; McKenzie, J.E.; Bossuyt, P.M.; Boutron, I.; Hoffmann, T.C.; Mulrow, C.D.; Shamseer, L.; Tetzlaff, J.M.; Akl, E.A.; Brennan, S.E.; et al. The PRISMA 2020 statement: An updated guideline for reporting systematic reviews. *BMJ* **2021**, *372*, n71. [CrossRef]
29. Tao, X.; Chen, Y.; Zhen, K.; Ren, S.; Lv, Y.; Yu, L. Effect of continuous aerobic exercise on endothelial function: A systematic review and meta-analysis of randomized controlled trials. *Front. Physiol.* **2023**, *14*, 1043108. [CrossRef] [PubMed]
30. You, Q.; Yu, L.; Li, G.; He, H.; Lv, Y. Effects of different intensities and durations of aerobic exercise on vascular endothelial function in middle-aged and elderly people: A meta-analysis. *Front. Physiol.* **2022**, *12*, 803102. [CrossRef]
31. Chen, Y.; Su, Q.; Yang, J.; Li, G.; Zhang, S.; Lv, Y.; Yu, L. Effects of rest interval and training intensity on jumping performance: A systematic review and meta-analysis investigating post-activation performance enhancement. *Front. Physiol.* **2023**, *14*, 1202789. [CrossRef] [PubMed]
32. Adams-Campbell, L.L.; Hicks, J.; Makambi, K.; Randolph-Jackson, P.; Mills, M.; Isaacs, C.; Dash, C. An 8-week exercise study to improve cancer treatment related fatigue and QOL among African American breast cancer patients undergoing radiation treatment: A pilot randomized clinical trial. *J. Natl. Med. Assoc.* **2023**, *115*, 199–206. [CrossRef] [PubMed]
33. Aydin, M.; Kose, E.; Odabas, I.; Meric Bingul, B.; Demirci, D.; Aydin, Z. The Effect of Exercise on Life Quality and Depression Levels of Breast Cancer Patients. *Asian Pac. J. Cancer Prev.* **2021**, *22*, 725–732. [CrossRef]
34. Cantarero-Villanueva, I.; Fernández-Lao, C.; Del Moral-Avila, R.; Fernández-de-Las-Peñas, C.; Feriche-Fernández-Castanys, M.B.; Arroyo-Morales, M. Effectiveness of core stability exercises and recovery myofascial release massage on fatigue in breast cancer survivors: A randomized controlled clinical trial. *Evid. Based Complement. Alternat. Med.* **2012**, *2012*, 620619. [CrossRef]
35. Cantarero-Villanueva, I.; Fernández-Lao, C.; Cuesta-Vargas, A.I.; Del Moral-Avila, R.; Fernández-de-Las-Peñas, C.; Arroyo-Morales, M. The effectiveness of a deep water aquatic exercise program in cancer-related fatigue in breast cancer survivors: A randomized controlled trial. *Arch. Phys. Med. Rehabil.* **2013**, *94*, 221–230. [CrossRef]
36. Cešeiko, R.; Eglītis, J.; Srebnijs, A.; Timofejevs, M.; Purmalis, E.; Erts, R.; Vētra, A.; Tomsone, S. The impact of maximal strength training on quality of life among women with breast cancer undergoing treatment. *Exp. Oncol.* **2019**, *41*, 166–172. [CrossRef]
37. Courneya, K.S.; Segal, R.J.; Mackey, J.R.; Gelmon, K.; Reid, R.D.; Friedenreich, C.M.; Ladha, A.B.; Proulx, C.; Vallance, J.K.; Lane, K.; et al. Effects of aerobic and resistance exercise in breast cancer patients receiving adjuvant chemotherapy: A multicenter randomized controlled trial. *J. Clin. Oncol.* **2007**, *25*, 4396–4404. [CrossRef] [PubMed]
38. De Luca, V.; Minganti, C.; Borrione, P.; Grazioli, E.; Cerulli, C.; Guerra, E.; Bonifacino, A.; Parisi, A. Effects of concurrent aerobic and strength training on breast cancer survivors: A pilot study. *Public Health* **2016**, *136*, 126–132. [CrossRef]
39. Dieli-Conwright, C.M.; Courneya, K.S.; Demark-Wahnefried, W.; Sami, N.; Lee, K.; Sweeney, F.C.; Stewart, C.; Buchanan, T.A.; Spicer, D.; Tripathy, D.; et al. Aerobic and resistance exercise improves physical fitness, bone health, and quality of life in overweight and obese breast cancer survivors: A randomized controlled trial. *Breast Cancer Res.* **2018**, *20*, 124. [CrossRef]
40. Gal, R.; Monninkhof, E.M.; van Gils, C.H.; Groenwold, R.H.H.; Elias, S.G.; van den Bongard, D.H.J.G.; Peeters, P.H.M.; Verkooijen, H.M.; May, A.M. Effects of exercise in breast cancer patients: Implications of the trials within cohorts (TwiCs) design in the UMBRELLA Fit trial. *Breast Cancer Res. Treat.* **2021**, *190*, 89–101. [CrossRef]

41. Hagstrom, A.D.; Marshall, P.W.; Lonsdale, C.; Cheema, B.S.; Fiatarone Singh, M.A.; Green, S. Resistance training improves fatigue and quality of life in previously sedentary breast cancer survivors: A randomised controlled trial. *Eur. J. Cancer Care* **2016**, *25*, 784–794. [CrossRef] [PubMed]
42. Koevoets, E.W.; Schagen, S.B.; de Ruiter, M.B.; Geerlings, M.I.; Witlox, L.; van der Wall, E.; Stuiver, M.M.; Sonke, G.S.; Velthuis, M.J.; Jobsen, J.J.; et al. Effect of physical exercise on cognitive function after chemotherapy in patients with breast cancer: A randomized controlled trial (PAM study). *Breast Cancer Res.* **2022**, *24*, 36. [CrossRef]
43. Mock, V.; Frangakis, C.; Davidson, N.E.; Ropka, M.E.; Pickett, M.; Poniatowski, B.; Stewart, K.J.; Cameron, L.; Zawacki, K.; Podewils, L.J.; et al. Exercise manages fatigue during breast cancer treatment: A randomized controlled trial. *Psychooncology* **2005**, *14*, 464–477. [CrossRef] [PubMed]
44. Moraes, R.F.; Ferreira-Júnior, J.B.; Marques, V.A.; Vieira, A.; Lira, C.A.B.; Campos, M.H.; Freitas-Junior, R.; Rahal, R.M.S.; Gentil, P.; Vieira, C.A. Resistance Training, Fatigue, Quality of Life, Anxiety in Breast Cancer Survivors. *J. Strength Cond. Res.* **2021**, *35*, 1350–1356. [CrossRef] [PubMed]
45. Mostafaei, F.; Azizi, M.; Jalali, A.; Salari, N.; Abbasi, P. Effect of exercise on depression and fatigue in breast cancer women undergoing chemotherapy: A randomized controlled trial. *Heliyon* **2021**, *7*, e07657. [CrossRef] [PubMed]
46. Nouri, R.; Braumann, K.M.; ChamPiri, B.M.; Schroeder, J.; Akochakian, M. Cancer related fatigue and upper limb disabilities cannot improve after 6 weeks resistance training with Thera-Band in breast cancer survivors. *Int. J. Appl. Exerc. Physiol.* **2018**, *7*, 76–84. [CrossRef]
47. Paulo, T.R.S.; Rossi, F.E.; Viezel, J.; Tosello, G.T.; Seidinger, S.C.; Simões, R.R.; De Freitas, R.; Freitas, I.F. The impact of an exercise program on quality of life in older breast cancer survivors undergoing aromatase inhibitor therapy: A randomized controlled trial. *Health Qual. Life Outcomes* **2019**, *17*, 17. [CrossRef] [PubMed]
48. Pinto, B.M.; Frierson, G.M.; Rabin, C.; Trunzo, J.J.; Marcus, B.H. Home-based physical activity intervention for breast cancer patients. *J. Clin. Oncol.* **2005**, *23*, 3577–3587. [CrossRef] [PubMed]
49. Pinto, B.; Stein, K.; Dunsiger, S. Peer mentorship to promote physical activity among cancer survivors: Effects on quality of life. *Psychooncology* **2015**, *24*, 1295–1302. [CrossRef]
50. Qiao, Y.; van Londen, G.J.; Brufsky, J.W.; Poppenberg, J.T.; Cohen, R.W.; Boudreau, R.M.; Glynn, N.W. Perceived physical fatigability improves after an exercise intervention among breast cancer survivors: A randomized clinical trial. *Breast Cancer* **2022**, *29*, 30–37. [CrossRef]
51. Rogers, L.Q.; Courneya, K.S.; Anton, P.M.; Verhulst, S.; Vicari, S.K.; Robbs, R.S.; McAuley, E. Effects of a multicomponent physical activity behavior change intervention on fatigue, anxiety, and depressive symptomatology in breast cancer survivors: Randomized trial. *Psychooncology* **2017**, *26*, 1901–1906. [CrossRef]
52. Santagnello, S.B.; Martins, F.M.; de Oliveira Junior, G.N.; de Freitas Rodrigues de Sousa, J.; Nomelini, R.S.; Murta, E.F.C.; Orsatti, F.L. Improvements in muscle strength, power, and size and self-reported fatigue as mediators of the effect of resistance exercise on physical performance breast cancer survivor women: A randomized controlled trial. *Support Care Cancer* **2020**, *28*, 6075–6084. [CrossRef] [PubMed]
53. Schmidt, M.E.; Wiskemann, J.; Armbrust, P.; Schneeweiss, A.; Ulrich, C.M.; Steindorf, K. Effects of resistance exercise on fatigue and quality of life in breast cancer patients undergoing adjuvant chemotherapy: A randomized controlled trial. *Int. J. Cancer* **2015**, *137*, 471–480. [CrossRef]
54. Sprod, L.K.; Hsieh, C.C.; Hayward, R.; Schneider, C.M. Three versus six months of exercise training in breast cancer survivors. *Breast Cancer Res. Treat.* **2010**, *121*, 413–419. [CrossRef] [PubMed]
55. Tjoe, J.A.; Piacentine, L.B.; Papanek, P.E.; Raff, H.; Richards, J.; Harkins, A.L.; Yin, J.; Ng, A.V. Team triathlon effects on physiological, psychological, and immunological measures in women breast cancer survivors. *Support Care Cancer* **2020**, *28*, 6095–6104. [CrossRef]
56. Travier, N.; Velthuis, M.J.; Steins Bisschop, C.N.; van den Buijs, B.; Monninkhof, E.M.; Backx, F.; Los, M.; Erdkamp, F.; Bloemendal, H.J.; Rodenhuis, C.; et al. Effects of an 18-week exercise programme started early during breast cancer treatment: A randomised controlled trial. *BMC Med.* **2015**, *13*, 121. [CrossRef] [PubMed]
57. van Waart, H.; Stuiver, M.M.; van Harten, W.H.; Geleijn, E.; Kieffer, J.M.; Buffart, L.M.; de Maaker-Berkhof, M.; Boven, E.; Schrama, J.; Geenen, M.M.; et al. Effect of Low-Intensity Physical Activity and Moderate- to High-Intensity Physical Exercise During Adjuvant Chemotherapy on Physical Fitness, Fatigue, and Chemotherapy Completion Rates: Results of the PACES Randomized Clinical Trial. *J. Clin. Oncol.* **2015**, *33*, 1918–1927. [CrossRef]
58. Liu, Y.C.; Hung, T.T.; Konara Mudiyanselage, S.P.; Wang, C.J.; Lin, M.F. Beneficial Exercises for Cancer-Related Fatigue among Women with Breast Cancer: A Systematic Review and Network Meta-Analysis. *Cancers* **2022**, *15*, 151. [CrossRef]
59. Hilfiker, R.; Meichtry, A.; Eicher, M.; Nilsson Balfe, L.; Knols, R.H.; Verra, M.L.; Taeymans, J. Exercise and other non-pharmaceutical interventions for cancer-related fatigue in patients during or after cancer treatment: A systematic review incorporating an indirect-comparisons meta-analysis. *Br. J. Sports Med.* **2018**, *52*, 651–658. [CrossRef]
60. Gleeson, M.; Bishop, N.C.; Stensel, D.J.; Lindley, M.R.; Mastana, S.S.; Nimmo, M.A. The anti-inflammatory effects of exercise: Mechanisms and implications for the prevention and treatment of disease. *Nat. Rev. Immunol.* **2011**, *11*, 607–615. [CrossRef]
61. Fairey, A.S.; Courneya, K.S.; Field, C.J.; Bell, G.J.; Jones, L.W.; Martin, B.S.; Mackey, J.R. Effect of exercise training on C-reactive protein in postmenopausal breast cancer survivors: A randomized controlled trial. *Brain Behav. Immun.* **2005**, *19*, 381–388. [CrossRef]

62. Kiecolt-Glaser, J.K.; Bennett, J.M.; Andridge, R.; Peng, J.; Shapiro, C.L.; Malarkey, W.B.; Emery, C.F.; Layman, R.; Mrozek, E.E.; Glaser, R. Yoga's impact on inflammation, mood, and fatigue in breast cancer survivors: A randomized controlled trial. *J. Clin. Oncol.* **2014**, *32*, 1040–1049. [CrossRef] [PubMed]
63. Hutnick, N.A.; Williams, N.I.; Kraemer, W.J.; Orsega-Smith, E.; Dixon, R.H.; Bleznak, A.D.; Mastro, A.M. Exercise and lymphocyte activation following chemotherapy for breast cancer. *Med. Sci. Sports Exerc.* **2005**, *37*, 1827–1835. [CrossRef]
64. Ryan, J.L.; Carroll, J.K.; Ryan, E.P.; Mustian, K.M.; Fiscella, K.; Morrow, G.R. Mechanisms of cancer-related fatigue. *Oncologist* **2007**, *12*, 22–34. [CrossRef]
65. Petersen, A.M.; Pedersen, B.K. The anti-inflammatory effect of exercise. *J. Appl. Physiol.* **2005**, *98*, 1154–1162. [CrossRef]
66. Galvão, D.A.; Taaffe, D.R.; Spry, N.; Joseph, D.; Newton, R.U. Combined resistance and aerobic exercise program reverses muscle loss in men undergoing androgen suppression therapy for prostate cancer without bone metastases: A randomized controlled trial. *J. Clin. Oncol.* **2010**, *28*, 340–347. [CrossRef]
67. Patel, J.G.; Bhise, A.R. Effect of Aerobic Exercise on Cancer-related Fatigue. *Indian J. Palliat. Care* **2017**, *23*, 355–361. [CrossRef] [PubMed]
68. Phillips, S.M.; McAuley, E. Physical activity and fatigue in breast cancer survivors: A panel model examining the role of self-efficacy and depression. *Cancer Epidemiol. Biomark. Prev.* **2013**, *22*, 773–781. [CrossRef] [PubMed]
69. McAuley, E.; White, S.M.; Rogers, L.Q.; Motl, R.W.; Courneya, K.S. Physical activity and fatigue in breast cancer and multiple sclerosis: Psychosocial mechanisms. *Psychosom. Med.* **2010**, *72*, 88–96. [CrossRef] [PubMed]
70. Hartescu, I.; Morgan, K.; Stevinson, C.D. Increased physical activity improves sleep and mood outcomes in inactive people with insomnia: A randomized controlled trial. *J. Sleep Res.* **2015**, *24*, 526–534. [CrossRef]
71. Cramp, F.; Byron-Daniel, J. Exercise for the management of cancer-related fatigue in adults. *Cochrane Database Syst. Rev.* **2012**, *11*, Cd006145. [CrossRef]
72. Kelley, G.A.; Kelley, K.S. Exercise and cancer-related fatigue in adults: A systematic review of previous systematic reviews with meta-analyses. *BMC Cancer* **2017**, *17*, 693. [CrossRef] [PubMed]
73. Steindorf, K.; Schmidt, M.E.; Klassen, O.; Ulrich, C.M.; Oelmann, J.; Habermann, N.; Beckhove, P.; Owen, R.; Debus, J.; Wiskemann, J.; et al. Randomized, controlled trial of resistance training in breast cancer patients receiving adjuvant radiotherapy: Results on cancer-related fatigue and quality of life. *Ann. Oncol.* **2014**, *25*, 2237–2243. [CrossRef] [PubMed]
74. Yang, T.Y.; Chen, M.L.; Li, C.C. Effects of an aerobic exercise programme on fatigue for patients with breast cancer undergoing radiotherapy. *J. Clin. Nurs.* **2015**, *24*, 202–211. [CrossRef] [PubMed]
75. Mijwel, S.; Jervaeus, A.; Bolam, K.A.; Norrbom, J.; Bergh, J.; Rundqvist, H.; Wengström, Y. High-intensity exercise during chemotherapy induces beneficial effects 12 months into breast cancer survivorship. *J. Cancer Surviv.* **2019**, *13*, 244–256. [CrossRef] [PubMed]
76. van Vulpen, J.K.; Peeters, P.H.; Velthuis, M.J.; van der Wall, E.; May, A.M. Effects of physical exercise during adjuvant breast cancer treatment on physical and psychosocial dimensions of cancer-related fatigue: A meta-analysis. *Maturitas* **2016**, *85*, 104–111. [CrossRef] [PubMed]
77. Juvet, L.K.; Thune, I.; Elvsaas, I.; Fors, E.A.; Lundgren, S.; Bertheussen, G.; Leivseth, G.; Oldervoll, L.M. The effect of exercise on fatigue and physical functioning in breast cancer patients during and after treatment and at 6 months follow-up: A meta-analysis. *Breast* **2017**, *33*, 166–177. [CrossRef] [PubMed]
78. Ormel, H.L.; van der Schoot, G.G.F.; Sluiter, W.J.; Jalving, M.; Gietema, J.A.; Walenkamp, A.M.E. Predictors of adherence to exercise interventions during and after cancer treatment: A systematic review. *Psychooncology* **2018**, *27*, 713–724. [CrossRef] [PubMed]
79. Meneses-Echávez, J.F.; González-Jiménez, E.; Ramírez-Vélez, R. Effects of supervised exercise on cancer-related fatigue in breast cancer survivors: A systematic review and meta-analysis. *BMC Cancer* **2015**, *15*, 77. [CrossRef] [PubMed]
80. Sweegers, M.G.; Altenburg, T.M.; Brug, J.; May, A.M.; van Vulpen, J.K.; Aaronson, N.K.; Arbane, G.; Bohus, M.; Courneya, K.S.; Daley, A.J.; et al. Effects and moderators of exercise on muscle strength, muscle function and aerobic fitness in patients with cancer: A meta-analysis of individual patient data. *Br. J. Sports Med.* **2019**, *53*, 812. [CrossRef] [PubMed]
81. Zhou, H.J.; Wang, T.; Xu, Y.Z.; Chen, Y.N.; Deng, L.J.; Wang, C.; Chen, J.X.; Tan, J.B. Effects of exercise interventions on cancer-related fatigue in breast cancer patients: An overview of systematic reviews. *Support Care Cancer* **2022**, *30*, 10421–10440. [CrossRef]
82. Danhauer, S.C.; Mihalko, S.L.; Russell, G.B.; Campbell, C.R.; Felder, L.; Daley, K.; Levine, E.A. Restorative yoga for women with breast cancer: Findings from a randomized pilot study. *Psychooncology* **2009**, *18*, 360–368. [CrossRef] [PubMed]
83. Li, G.; You, Q.; Hou, X.; Zhang, S.; Du, L.; Lv, Y.; Yu, L. The effect of exercise on cognitive function in people with multiple sclerosis: A systematic review and meta-analysis of randomized controlled trials. *J. Neurol.* **2023**, *270*, 2908–2923. [CrossRef] [PubMed]
84. Campbell, K.L.; Winters-Stone, K.M.; Wiskemann, J.; May, A.M.; Schwartz, A.L.; Courneya, K.S.; Zucker, D.S.; Matthews, C.E.; Ligibel, J.A.; Gerber, L.H.; et al. Exercise Guidelines for Cancer Survivors: Consensus Statement from International Multidisciplinary Roundtable. *Med. Sci. Sports Exerc.* **2019**, *51*, 2375–2390. [CrossRef] [PubMed]
85. Rao, A.V.; Cohen, H.J. Fatigue in older cancer patients: Etiology, assessment, and treatment. *Semin. Oncol.* **2008**, *35*, 633–642. [CrossRef] [PubMed]

86. Courneya, K.S.; Karvinen, K.H. Exercise, aging, and cancer. *Appl. Physiol. Nutr. Metab.* **2007**, *32*, 1001–1007. [CrossRef]
87. Montaño-Rojas, L.S.; Romero-Pérez, E.M.; Medina-Pérez, C.; Reguera-García, M.M.; de Paz, J.A. Resistance Training in Breast Cancer Survivors: A Systematic Review of Exercise Programs. *Int. J. Environ. Res. Public Health* **2020**, *17*, 6511. [CrossRef]

life

MDPI

Review

Comparing the Effect of Isoinertial Flywheel Training and Traditional Resistance Training on Maximal Strength and Muscle Power in Healthy People: A Systematic Review and Meta-Analysis

Zhongzhong Hu [1], Yuhang Liu [2], Keke Huang [1], Hao Huang [1], Feng Li [1] and Xiaoyi Yuan [2,*]

[1] School of Sports Science, Wenzhou Medical University, Wenzhou 325035, China; bsu2014hzz@163.com (Z.H.); huangkekerunning@163.com (K.H.); huanghaomedical@163.com (H.H.); lifengball@163.com (F.L.)

[2] China Athletics College, Beijing Sport University, Beijing 100084, China; 15031958573@163.com

* Correspondence: bsu_yuan@bsu.edu.cn

Abstract: Background: This systematic review and meta-analysis aimed to analyze whether isoinertial flywheel training (FWT) is superior to traditional resistance training (TRT) in enhancing maximal strength and muscle power in healthy individuals. Methods: Electronic searches were conducted in the Web of Science, PubMed, Cochrane Library, SPORTDiscus, and Scopus databases up to 21 April 2024. Outcomes were analyzed as continuous variables using either a random or fixed effects model to calculate the standardized mean difference (SMD) and 95% confidence intervals (CI). Results: A total of sixteen articles, involving 341 subjects, met the inclusion criteria and were included in the statistical analyses. The pooled results indicate no statistically significant differences between FWT and TRT in developing maximal strength in healthy individuals (SMD = 0.24, 95% CI [−0.26, 0.74], $p = 0.35$). Additionally, the pooled outcomes showed a small-sized effect in muscle power with FWT (SMD = 0.47, 95% CI [0.10, 0.84]), which was significantly higher than that with TRT ($p = 0.01$) in healthy individuals. Subgroup analysis revealed that when the total number of FWT sessions is between 12 and 18 (1–3 times per week), it significantly improves muscle power (SMD = 0.61, 95% CI [0.12, 1.09]). Significant effects favoring FWT for muscle power were observed in both well-trained (SMD = 0.58, 95% CI [0.04, 1.13]) and untrained individuals (SMD = 1.40, 95% CI [0.23, 2.57]). In terms of exercise, performing flywheel training with squat and lunge exercises significantly enhances muscle power (SMD = 0.43; 95% CI: 0.02–0.84, and $p = 0.04$). Interestingly, FWT was superior to weight stack resistance training (SMD = 0.61, 95% CI [0.21, 1.00]) in enhancing muscle power, while no significant differences were found compared to barbell free weights training (SMD = 0.36, 95% CI [−0.22, 0.94]). Conclusions: This meta-analysis confirms the superiority of FWT compared to TRT in promoting muscle power in both healthy untrained and well-trained individuals. Squats and lunges for FWT are more suitable for improving lower limb explosive power. It is recommended that coaches and trainers implement FWT for six weeks, 2–3 times per week, with at least a 48 h interval between each session. Although FWT is not superior to free weights training, it is advisable to include FWT in sport periodization to diversify the training stimuli for healthy individuals.

Keywords: eccentric overload training; isoinertial flywheel training; maximal strength; muscle power

Citation: Hu, Z.; Liu, Y.; Huang, K.; Huang, H.; Li, F.; Yuan, X. Comparing the Effect of Isoinertial Flywheel Training and Traditional Resistance Training on Maximal Strength and Muscle Power in Healthy People: A Systematic Review and Meta-Analysis. *Life* **2024**, *14*, 908. https://doi.org/10.3390/life14070908

Academic Editor: Panagiotis Georgianos

Received: 24 June 2024
Revised: 11 July 2024
Accepted: 18 July 2024
Published: 21 July 2024

1. Introduction

The maintenance and improvement of muscle strength and power are critical goals of physical training interventions across various populations, with resistance training emerging as the most prevalent method for achieving these outcomes [1,2]. Over the years, numerous methods to enhance strength were suggested, including the use of free weights, weight stacks, resistance bands, flywheels, and pneumatic resistance machines [3]. Traditional resistance training (TRT) involving free weights and weight stack machines, which rely on gravity-dependent loads, were demonstrated to elicit desirable structural and neural adaptations in healthy individuals [4]. In conventional setups, the external load provided by resistance equipment remains static throughout the entire range of motion. Eccentric muscle contractions, however, allow for higher force production than concentric contractions [5,6], the relative load during the eccentric phase in constant resistance training is inadvertently lower than in the concentric phase. This discrepancy leads to suboptimal loading, resulting in reduced motor unit recruitment and firing rates during the eccentric phase [7], potentially causing diminished sarcoplasmic calcium release and, consequently, a lesser stimulus for myocellular adaptation [8]. As a result, this traditional approach may not provide an optimal stimulus for the critical eccentric phase.

Eccentric overload training (EOT) can reduce skeletal muscle resistance in the weakest areas of motion, provide greater resistance in stronger areas, and align more closely with human strength curves to enable muscles to function over a broader range [9–11]. While EOT can be implemented using gravity-dependent (GD) devices, these often require third-party assistance, posing a limitation in various settings. Historically, practitioners utilized weight releasers in traditional training methods to achieve supramaximal eccentric loads, although this approach still presents certain constraints. This form of resistance training effectively reduces the mechanical disadvantage of the sticking point commonly encountered in free weight training [12,13]. Recognizing these limitations, several innovative methods emerged, including the adoption of non-gravity-dependent technology. Specifically, isoinertial flywheel devices, which harness the inertia of a rotating wheel and the subsequent stored kinetic energy, offer a higher eccentric load compared to traditional weight training methods [14,15]. Although no differences were found in muscle fatigue levels, flywheel training (FWT) induced greater physiological stress than barbell squat training. This was observed through a greater decrease in muscle oxygen saturation and a longer reoxygenation period [16].

Scientific literature acknowledges that FWT induces several morphological and neural-adaptive changes in the human body. These include increases in peak power output, muscle cross-sectional area, musculotendinous stiffness, as well as improvements in motor unit recruitment, rate coding (firing frequency), synchronous motor unit activity, and neuromuscular inhibition [17,18]. In contrast to TRT, this form of accentuated eccentric training induces a prolonged eccentric strain, which may lead to superior adaptations. Prolonged eccentric training appears to increase eccentric kinetic energy and enhance performance more effectively than traditional methodologies [19]. Inertial technology emerged as an alternative that enables accentuating eccentric overload in more specific sports actions, such as changing direction. This is essential for player optimization, reducing the risk of injuries, and aiding in injury rehabilitation [20]. While studies confirm that FWT can yield more acute [21–23] and long-term [24–26] training effects on strength performance than traditional constant resistance training, some authors report no significant difference between the two methods [27,28]. A limitation of existing studies lies in the use of notably different protocols and execution methodologies. For instance, variations in training methods, targeted muscle groups, sets and repetitions performed, measurement tools, eccentric load applied, participants' age, and training experience differ significantly among studies. Nonetheless, our analysis builds on previous meta-analyses by incorporating a larger number of studies, more recently published data, and comparative analyses on the effects of FWT on different training levels of participants, and inertial flywheel training compared to free weights or weight stack training. The primary aim of this meta-analysis was to compare the effects of FWT versus TRT on muscle power and maximal strength by examining and compiling relevant studies.

2. Methods

This meta-analysis was conducted in accordance with the Preferred Reporting Items for Systematic Reviews and Meta-analyses (PRISMA) statement guidelines [29]. Prior to the search, a review protocol was registered at PROSPERO (ID = CRD42023491903).

2.1. Search Strategy

The electronic databases PubMed, SPORTDiscus, Web of Science, Cochrane Library, and Scopus were searched for randomized controlled trials on flywheel training from their inception to 21 April 2024. Search terms included: 'eccentric overload training', 'flywheel training', 'flywheel resistance training', 'flywheel exercise', 'isoinertial training/inertial training', 'isoinertial exercise/inertial exercise', 'strength', and 'power'. Boolean operators 'AND' and 'OR' were used to combine key search terms. Using the WOS database for example, (AB = (flywheel training OR flywheel resistance training OR flywheel exercise OR isoinertial training OR inertial training OR isoinertial exercise OR inertial exercise)) AND (AB = (strength OR power)). When applicable, filters were used during the initial literature search to identify relevant articles. A hand-search of the reference lists of relevant articles was also conducted for other potentially relevant references.

2.2. Inclusion/Exclusion Criteria

To rate studies for eligibility, a participants, intervention, comparators, study outcomes, and study design (PICOS) approach was used [30]. An article was eligible for inclusion if it met all of the following criteria: (1) the original article was a randomized controlled trial (RCT); (2) participants were healthy, with no imposed limitations concerning gender, training status, sport specialty, or body composition; (3) the manuscript included an FWT intervention and a control or alternative intervention group aimed at evaluating training adaptations in strength and/or power; (4) the article stipulated that participants completed an FWT protocol lasting at least four weeks; and (5) the study provided data on at least one of the following outcome measures: strength (e.g., 1 RM, maximal voluntary contraction, and peak torque) and power (e.g., jump height, rate of force development, and peak power).

An article was excluded if it met any of the following criteria: (1) it was a non-randomized controlled trial; (2) it failed to meet the minimum requirements for the training protocol (e.g., duration or frequency); (3) the document was a literature review, abstract, editorial commentary, or letter to the editor; (4) it was not written in English; (5) means and standard deviations were not reported, and the authors did not respond to our inquiries; and (6) the study involved participants with any pathology or those receiving treatment for musculoskeletal injuries in the trained limb.

Titles and abstracts identified in the search were downloaded into EndNote 20, after which cross-references and duplicates were deleted. All publications potentially relevant for inclusion were independently assessed by two reviewers, with full texts obtained if necessary. Any discrepancies that arose were resolved during a consensus meeting, with the provision that a third reviewer was available if needed.

2.3. Study Coding and Data Extraction

Two reviewers independently extracted data using a specially designed standardized form, focusing on general study information, participant demographics, intervention characteristics, and outcome measures. If the necessary data were not explicitly available in tables or the text's results section, the first author of the systematic review proactively contacted the original authors to request the missing data. When the authors did not have access to their data, essential details, such as means and standard deviations for outcome measures, were meticulously extracted from figures and graphs using Web Plot Digitizer V4.7 software. To ensure accuracy, another reviewer then rigorously verified the validity of the data extraction.

Each article was read and coded by two investigators focusing on several variables: (a) descriptive information, such as participants' details (age, body mass, and height), physical activity status (trained or untrained), sex, and the total number of participants; (b) specifics of the program exercises, including the type of exercise (knee extension, squat, half-squat, leg press, deadlift, and bench press); (c) program variables, detailing the frequency of weekly sessions, duration of the training period, total number of sessions, number of sets per session, number of repetitions per set, and training intensity; and (d) outcome measurements, capturing measures of maximal muscle strength and/or muscle power. The investigators' mean agreement was quantified using an intraclass correlation coefficient (ICC), with the coding agreement assessed by comparing the number of variables on which they aligned versus the total coded. A mean agreement of 0.90 is upheld as an appropriate level of reliability for such coding procedures [31]. Any discrepancies in coding between investigators were meticulously scrutinized and resolved before proceeding with the analysis.

2.4. Quality Assessment

Two investigators conducted independent quality assessments of the included studies, with any disagreements resolved through a consensus meeting mediated by a third party. We used the 2019 Cochrane Risk of Bias 2 tool (RoB2) to assess the risk of bias across five domains: randomization process, deviations from intended interventions, missing outcome data, measurement of the outcome, and selection of the reported result [32]. Each of the 16 studies included in the quantitative analysis was longitudinal. Two independent authors assessed the quality using checklists. If there was a disagreement regarding the risk of bias assessment findings, a third reviewer was consulted to evaluate the data and make the final decision.

2.5. Statistics and Data Analysis

Stata 17.0 and Reviewer Manager 5.4 software were instrumental for various tasks, including data merging, subgroup analysis, forest plot generation, heterogeneity analysis, meta-regression, and assessing publication bias. For the primary outcome focusing on muscle strength and power, we calculated intervention effects using standardized mean differences (SMD) with 95% confidence intervals (CI), which is appropriate due to the continuous nature of the data. Effect sizes were stratified as small (0.2), medium (0.5), or large (0.8 or greater) [33]. We employed the I^2 test to examine the heterogeneity of each trial, with benchmarks set at 25%, 50%, and 75% for low, medium, and high statistical heterogeneity, respectively. The chi-squared and I^2 statistics were pivotal in describing the level of heterogeneity or homogeneity among the comparisons, with a p-value threshold of less than 0.05 indicating significant heterogeneity [34]. In cases where the heterogeneity test showed no significant differences, a fixed-effects model was adopted for the meta-analysis. Conversely, a random effects model was applied in the presence of high heterogeneity. We conducted a detailed subgroup analysis to identify and analyze potential sources of heterogeneity.

3. Results

3.1. Study Selection

A search of electronic databases, along with scanning the reference lists, yielded 1367 relevant studies. After removing duplicates, 1019 titles and abstracts were screened. From these, 934 records were excluded based on their titles and abstracts, and 6 records were excluded due to the unavailability of full text. This led to the selection of 79 studies, which were then carefully screened for eligibility. During this process, two additional records were identified through meticulous examination of reference lists and citations of pertinent articles. A total of 65 studies were excluded for the following reasons: (1) 3 studies were published in German and 1 study was published in Korean; (2) the participants in 23 studies had diseases or were injured; (3) 10 studies performed interventions during simulated microgravity; (4) 8 studies focused on acute effects; (5) 6 studies did not have a control group; (6) the control group in 5 studies did not use free weights or weight stack training as the intervention; (7) the outcome measures in 4 studies did not include maximal strength and power; (8) 3 studies were not RCT designs; and (9) the dropout rate was greater than 15% in 2 studies. Ultimately, 16 studies met the inclusion criteria and were included in the meta-analysis. The detailed flowchart in Figure 1 illustrates the systematic selection process of the studies.

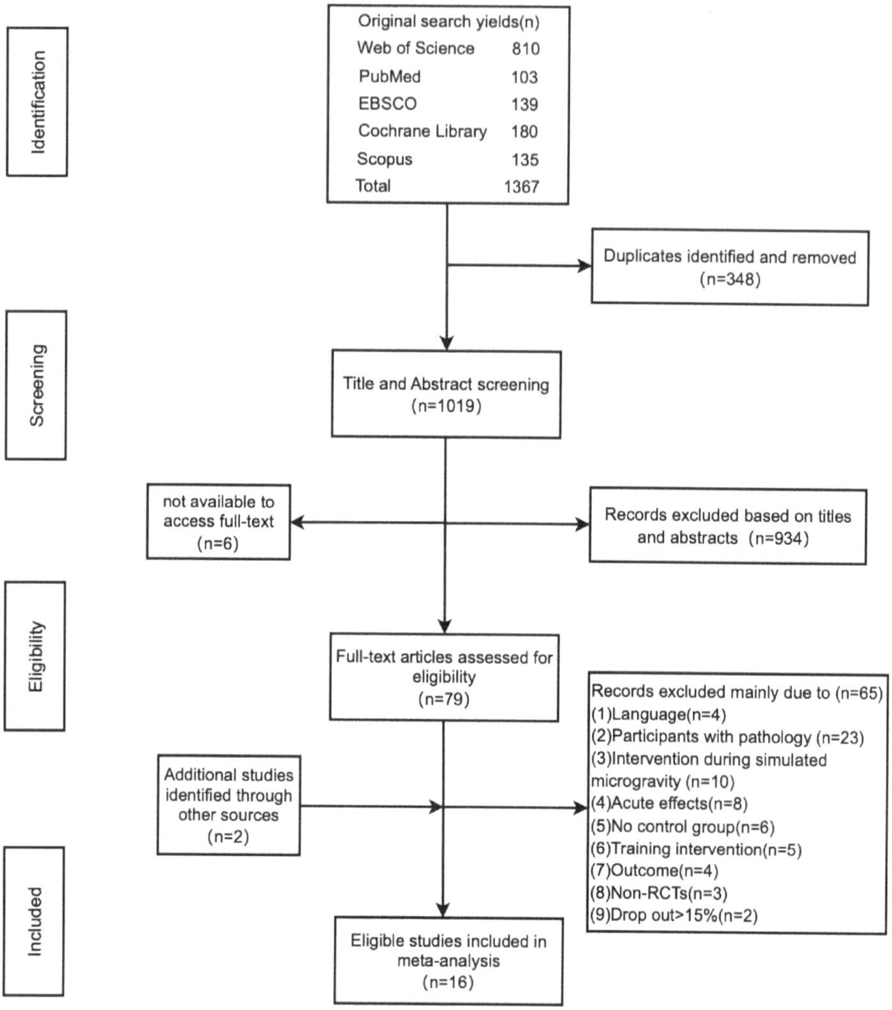

Figure 1. Flow chart illustrating the selection process for all included and excluded studies.

3.2. Descriptive Characteristics of the Studies

The main characteristics of the studies included in the review, encompassing participants, interventions, and results, are depicted in Table 1. Following an adjustment for dropouts, the total number of participants in the 16 studies was 341. Of these 341 participants, 171 undertook FWT, while the remaining 170 engaged in free weights training or weight stack training. The estimated average ages of the experimental and control groups were 32.18 ± 16.68 and 34.01 ± 18.51, respectively. Notably, the distribution of genders across the studies was imbalanced, with only three studies incorporating female participants, culminating in a demographic of just 32 women and 309 men. The participants in three studies were novices in resistance training, lacking or having scant experience in this discipline. Conversely, thirteen studies involved subjects with prior strength training experience: five studies included seasoned athletes engaged in professional or semi-professional leagues, and eight involved individuals who trained recreationally (two with college students active in sports, three involving strength-trained individuals, and three focusing on junior athletes).

Table 1. Basic characteristics of the included studies.

Study (Year)	Characteristics of Participants					Intervention					Results
	N (M/F)	Height (cm)	Weight (kg)	Age (y)	Training Experience	Exercise (Experimental Group Equipment)	Set × Reps	Frequency × Duration = Sessions	Control Group Intensity	Experimental Group Intensity	
[35]	12/12	—	—	69.8 ± 1.3	untrained	knee extension (YoYo flywheel device)	1–4 × 8–12	3 × 12 = 36	80%1 RM	—	MVC ↑ 8%; PT ↑ 28.0% (*)
[36]	15/0	182.8 ± 7.7	91.0 ± 13.8	39.3 ± 8.6	untrained	knee extension (YoYo flywheel device)	4 × 7	2–3 × 5 = 12	7 RM	7 RM	Con PP ↑ 9.0%; ECC PP 12.0%; MVC ↑ 11.6% (*)
[37]	17/0	185.2 ± 8.1	90.0 ± 15.8	39.1 ± 6.6	untrained	knee extension (YoYo flywheel device)	4 × 7	2–3 × 5 = 12	7 RM	7 RM	MVC ↑ 8.1% (*)
[38]	23/0	176.8 ± 3.34	76.8 ± 7.83	22.5 ± 2.5	RT (physically active males)	front step exercise (inertial flywheel device)	5–7 × 8	3 × 6 = 18	8 RM	8 RM	MVC ↑ 11.0%
[39]	29/0	185.0 ± 5.9	83.9 ± 3.9	21.7 ± 2.7	WT (professional handball players)	leg-press (YoYo flywheel device)	4 × 7	2–3 × 6 = 15	7 RM	7 RM	CMJ height ↑ 9.8% (*); PP ↑ 12.9% (*); 1 RM ↑ 12.2%
[40]	8/8	173.0 ± 13.0	79.0 ± 22.0	26.0 ± 4.0	RT (recreationally active individuals)	unilateral knee extension (YoYo flywheel device)	4 × 7	2–3 × 8 = 20	8–12 RM	7 RM	PP ↑ 29.2%; 1 RM ↑ 25.3%
[41]	40/0	180.0 ± 11.0	77.0 ± 5.0	23 ± 4	WT (Italian fourth-division soccer players)	squat (Desmotec flywheel device)	4–6 × 8	1 × 8 = 8	80%1 RM	8 RM	CMJ height ↑ 10%; BS 1 RM ↑ 7%
[42]	32/0	177.6 ± 5.4	75.9 ± 7.6	21.0 ± 1.4	RT (amateur soccer student players)	half squat (Desmotec flywheel device)	3–4 × 5 or 3–6 × 6	2 × 8 = 16	—	5 or 6 RM	CMJ height↑ 4.0%; 1 RM ↑ 9.7%
[43]	8/0	193.5 ± 8.0	87.4 ± 11.7	21.3 ± 3.5	WT (professional basketball players)	half squat (inertial flywheel device)	4–6 × 10	1 × 6 = 6	14 RM	10 RM	CMJ height ↑ (*)
[44]	24/0	190.6 ± 5.9	77.2 ± 7.0	17.6 ± 0.6	RT (junior basketball players)	half squat, Romanian deadlift (isoinertial flywheel device)	2–4 × 8	1–2 × 8 = 12	80%1 RM	8 RM	CMJ height ↑ 11.7% (*); MVC ↑ 18.7%
[45]	16/0	—	93.0 ± 13.1	18.0 ± 1.0	RT (academy rugby union players)	squat, Romanian deadlift, Bulgarian split squat (kbox flywheel device)	4–5 × 6 or 8	2 × 4 = 8	6 RM or 8 RM	6 RM or 8 RM	CMJ PP ↑ 4.0%; CMJ height ↑ 4.9%

Table 1. *Cont.*

Study (Year)	Characteristics of Participants					Intervention						Results
	N (M/F)	Height (cm)	Weight (kg)	Age (y)	Training Experience	Exercise (Experimental Group Equipment)	Set × Reps	Frequency × Duration = Sessions	Control Group Intensity	Experimental Group Intensity	
[46]	16/0	174.4 ± 7.8	64.5 ± 8.6	15.5 ± 1.2	RT (junior tennis players)	chest press, shoulder press, row, closed stance, and chest crossover (isoinertial flywheel device)	3 × 6 or 8	2 × 8 = 16	50–70% 1 RM	RPE 5–7	CMJ height ↑ 9.7% (*)
[47]	22/0	178.0 ± 1.8	71.5 ± 6.9	21.8 ± 2.7	WT (elite collegiate long-distance runners)	squat (kbox flywheel device)	4 × 7	3 × 6 = 18	85%1 RM	7 RM	CMJ height ↑ 12.0 (*)
[48]	34/0	174.0 ± 7.3	70.5 ± 13.3	16.0 ± 1.4	RT (junior handball players)	lunge, acceleration, squat, single leg hop, and crossover step (isoinertial flywheel device)	3 × 8 or 12	2 × 8 = 16	RPE 6–9	RPE 6–9	UCMJD height ↑ 21.9%
[49]	18/0	184.1 ± 9.7	78.9 ± 10.0	18.6 ± 0.8	WT (elite hockey players)	bilateral/unilateral squat, leg curl, and leg press (isoinertial flywheel device)	3–4 × 6 or 4 × 7	1–2 × 8 = 14	4–12 RM	6 or 7 RM	CMJ height ↑ 5.7%
[50]	11/12	170.0 ± 2.0	73.8 ± 15.9	24.15 ± 3.9	RT (physically active adults)	squat, bench press, deadlift, and row (flywheel training platform)	3 × 4–12	3 × 10 = 30	—	—	MVIT ↑ 11.4

N = number, M = male, F = female, RT = recreationally trained, WT = well-trained, CMJ = countermovement jump, UCMJD = unilateral countermovement jump with dominant leg, PP = peak power, BS = back squat, PT = peak torque, MVC = maximal voluntary contraction, MVIT = Maximal voluntary isometric torque, RM = repetition maximum, RPE = rate of perceived exertion, Con = concentric, and ECC = eccentric, ↑ Statistically significant within-group differences ($p < 0.05$), * statistically significant difference between EOT and CON groups ($p < 0.05$).

In the experimental groups, all studies utilized inertial flywheel devices for eccentric overload (EO). Regarding the control groups, eight studies implemented free weights training, while eight opted for resistance training with a weight stack machine. The duration of training interventions varied from 4 to 12 weeks, with participants undertaking an average of 2.3 ± 0.7 sessions weekly, culminating in 17.5 ± 11.5 sessions per study. Among the studies, there was variation in the total number of sets (3–7) and repetitions (4–12) per session. Inertia was a key variable, with eleven studies employing a range from 0.0291 to 0.145 kg·m^2; however, five studies did not disclose the inertia used. The most commonly used load intensity for the control group was 80% 1 RM or 7 RM.

3.3. Quality of the Selected Studies

Assessing the risk of bias was conducted using a revised version of the Cochrane Risk of Bias (RoB 2) tool, evaluating individual studies across six different domains of bias (Figure 2A). Only one study [45] reported randomization sequence generation. All sixteen studies included in the meta-analysis were rated as having "some concerns" for risk of bias (Figure 2B), likely because it is impossible to blind subjects to experimental and control groups in studies of this nature. Nevertheless, it is important to acknowledge that blinding participants is a notably challenging criterion to meet in this context. Adding to the credibility of these assessments, the studies demonstrated significantly high inter-rater reliability (ICC = 0.95).

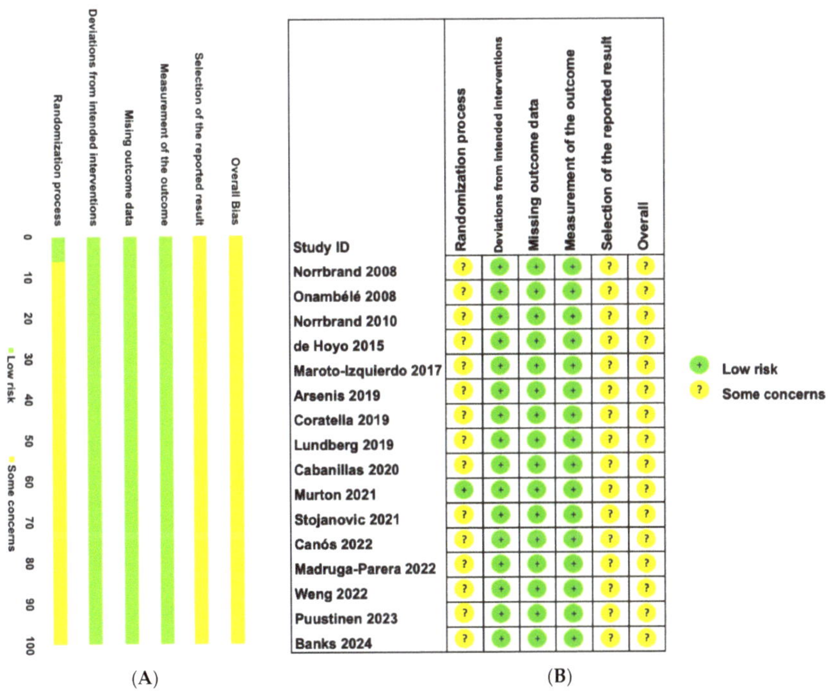

Figure 2. Summary of the risk of bias of studies included in this meta-analysis [35–50]. (**A**) Summary of 16 studies in six different domains of bias. (**B**) Details of 16 studies in six different domains of bias.

3.4. Publication Bias and Sensitivity

A funnel plot visually represents each individual study's effect by considering the study size in relation to the difference observed between pre- and post-tests. A symmetrical funnel plot, centered around the mean effect of the collective studies, indicates that the identification and selection processes are likely free from bias [51]. The corresponding funnel plots are illustrated in Figure 3A,B. The visual inspection of the funnel plot indicates a symmetrical distribution pattern of the effects, illustrating the absence of publication bias. This is corroborated by Egger's regression outcome, which also indicates that the distribution pattern of the effect in the funnel plot is symmetrical (maximal muscle strength: t = 2.06, p = 0.073; muscle power: t = 2.13, p = 0.059).

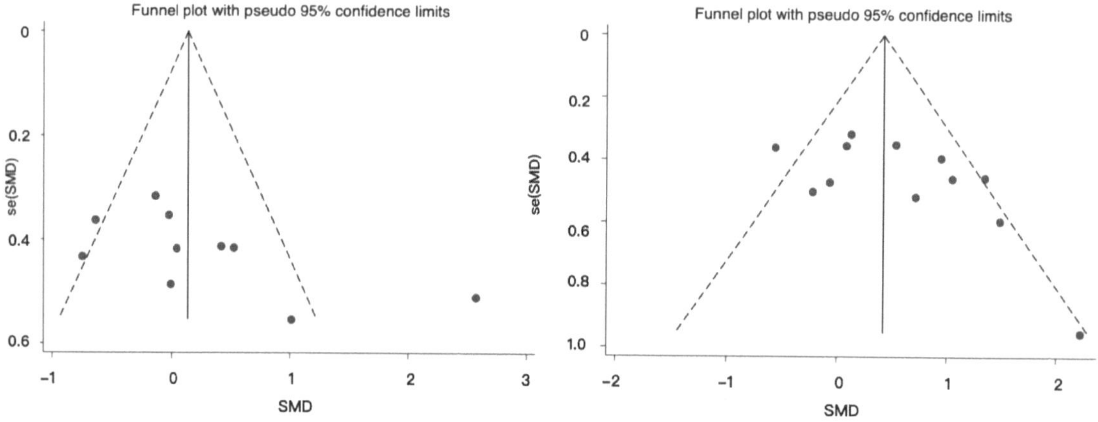

(**A**) (**B**)

Figure 3. Funnel plot illustrating the symmetrical distribution of the effects across the included studies. (**A**) Funnel plot for studies about maximal muscle strength. (**B**) Funnel plot for studies about muscle power.

In a separate sensitivity analysis, we assessed the contribution of each study to the overall improvement in maximal muscle strength (Figure 4A) and muscle power (Figure 4B) detected in this meta-analysis. This was achieved by successively omitting the results of each study from the comparisons made with the data from the remaining studies. In each scenario where the results of one study were omitted, no significant differences were detected, indicating the robust contribution of all the studies to the observed gains in maximal muscle strength and muscle power.

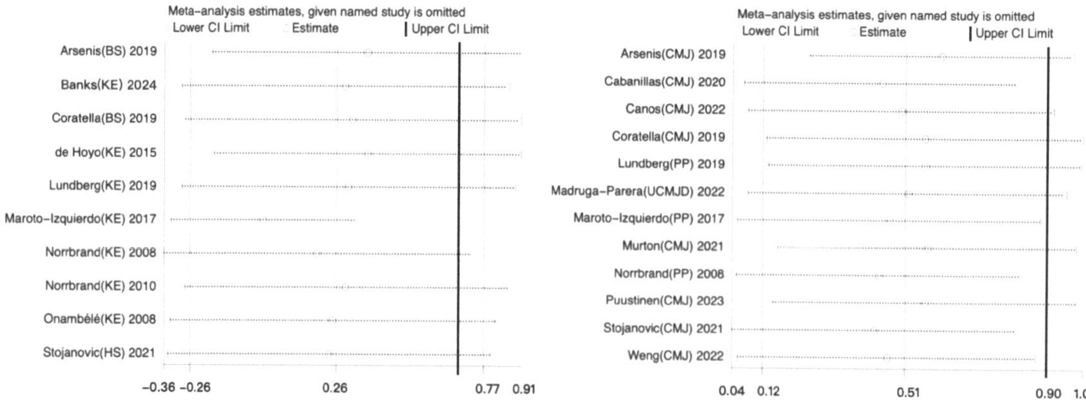

(**A**) (**B**)

Figure 4. Sensitivity analysis for this meta-analysis [35–50]. (**A**) Sensitivity analysis for studies about maximal muscle strength. (**B**) Sensitivity analysis for studies about muscle power.

3.5. *Main Analysis*

3.5.1. Meta-Analysis Results on Muscle Power

A total of twelve reports were included in the meta-analysis. The data presented in Figure 5 reveal significant differences between FWT and TRT in improving the muscle power of healthy subjects (ES = 0.47; 95% CI: 0.10–0.84, $p = 0.01$). However, our analysis detected moderate statistical heterogeneity ($I^2 = 54\%$).

Study or Subgroup	FWT Mean	FWT SD	FWT Total	TRT Mean	TRT SD	TRT Total	Weight	Std. Mean Difference IV, Random, 95% CI
Arsenis(CMJ) 2019	44.73	5.1	16	47.9	6.18	16	10.2%	-0.55 [-1.25, 0.16]
Cabanillas(CMJ) 2020	43.27	1.65	4	38.05	2.89	4	3.0%	1.93 [0.01, 3.85]
Canos(CMJ) 2022	36.1	6	8	32.34	4.46	8	7.3%	0.67 [-0.34, 1.69]
Coratella(CMJ) 2019	43.5	6.9	20	42.6	7.7	20	11.1%	0.12 [-0.50, 0.74]
Lundberg(PP) 2019	158.1	61.7	16	153	64.5	16	10.3%	0.08 [-0.61, 0.77]
Madruga-Parera(UCMJD) 2022	19.15	3.72	17	17.24	3.45	17	10.4%	0.52 [-0.17, 1.20]
Maroto-Izquierdo(PP) 2017	1009.5	67.4	15	954.9	45.1	14	9.5%	0.92 [0.15, 1.69]
Murton(CMJ) 2021	38.45	3.64	8	39.45	5.36	8	7.5%	-0.21 [-1.19, 0.78]
Norrbrand(PP) 2008	345	123	7	212	43	8	6.2%	1.40 [0.23, 2.57]
Puustinen(CMJ) 2023	43.8	3.2	9	44.1	5.6	9	8.0%	-0.06 [-0.99, 0.86]
Stojanovic(CMJ) 2021	59.29	2.97	12	55.22	3.07	12	8.3%	1.30 [0.41, 2.20]
Weng(CMJ) 2022	62.1	5.55	12	56.4	5.3	10	8.2%	1.01 [0.11, 1.91]
Total (95% CI)			**144**			**142**	**100.0%**	**0.47 [0.10, 0.84]**

Heterogeneity: Tau² = 0.22; Chi² = 23.91, df = 11 (P = 0.01); I² = 54%
Test for overall effect: Z = 2.50 (P = 0.01)

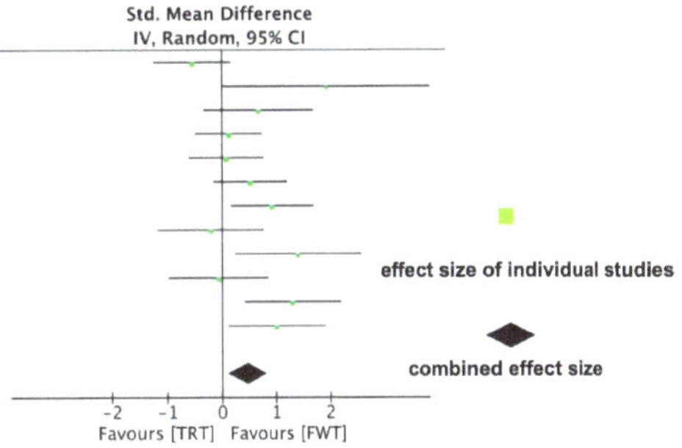

Figure 5. Forest plot with meta-analysis of standardized mean difference showing comparison of flywheel training (FWT) versus traditional resistance training (TRT) on muscle power [36,39,41–49].

3.5.2. Meta-Analysis Results on Maximal Strength

Ten of the included reports examined the effects of FWT versus TRT on maximal muscle strength, with measures encompassing 1 RM, peak torque, and maximal voluntary contraction. The analysis (Figure 6) revealed no significant differences in outcomes between the two training modalities (ES = 0.24; 95% CI: −0.26–0.74, $p = 0.35$), although a high degree of heterogeneity was observed ($I^2 = 74\%$).

Study or Subgroup	FWT Mean	FWT SD	FWT Total	TRT Mean	TRT SD	TRT Total	Weight	Std. Mean Difference IV, Random, 95% CI
Arsenis(BS) 2019	159.06	19.42	16	172.18	21.52	16	10.8%	-0.62 [-1.34, 0.09]
Banks(KE) 2024	437.2	126.9	11	432	121.5	12	10.1%	0.04 [-0.78, 0.86]
Coratella(BS) 2019	100.1	16.94	20	102.41	16.17	20	11.4%	-0.14 [-0.76, 0.48]
de Hoyo(KE) 2015	1096	247.4	11	1244.4	139.1	12	9.9%	-0.72 [-1.57, 0.13]
Lundberg(KE) 2019	23.1	11.3	16	23.4	11	16	10.9%	-0.03 [-0.72, 0.67]
Maroto-Izquierdo(KE) 2017	290.3	12.5	15	262	9.1	14	8.9%	2.50 [1.50, 3.51]
Norrbrand(KE) 2008	599	156	7	482	63	8	8.4%	0.95 [-0.14, 2.04]
Norrbrand(KE) 2010	622	91	9	623	103	8	9.2%	-0.01 [-0.96, 0.94]
Onambélé(KE) 2008	122	14	12	116	14.2	12	10.2%	0.41 [-0.40, 1.22]
Stojanovic(HS) 2021	109.83	7.81	12	105.25	9.36	12	10.1%	0.51 [-0.30, 1.33]
Total (95% CI)			**129**			**130**	**100.0%**	**0.24 [-0.26, 0.74]**

Heterogeneity: Tau² = 0.47; Chi² = 34.09, df = 9 (P < 0.0001); I² = 74%
Test for overall effect: Z = 0.94 (P = 0.35)

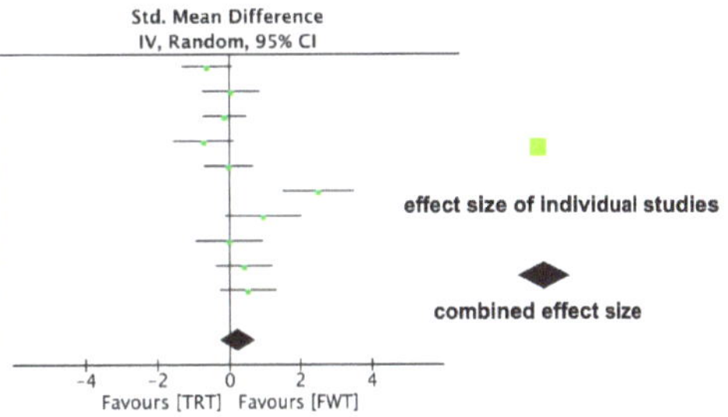

Figure 6. Forest plot with meta-analysis of standardized mean difference showing comparison of flywheel training (FWT) versus traditional resistance training (TRT) on maximal strength [35–42,44,50].

3.6. *Subgroup Analysis*

3.6.1. Subject-Related Moderating Variables

The impact of strength training experience on the differential effects of FWT versus TRT on maximal muscle strength and muscle power is illustrated in Figure 7. A univariate subgroup analysis indicated that strength training experience did not significantly influence the FWT/TRT effects on maximal muscle strength ($p = 0.35$, Figure 7B), while it did play a crucial role in modulating the effects on muscle power ($p = 0.01$, Figure 7A). There were significant and moderate-sized effects in favor of FWT over TRT for muscle power among well-trained individuals (SMD = 0.58, $p = 0.04$), alongside large-sized effects observed for muscle power in untrained individuals (SMD = 1.40, $p = 0.02$). However, for maximal muscle strength, there were no notable differences between FWT and TRT in both untrained and trained participants ($p > 0.05$).

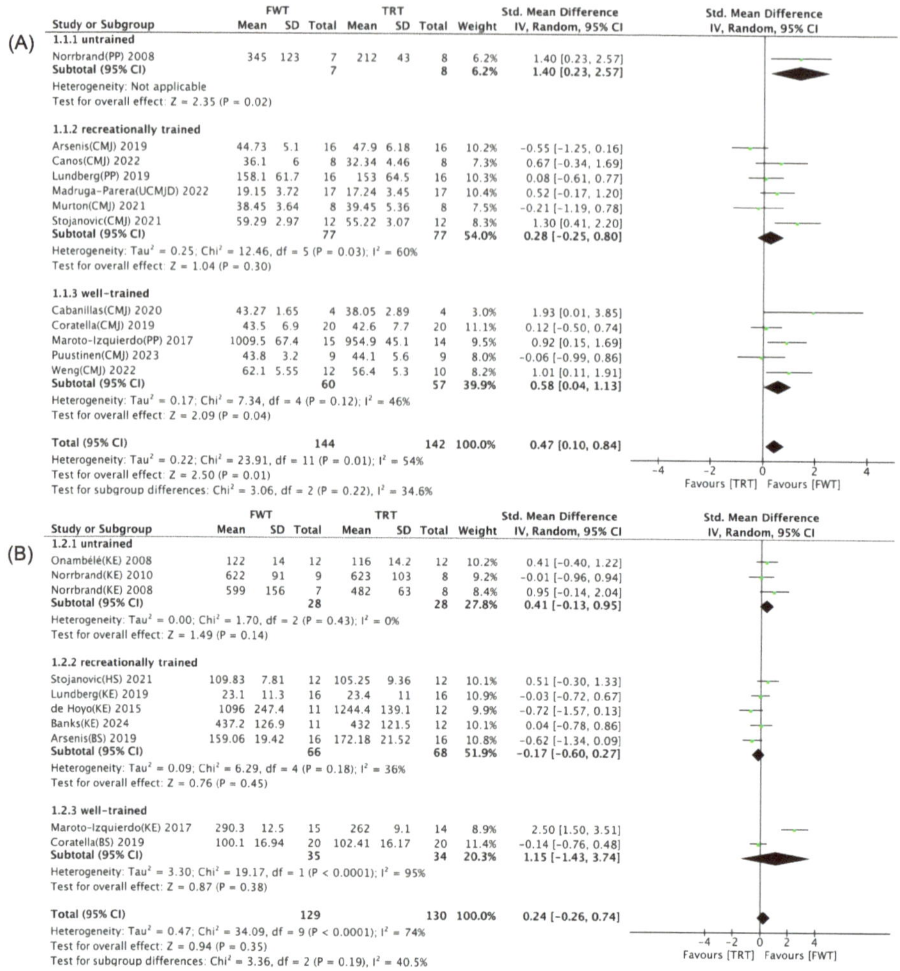

Figure 7. Forest plot with subgroup analysis of strength training experience [35–50]. (**A**) Effect of strength training experience on muscle power. (**B**) Effect of strength training experience on maximal muscle strength. ■ effect size of individual studies. ◆ combined effect size.

3.6.2. Training-Related Programming Parameters

The effects of training-related programming parameters for FWT/TRT on maximal muscle strength and muscle power are illustrated in Figures 8–10. Univariate subgroup analyses highlighted that the total number of training sessions, the type of control group intervention, and the selected exercise significantly influenced the impact of FWT versus TRT on muscle power, with notable distinctions based on these variables ($p < 0.05$). As shown in Figure 8A, a significant and moderate enhancement of muscle power was noted with FWT for those undertaking 12–18 training sessions (SMD = 0.61; 95% CI: 0.12–1.09, $p = 0.01$), contrasting with a lack of such improvement for schedules with fewer than 12 or more than 18 sessions ($p > 0.05$). Furthermore, muscle power significantly increased (Figure 9A) when comparing FWT with weight stack training (SMD = 0.61; 95% CI: 0.21–1.00, $p = 0.003$), unlike its free weight training counterpart ($p = 0.22$). According to Figure 10A, performing flywheel training with squat and lunge exercises significantly enhances muscle power (SMD = 0.43; 95% CI: 0.02–0.84, $p = 0.04$). However, when using knee extension exercises, there is no significant difference between FWT and TRT ($p = 0.06$).

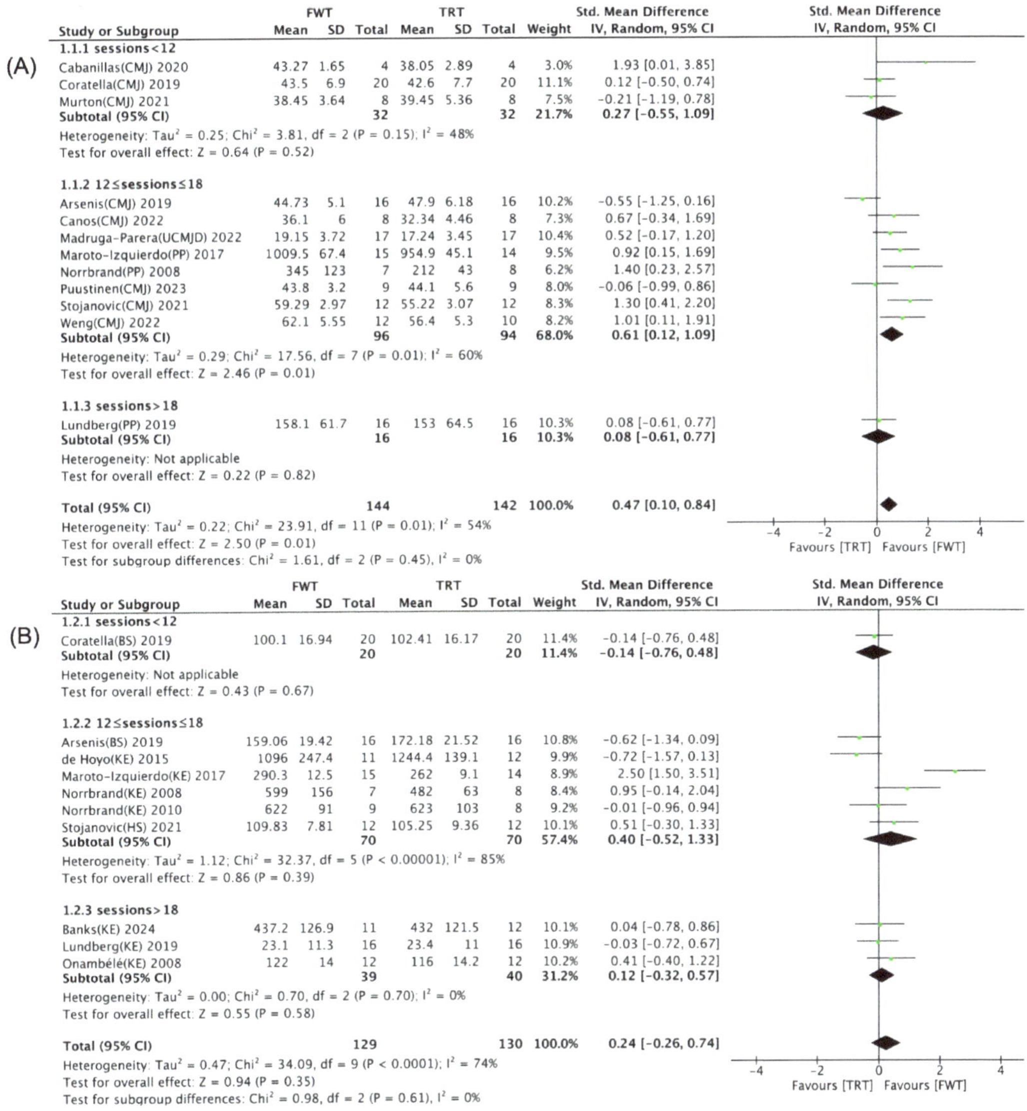

Figure 8. Forest plot with subgroup analysis of total number of training sessions [35–50]. (**A**) Effect of total number of training sessions on muscle power. (**B**) Effect of total number of training sessions on maximal muscle strength. ■ effect size of individual studies. ◆ combined effect size.

(A)

Study or Subgroup	FWT Mean	FWT SD	FWT Total	TRT Mean	TRT SD	TRT Total	Weight	Std. Mean Difference IV, Random, 95% CI
1.1.1 weight-stack machine								
Canos(CMJ) 2022	36.1	6	8	32.34	4.46	8	7.3%	0.67 [−0.34, 1.69]
Lundberg(PP) 2019	158.1	61.7	16	153	64.5	16	10.3%	0.08 [−0.61, 0.77]
Madruga-Parera(UCMJD) 2022	19.15	3.72	17	17.24	3.45	17	10.4%	0.52 [−0.17, 1.20]
Maroto-Izquierdo(PP) 2017	1009.5	67.4	15	954.9	45.1	14	9.5%	0.92 [0.15, 1.69]
Norrbrand(PP) 2008	345	123	7	212	43	8	6.2%	1.40 [0.23, 2.57]
Subtotal (95% CI)			**63**			**63**	**43.6%**	**0.61 [0.21, 1.00]**
Heterogeneity: Tau² = 0.03; Chi² = 4.70, df = 4 (P = 0.32); I² = 15%								
Test for overall effect: Z = 2.99 (P = 0.003)								
1.1.2 free weights(barbell)								
Arsenis(CMJ) 2019	44.73	5.1	16	47.9	6.18	16	10.2%	−0.55 [−1.25, 0.16]
Cabanillas(CMJ) 2020	43.27	1.65	4	38.05	2.89	4	3.0%	1.93 [0.01, 3.85]
Coratella(CMJ) 2019	43.5	6.9	20	42.6	7.7	20	11.1%	0.12 [−0.50, 0.74]
Murton(CMJ) 2021	38.45	3.64	8	39.45	5.36	8	7.5%	−0.21 [−1.19, 0.78]
Puustinen(CMJ) 2023	43.8	3.2	9	44.1	5.6	9	8.0%	−0.06 [−0.99, 0.86]
Stojanovic(CMJ) 2021	59.29	2.97	12	55.22	3.07	12	8.3%	1.30 [0.41, 2.20]
Weng(CMJ) 2022	62.1	5.55	12	56.4	5.3	10	8.2%	1.01 [0.11, 1.91]
Subtotal (95% CI)			**81**			**79**	**56.4%**	**0.36 [−0.22, 0.94]**
Heterogeneity: Tau² = 0.37; Chi² = 17.24, df = 6 (P = 0.008); I² = 65%								
Test for overall effect: Z = 1.22 (P = 0.22)								
Total (95% CI)			**144**			**142**	**100.0%**	**0.47 [0.10, 0.84]**
Heterogeneity: Tau² = 0.22; Chi² = 23.91, df = 11 (P = 0.01); I² = 54%								
Test for overall effect: Z = 2.50 (P = 0.01)								
Test for subgroup differences: Chi² = 0.48, df = 1 (P = 0.49), I² = 0%								

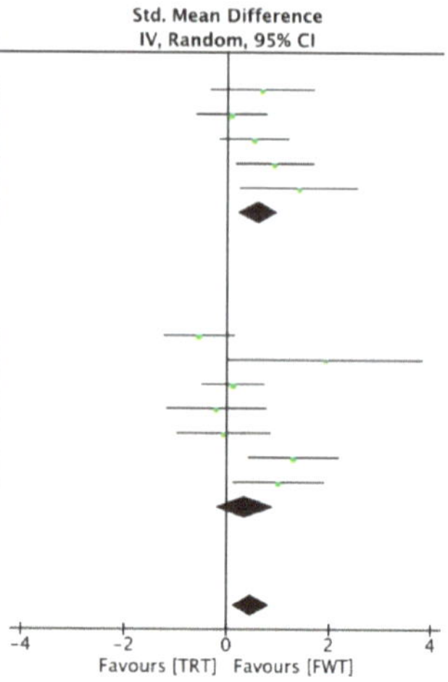

(B)

Study or Subgroup	FWT Mean	FWT SD	FWT Total	TRT Mean	TRT SD	TRT Total	Weight	Std. Mean Difference IV, Random, 95% CI
1.2.1 weight-stack machine								
de Hoyo(KE) 2015	1096	247.4	11	1244.4	139.1	12	9.9%	−0.72 [−1.57, 0.13]
Lundberg(KE) 2019	23.1	11.3	16	23.4	11	16	10.9%	−0.03 [−0.72, 0.67]
Maroto-Izquierdo(KE) 2017	290.3	12.5	15	262	9.1	14	8.9%	2.50 [1.50, 3.51]
Norrbrand(KE) 2008	599	156	7	482	63	8	8.4%	0.95 [−0.14, 2.04]
Norrbrand(KE) 2010	622	91	9	623	103	8	9.2%	−0.01 [−0.96, 0.94]
Onambélé(KE) 2008	122	14	12	116	14.2	12	10.2%	0.41 [−0.40, 1.22]
Subtotal (95% CI)			**70**			**70**	**57.5%**	**0.49 [−0.35, 1.32]**
Heterogeneity: Tau² = 0.87; Chi² = 26.59, df = 5 (P < 0.0001); I² = 81%								
Test for overall effect: Z = 1.14 (P = 0.25)								
1.2.2 free weights(barbell)								
Arsenis(BS) 2019	159.06	19.42	16	172.18	21.52	16	10.8%	−0.62 [−1.34, 0.09]
Banks(KE) 2024	437.2	126.9	11	432	121.5	12	10.1%	0.04 [−0.78, 0.86]
Coratella(BS) 2019	100.1	16.94	20	102.41	16.17	20	11.4%	−0.14 [−0.76, 0.48]
Stojanovic(HS) 2021	109.83	7.81	12	105.25	9.36	12	10.1%	0.51 [−0.30, 1.33]
Subtotal (95% CI)			**59**			**60**	**42.5%**	**−0.09 [−0.53, 0.36]**
Heterogeneity: Tau² = 0.06; Chi² = 4.38, df = 3 (P = 0.22); I² = 31%								
Test for overall effect: Z = 0.38 (P = 0.70)								
Total (95% CI)			**129**			**130**	**100.0%**	**0.24 [−0.26, 0.74]**
Heterogeneity: Tau² = 0.47; Chi² = 34.09, df = 9 (P < 0.0001); I² = 74%								
Test for overall effect: Z = 0.94 (P = 0.35)								
Test for subgroup differences: Chi² = 1.41, df = 1 (P = 0.23), I² = 29.2%								

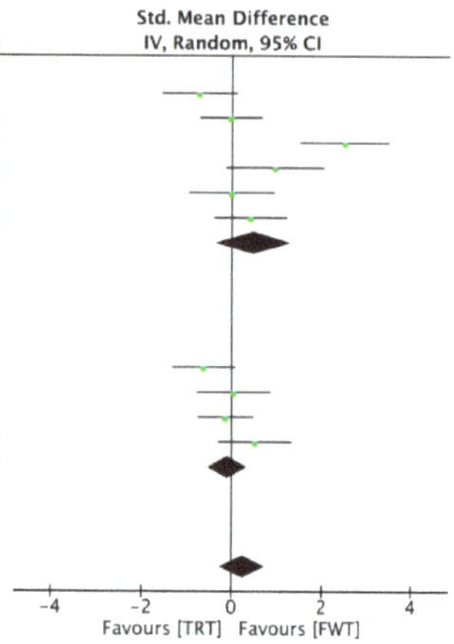

Figure 9. Forest plot with subgroup analysis of control group's intervention [35–50]. (**A**) Effect of control group's intervention on muscle power. (**B**) Effect of control group's intervention on maximal muscle strength. ■ effect size of individual studies. ◆ combined effect size.

Regarding the total number of training sessions (Figure 8B), there were no significant differences between the two training modalities in maximal muscle strength, regardless of whether participants undertook 12–18 sessions or fewer than 12 or more than 18 sessions ($p > 0.05$). In terms of the control group's intervention (Figure 9B), no significant differences were detected in maximal muscle strength when performing weight stack or free weights training as the control intervention ($p > 0.05$). Regardless of whether single-joint or multi-joint training exercises are selected, there is no statistically significant difference between FWT and TRT in enhancing maximum muscle strength ($p > 0.05$). However, as shown in Figure 10B, selecting knee extension exercises during FWT is more conducive to maximum strength gains (SMD = 0.41; 95% CI: −0.29–1.12), while choosing squat exercises during TRT yields better results for maximum strength (SMD = −0.11; 95% CI: −0.71–0.49).

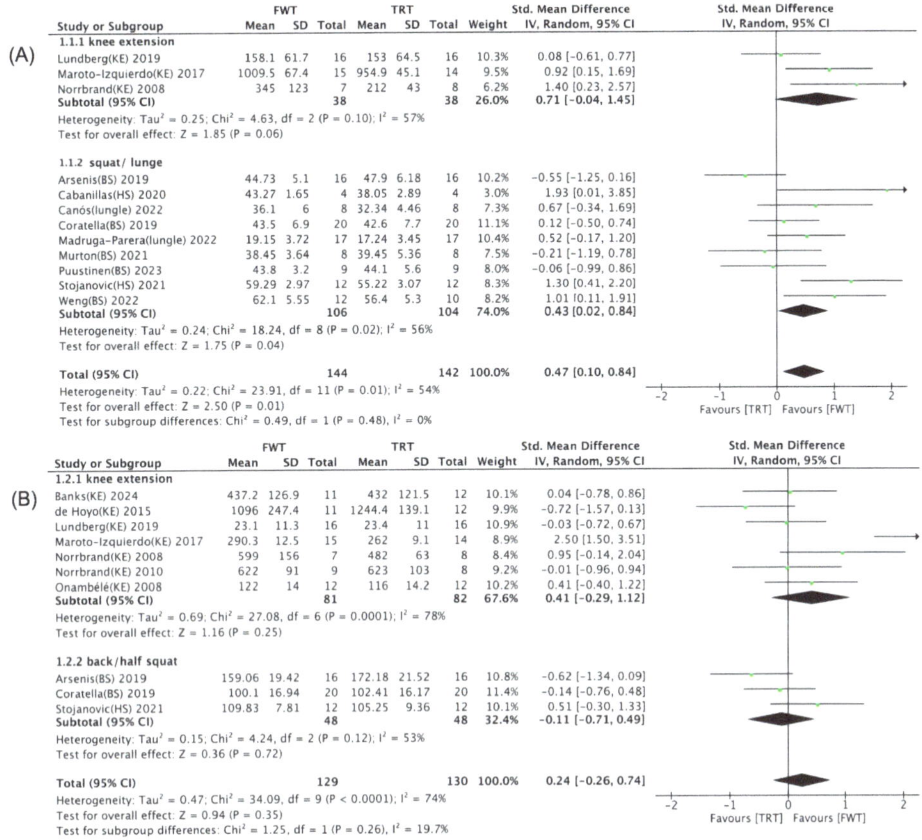

Figure 10. Forest plot with subgroup analysis of selected exercise [35–50]. (**A**) Effect of selected exercise on muscle power. (**B**) Effect of selected exercise on maximal muscle strength. ■ effect size of individual studies. ◆ combined effect size.

4. Discussion

4.1. Main Analysis

This study systematically synthesized and quantified existing evidence, comparing the effects of FWT versus TRT on muscle power and maximal strength. Sixteen studies meeting the inclusion criteria were analyzed, with twelve studies assessing muscle power and ten studies focusing on maximal muscle strength. This meta-analysis included only studies involving healthy individuals, both untrained and trained, further subdivided into well-trained and recreationally trained categories. The results indicate a small but significant increase in muscle power with FWT (ES = 0.47; 95% CI: 0.10–0.84, $p = 0.01$) compared to TRT. From a molecular perspective, FWT appears to increase mRNA levels of genes expressed predominantly in fast glycolytic fibers, potentially inducing a faster muscle phenotype. This adaptation likely optimizes muscle for explosive, high-speed actions [52]. Furthermore, our analysis revealed no significant difference between FWT and TRT concerning maximal strength ($p > 0.05$). Despite extensive research elucidating the benefits associated with FWT, discrepancies persist regarding its efficacy on maximal strength. A previous meta-analysis encompassing seven studies on FWT concluded that it does not offer clear advantages over TRT in enhancing maximal strength [28]. Conversely, De Keijzer provides a contrasting perspective, affirming the effectiveness of FWT in boosting maximal strength across both healthy and athletic cohorts. De Keijzer's synthesis of 11 pertinent reviews posits FWT as a viable alternative to conventional resistance training, citing improvements in muscular strength, power, and jump performance among diverse population groups [53]. Nevertheless, variables across studies, such as exercise selection, instructional nuances (e.g., delayed eccentric action), training background, frequency, and duration, necessitate careful consideration in interpreting these results.

4.2. Subgroup Analysis

A limitation of existing studies lies in the use of notably different protocols and execution methodologies. For instance, variations in training methods, targeted muscle groups, sets and repetitions performed, and training experience differ significantly among selected studies. However, our study overcame this limitation by conducting subgroup analysis. By categorizing and separately analyzing different variables, such as total number of training sessions, participants' strength training experience, training methods, and selected exercise, we were able to more accurately assess the impact of these factors on training outcomes.

4.2.1. Total Number of Training Sessions

FWT is a unique training mode gaining popularity in the research community. However, the optimal duration of an FWT program for athletes to develop sufficient neural and muscular adaptations, thereby enhancing maximal strength and power, remains elusive. Our meta-analysis, encompassing multiple studies with varied protocols, indicates that 12–18 sessions of FWT result in significantly greater power gains compared to TRT. These findings not only corroborate those reported by Sanchez et al. [54], but also demonstrate increased statistical power (from 103 to 208 participants) and a heightened magnitude of strength performance change, from 0.21 (small) to 0.61 (moderate). FWT was documented to cause subcellular damage to the contractile and structural components of skeletal muscle, inducing local and systemic inflammatory responses [55–57]. Consequently, an overly brief intervention period might involve a recovery process, potentially leading to deteriorated sport performance due to fatigue [58]. Moreover, participants typically require two or three familiarization sessions to acclimate to the training apparatus and techniques. FWT programs encompassing 12–18 sessions yielded favorable outcomes [37,39,46] as they afford sufficient recovery and over-compensation periods, which is advantageous for muscle performance. Conversely, excessively prolonged interventions may trigger physiological and neural adaptations, or even training burnout, hindering further improvement or even causing regression in sports capabilities [59]. Notably, eccentrically induced muscle damage can alter resting metabolic rates for up to 48 h post-exercise [60–62]. Buonsenso et al. also suggested that 2–3 times per week of FWT is an optimal frequency to improve jump performance [18]. Thus, we recommend coaches and trainers allocate a 6-week training block, comprising 2–3 weekly sessions, each followed by 48 h of recovery, to foster optimal muscle performance through FWT. To avoid training adaptations resulting from long-term FWT, it is recommended to increase the moment of inertia or shorten rest intervals to enhance training intensity when the total number of sessions exceeds 18.

4.2.2. Strength Training Experience

When subjects were stratified according to training experience, untrained and well-trained individuals achieved significantly greater power gains with FWT than with TRT. It should be highlighted that subjects labeled as "well-trained" in our study are typically considered "elite athletes," participating in professional or semi-professional leagues. Suarez-Arrones et al. found similar results, applying FWT to elite soccer players throughout an entire competitive season, which significantly improved half squat power output [63]. Although only one study on the untrained population met the criteria for inclusion in the subgroup analysis, Sáez-Michea demonstrated that strength training with the isoinertial method effectively improves CMJ jump ability, running velocity, and dynamic postural balance in healthy untrained adults [64]. Due to the limited number of studies, further high-quality investigations are necessary to confirm current findings. In addition, the power increment observed for recreationally trained subjects undertaking an FWT program versus a traditional program did not vary significantly (ES = 0.28; 95% CI: −0.25–0.80, $p = 0.30$). In our analysis, "recreationally trained" refers to participants other than well-trained athletes with more than one year of strength training experience. As consistently reported in the literature, there is no significant increase in eccentric hamstring strength following a six-week flywheel leg curl protocol [65]. These results are inconsistent with those of Allen, who noted that diverse FWT interventions can effectively improve strength, power, and jump performance in male soccer players of varying levels [66]. Differences in maximum neural activation and recovery ability between training sessions of athletes of different technical levels may explain the differences in power outcomes. Another explanation could be that well-trained athletes, having more familiarity with flywheel tempo, engage more actively in both concentric and eccentric actions, thereby achieving greater power gains. Indeed, the enhanced eccentric overload generated during isoinertial FWT is generally more pronounced in individuals with prior strength experience, underscoring the necessity of proper technique to optimize this training method's benefits [67]. Thus, it is possible to conclude that well-trained athletes may be able to reap greater benefits from FWT than recreationally trained individuals.

In our analysis, we also examined the effects of TRT versus FWT on maximal muscle strength among subjects with different training backgrounds. However, no significant differences were detected between TRT and FWT in either trained or untrained populations. A previous review confirmed that FWT is a valid strategy to improve strength; however, differences with traditional training programs were not clearly established, making it impossible to state that FWT is superior to TRT methodologies [68]. Flywheel eccentric overload training protocols did not improve lower-body one-repetition maximum (1 RM) more effectively than traditional training methods, but the evidence is insufficient due to decreased compliance with the intervention, which was connected to the effects of delayed onset muscle soreness [69]. Interestingly, Sagelv et al. reported improvement in maximal squat strength in amateur soccer players, with more significant gains after a traditional squat protocol than with a FWT program [70]. However, the dropout rate of 18.75% in this study may reduce the credibility of the results. Such discrepancies in reported outcomes may stem from variations in exercise selection and loading parameters used in each study. Our meta-analysis revealed that FWT is a favorable strategy for improving power in both elite athletes and untrained individuals, presenting clear implications for coaches and sports science specialists. However, it is pertinent to acknowledge that this meta-analysis encompassed a limited number of studies. Therefore, conducting further robust research is essential for a more comprehensive understanding of performance adaptations following FWT interventions.

4.2.3. Control Group's Intervention

In our meta-analysis, the control group interventions included barbell free weights training and weight stack training. Based on the available data, inertial flywheel resistance training was superior to weight stack resistance training in enhancing muscle power, while no significant differences were detected when comparing it to barbell free weights training. Our findings confirm those of the study by Alkner, which concluded that quadriceps muscle activation was superior in flywheel exercise compared to weight stack knee extension exercise [71]. This outcome aligns with the findings of Nunez Sanchez, who noted that FWT provided additional benefits to muscle strength compared with knee extension machines [54]. In contrast, Raya-Gonzalez identified no differences between inertial flywheel and barbell free weights training in muscle strength [68]. This discrepancy may be due to isolated single-joint movements not fully exploiting the stretch-shortening cycle, whereas multi-joint exercises promote the recruitment of a greater number of motor units. Both free weight training and FWT share the benefit of activating more muscle groups, which is not possible with weight stack training. Allen suggested that enhanced utilization of elastic potential energy during the stretch-shortening cycle (SSC) and increased muscle–tendon unit stiffness from FWT could significantly boost jump performance [66]. Similarly, strength augmentations potentially arise from neural adaptations, such as heightened neural drive, modified motor unit firing rates, and improved motor unit synchronization, all of which might be amplified by multi-joint movements [62,72]. However, some meta-analyses generated contradictory results, showing more significant improvements in maximal strength and power output with inertial flywheel resistance training than with gravity-dependent resistance training [52,66,73,74]. These differences could be due to heterogeneity between participants or differences in training volume, rest intervals, and the inertia of the flywheels used in each study. A range of inertial intensities (0.025–0.11 kg·m^2) are generally recommended to induce chronic adaptations and enhance athletic performance [41,75]. It is found that higher inertial intensities may be preferable for developing force, while lower inertial intensities could be used for power purposes [76]. Sabido suggested prescribing 3 min rest intervals when performing flywheel squat exercises regardless of the inertial load; conversely, when using 2 min rest intervals, the inertial load should be light [77].

4.2.4. Selected Exercise

This study aligns with the findings of Loren Z., [78] demonstrating that flywheel inertial resistance is particularly suitable for lower limb exercises, such as squats and lunges, because it increases the demand on the hip extensors and ankle plantar flexors while reducing the mechanical demand on the knee extensors. Based on the number of joints involved during exercise, training movements can be classified into single-joint and multi-joint exercises. In our meta-analysis, most researchers selected knee extension as the preferred single-joint exercise, while squats and lunges were the main multi-joint exercises. Anatomically, squats engage multiple muscle groups, including the quadriceps, gluteus maximus, hamstrings, adductors, lower back, core, and calf muscles. Strengthening these muscle groups inevitably enhances lower limb explosive power. Biomechanically, squats require practitioners to control the eccentric lowering phase and quickly overcome resistance during the upward phase, effectively utilizing the elastic potential energy stored during the SSC [79], thereby promoting the growth of lower limb explosive power. Leg extension is a classic method for isolating and training the quadriceps. Studies [40] found that 8 weeks (2–3 days per week) of knee extension using flywheel training technology increased quadriceps femoris muscle hypertrophy by 8%, with similar increases in 1 RM and peak power observed in weight stack resistance exercise. This further confirms the findings of this study.

5. Limitations

Recognizing the limitations of this meta-analysis, it is important to note that numerous studies were excluded due to incomplete data or strict inclusion criteria, resulting in a limited number of studies available for subgroup analysis. Most of the investigations included in this review were conducted on male elite athletes, youth athletes, or recreationally trained individuals, thereby limiting the applicability of the conclusions to female and elderly populations. Further research into the effects of FWT on the physical performance of women and older individuals is essential. Another potential shortcoming of this review is the likelihood of selection bias. Our selection was confined to full-text papers published in English, potentially introducing a language bias. Additionally, the diversity in exercise interventions (such as intervention duration, exercises employed, volume, and intensity) possibly significantly influenced the outcomes observed in the meta-analysis. Compounding this, specific details of the training protocols, such as inertial loads, were not consistently documented in the studies reviewed. Despite these constraints, this meta-analysis provides a comprehensive overview of existing research and elucidates, based on scientific literature, the advantages of employing FWT to enhance muscle power.

6. Future Research Perspectives

Future studies are necessary to conduct multi-center, large-sample, and long-term randomized controlled trials to provide more reliable evidence-based support for clinical practice. It is recommended that specific characteristics of FWT (volume, intensities, duration, rest interval, and exercise selection) be clearly identified to ascertain the dose–response relationship that maximizes improvements across various demographics, such as different age groups (e.g., adults vs. elders), genders (e.g., male vs. female), types of sports (e.g., volleyball vs. basketball), and athletic levels (e.g., amateur vs. professional). Additionally, exploring the impact of diverse tempo strategies, including accelerated and delayed movement velocity during concentric and eccentric phases, on training outcomes is crucial. Future research also needs to explore load quantification and monitoring in FWT, particularly clarifying the relationship between inertia power and velocity power. Another research direction worth pursuing is the distinct effects of training with different equipment, such as vertical flywheel devices, horizontal flywheel devices, and seated leg curl flywheel devices.

7. Conclusions

In conclusion, the results of this systematic review and meta-analysis indicate that isoinertial FWT is an effective tool for enhancing performance aspects closely tied to muscle power, such as countermovement jump and peak power output, in both healthy untrained and well-trained individuals. Moreover, choosing squats and lunges for FWT is more suitable for improving lower limb explosive power. It is recommended that coaches and trainers plan for a six-week period, conducting sessions 2–3 times a week, with an interval of at least 48 h between each session. It is necessary to increase the moment of inertia or shorten rest intervals to enhance training intensity when the total number of sessions exceeds 18. Furthermore, this meta-analysis found that inertial flywheel resistance training is superior to weight stack resistance training in enhancing muscle power, while no significant differences were found when compared to barbell free weights training.

Author Contributions: Z.H. and F.L. were instrumental in the conception and design of the study. K.H. and H.H. were responsible for organizing the database. Statistical analyses were carried out by Z.H. and Y.L., who also drafted the manuscript. Their efforts were supplemented by K.H. and H.H. who assisted in interpreting the data and further enriched the study with their additional suggestions. The manuscript was thoroughly revised and finalized by X.Y. All authors have read and agreed to the published version of the manuscript.

Funding: This investigation did not receive any grants and was not performed with financial support from any funding agency.

Institutional Review Board Statement: Not applicable.

Informed Consent Statement: Not applicable.

Data Availability Statement: The original contributions presented in the study are included in the article, further inquiries can be directed to the corresponding author.

Acknowledgments: We would like to express our gratitude to the authors of the included studies for generously providing the essential information required to conduct our systematic review with meta-analysis.

Conflicts of Interest: The authors declare that the research was conducted in the absence of any commercial or financial relationships that could be construed as a potential conflict of interest.

References

1. Schumann, M.; Feuerbacher, J.F.; Sunkeler, M.; Freitag, N.; Ronnestad, B.R.; Doma, K.; Lundberg, T.R. Compatibility of Concurrent Aerobic and Strength Training for Skeletal Muscle Size and Function: An Updated Systematic Review and Meta-Analysis. *Sports Med.* **2022**, *52*, 601–612. [CrossRef]
2. Suchomel, T.J.; Nimphius, S.; Bellon, C.R.; Stone, M.H. The Importance of Muscular Strength: Training Considerations. *Sports Med.* **2018**, *48*, 765–785. [CrossRef] [PubMed]

3. Ratamess, N.A.; Alvar, B.A.; Evetoch, T.E.; Housh, T.J.; Ben Kibler, W.; Kraemer, W.J.; Triplett, N.T. Progression Models in Resistance Training for Healthy Adults. *Med. Sci. Sports Exerc.* **2009**, *41*, 687–708. [CrossRef]
4. Morris, S.J.; Oliver, J.L.; Pedley, J.S.; Haff, G.G.; Lloyd, R.S. Comparison of Weightlifting, Traditional Resistance Training and Plyometrics on Strength, Power and Speed: A Systematic Review with Meta-Analysis. *Sports Med.* **2022**, *52*, 1533–1554. [CrossRef] [PubMed]
5. Roig, M.; O'Brien, K.; Kirk, G.; Murray, R.; McKinnon, P.; Shadgan, B.; Reid, W.D. The effects of eccentric versus concentric resistance training on muscle strength and mass in healthy adults: A systematic review with meta-analysis. *Br. J. Sports Med.* **2009**, *43*, 556–568. [CrossRef]
6. Franchi, M.V.; Reeves, N.D.; Narici, M.V. Skeletal Muscle Remodeling in Response to Eccentric vs. Concentric Loading: Morphological, Molecular, and Metabolic Adaptations. *Front. Physiol.* **2017**, *8*, 447. [CrossRef] [PubMed]
7. Sogaard, K.; Christensen, H.; Jensen, B.R.; Finsen, L.; Sjogaard, G. Motor control and kinetics during low level concentric and eccentric contractions in man. *Electroencephalogr. Clin. Neurophysiol.* **1996**, *101*, 453–460. [CrossRef]
8. Gehlert, S.; Bloch, W.; Suhr, F. Ca^{2+}-dependent regulations and signaling in skeletal muscle: From electro-mechanical coupling to adaptation. *Int. J. Mol. Sci.* **2015**, *16*, 1066–1095. [CrossRef]
9. Hortobagyi, T.; Devita, P.; Money, J.; Barrier, J. Effects of standard and eccentric overload strength training in young women. *Med. Sci. Sports Exerc.* **2001**, *33*, 1206–1212. [CrossRef]
10. Suchomel, T.J.; Wagle, J.P.; Douglas, J.; Taber, C.B.; Harden, M.; Haff, G.G.; Stone, M.H. Implementing Eccentric Resistance Training-Part 2: Practical Recommendations. *J. Funct. Morphol. Kinesiol.* **2019**, *4*, 55. [CrossRef]
11. Patus, J. Accentuated Eccentric Loading is Superior to Traditional Loading for Improving Acute Countermovement Jump Performance in Adult, Resistance-Trained Males. *J. Sport. Rehabil.* **2021**, *30*, 956–960. [CrossRef] [PubMed]
12. van den Tillaar, R.; Andersen, V.; Saeterbakken, A.H. The Existence of a Sticking Region in Free Weight Squats. *J. Hum. Kinet.* **2014**, *42*, 63–71. [CrossRef] [PubMed]
13. Walker, S.; Blazevich, A.J.; Haff, G.; Tufano, J.J.; Newton, R.U.; Hakkinen, K. Greater Strength Gains after Training with Accentuated Eccentric than Traditional Isoinertial Loads in Already Strength-Trained Men. *Front. Physiol.* **2016**, *7*, 149. [CrossRef] [PubMed]
14. Beato, M.; Dello Iacono, A. Implementing Flywheel (Isoinertial) Exercise in Strength Training: Current Evidence, Practical Recommendations, and Future Directions. *Front. Physiol.* **2020**, *11*, 569. [CrossRef] [PubMed]
15. Monajati, A.; Larumbe-Zabala, E.; Goss-Sampson, M.; Naclerio, F. Injury Prevention Programs Based on Flywheel vs. Body Weight Resistance in Recreational Athletes. *J. Strength. Cond. Res.* **2021**, *35* (Suppl. 1), S188–S196. [CrossRef] [PubMed]
16. Timón, R.; Ponce-González, J.G.; González-Montesinos, J.L.; Olcina, G.; Pérez-Pérez, A.; Castro-Piñero, J. Inertial flywheel resistance training and muscle oxygen saturation. *J. Sports Med. Phys. Fit.* **2018**, *58*, 1618–1624. [CrossRef] [PubMed]
17. Friedmann-Bette, B.; Bauer, T.; Kinscherf, R.; Vorwald, S.; Klute, K.; Bischoff, D.; Muller, H.; Weber, M.A.; Metz, J.; Kauczor, H.U.; et al. Effects of strength training with eccentric overload on muscle adaptation in male athletes. *Eur. J. Appl. Physiol.* **2010**, *108*, 821–836. [CrossRef] [PubMed]
18. Buonsenso, A.; Centorbi, M.; Iuliano, E.; Di Martino, G.; Della Valle, C.; Fiorilli, G.; Calcagno, G.; di Cagno, A. A Systematic Review of Flywheel Training Effectiveness and Application on Sport Specific Performances. *Sports* **2023**, *11*, 76. [CrossRef]
19. Tesch, P.A.; Fernandez-Gonzalo, R.; Lundberg, T.R. Clinical Applications of Iso-Inertial, Eccentric-Overload (YoYo) Resistance Exercise. *Front. Physiol.* **2017**, *8*, 241. [CrossRef]
20. Xu, J.; Thompson, B.J.; Spencer, S.B.; Studenka, B.E.; Bressel, E. Effects of Flywheel Resistance Training on Muscle Function and Sport-specific Performance in Collegiate Club Water Polo Players. *Res. Q. Exerc. Sport* **2023**, *94*, 98–109. [CrossRef]
21. Tseng, K.W.; Chen, J.R.; Chow, J.J.; Tseng, W.C.; Condello, G.; Tai, H.L.; Fu, S.K. Post-activation Performance Enhancement after a Bout of Accentuated Eccentric Loading in Collegiate Male Volleyball Players. *Int. J. Environ. Res. Public Health* **2021**, *18*, 13110. [CrossRef] [PubMed]
22. Xie, H.; Zhang, W.; Chen, X.; He, J.; Lu, J.; Gao, Y.; Li, D.; Li, G.; Ji, H.; Sun, J. Flywheel eccentric overload exercises versus barbell half squats for basketball players: Which is better for induction of post-activation performance enhancement? *PLoS ONE* **2022**, *17*, e0277432. [CrossRef]
23. Maroto-Izquierdo, S.; Bautista, I.J.; Martin Rivera, F. Post-activation performance enhancement (PAPE) after a single bout of high-intensity flywheel resistance training. *Biol. Sport* **2020**, *37*, 343–350. [CrossRef] [PubMed]
24. Wagle, J.P.; Taber, C.B.; Cunanan, A.J.; Bingham, G.E.; Carroll, K.M.; DeWeese, B.H.; Sato, K.; Stone, M.H. Accentuated Eccentric Loading for Training and Performance: A Review. *Sports Med.* **2017**, *47*, 2473–2495. [CrossRef]
25. Pecci, J.; Munoz-Lopez, A.; Jones, P.A.; Sanudo, B. Effects of 6 weeks in-season flywheel squat resistance training on strength, vertical jump, change of direction and sprint performance in professional female soccer players. *Biol. Sport* **2023**, *40*, 521–529. [CrossRef]
26. Naczk, M.; Naczk, A.; Brzenczek-Owczarzak, W.; Arlet, J.; Adach, Z. Impact of inertial training on strength and power performance in young active men. *J. Strength. Cond. Res.* **2016**, *30*, 2107–2113. [CrossRef]
27. Toien, T.; Pedersen Haglo, H.; Unhjem, R.; Hoff, J.; Wang, E. Maximal strength training: The impact of eccentric overload. *J. Neurophysiol.* **2018**, *120*, 2868–2876. [CrossRef] [PubMed]

28. Vicens-Bordas, J.; Esteve, E.; Fort-Vanmeerhaeghe, A.; Bandholm, T.; Thorborg, K. Is inertial flywheel resistance training superior to gravity-dependent resistance training in improving muscle strength? A systematic review with meta-analyses. *J. Sci. Med. Sport* **2018**, *21*, 75–83. [CrossRef]
29. Takkouche, B.; Norman, G. PRISMA statement. *Epidemiology* **2011**, *22*, 128. [CrossRef]
30. Liberati, A.; Altman, D.G.; Tetzlaff, J.; Mulrow, C.; Gotzsche, P.C.; Ioannidis, J.P.A.; Clarke, M.; Devereaux, P.J.; Kleijnen, J.; Moher, D. The PRISMA Statement for Reporting Systematic Reviews and Meta-Analyses of Studies That Evaluate Health Care Interventions: Explanation and Elaboration. *Ann. Intern. Med.* **2009**, *151*, W65–W94. [CrossRef]
31. Koo, T.K.; Li, M.Y. A Guideline of Selecting and Reporting Intraclass Correlation Coefficients for Reliability Research. *J. Chiropr. Med.* **2016**, *15*, 155–163. [CrossRef] [PubMed]
32. Higgins, J.P.T.; Cochrane Collaboration. *Cochrane Handbook for Systematic Reviews of Interventions*, 2nd ed.; Wiley-Blackwell: Hoboken, NJ, USA, 2020.
33. Cohen, J. *Statistical Power Analysis for the Behavioral Sciences*, 2nd ed.; L. Erlbaum Associates: Hillsdale, NJ, USA, 1988; 567p.
34. Higgins, J.P.; Thompson, S.G.; Deeks, J.J.; Altman, D.G. Measuring inconsistency in meta-analyses. *BMJ* **2003**, *327*, 557–560. [CrossRef]
35. Onambélé, G.L.; Maganaris, C.N.; Mian, O.S.; Tam, E.; Rejc, E.; McEwan, I.M.; Narici, M.V. Neuromuscular and balance responses to flywheel inertial versus weight training in older persons. *J. Biomech.* **2008**, *41*, 3133–3138. [CrossRef] [PubMed]
36. Norrbrand, L.; Fluckey, J.D.; Pozzo, M.; Tesch, P.A. Resistance training using eccentric overload induces early adaptations in skeletal muscle size. *Eur. J. Appl. Physiol.* **2008**, *102*, 271–281. [CrossRef] [PubMed]
37. Norrbrand, L.; Pozzo, M.; Tesch, P.A. Flywheel resistance training calls for greater eccentric muscle activation than weight training. *Eur. J. Appl. Physiol.* **2010**, *110*, 997–1005. [CrossRef] [PubMed]
38. de Hoyo, M.; Pozzo, M.; Sanudo, B.; Carrasco, L.; Gonzalo-Skok, O.; Dominguez-Cobo, S.; Moran-Camacho, E. Effects of a 10-week in-season eccentric-overload training program on muscle-injury prevention and performance in junior elite soccer players. *Int. J. Sports Physiol. Perform.* **2015**, *10*, 46–52. [CrossRef] [PubMed]
39. Maroto-Izquierdo, S.; Garcia-Lopez, D.; de Paz, J.A. Functional and Muscle-Size Effects of Flywheel Resistance Training with Eccentric-Overload in Professional Handball Players. *J. Hum. Kinet.* **2017**, *60*, 133–143. [CrossRef] [PubMed]
40. Lundberg, T.R.; Garcia-Gutierrez, M.T.; Mandic, M.; Lilja, M.; Fernandez-Gonzalo, R. Regional and muscle-specific adaptations in knee extensor hypertrophy using flywheel versus conventional weight-stack resistance exercise. *Appl. Physiol. Nutr. Metab.* **2019**, *44*, 827–833. [CrossRef] [PubMed]
41. Coratella, G.; Beato, M.; Ce, E.; Scurati, R.; Milanese, C.; Schena, F.; Esposito, F. Effects of in-season enhanced negative work-based vs traditional weight training on change of direction and hamstrings-to-quadriceps ratio in soccer players. *Biol. Sport* **2019**, *36*, 241–248. [CrossRef]
42. Arsenis, S.; Gioftsidou, A.; Smilios, I.; Malliou, P.; Chatzinikolaou, A.O.; Gkogkos, D.; Beneka, A. The effects of flywheel training on the strength and power of lower limbs. *Sport Sci.* **2019**, *12*, 8–19.
43. Cabanillas, R.; Serna, J.; Muñoz-Arroyave, V.; Ramos, J.A.E. Effect of eccentric overload through isoinertial technology in basketball players. *Rev. Bras. Cineantropometria Desempenho Hum.* **2020**, *22*, e59831. [CrossRef]
44. Stojanovic, M.D.M.; Mikic, M.; Drid, P.; Calleja-Gonzalez, J.; Maksimovic, N.; Belegisanin, B.; Sekulovic, V. Greater Power but Not Strength Gains Using Flywheel Versus Equivolumed Traditional Strength Training in Junior Basketball Players. *Int. J. Environ. Res. Public Health* **2021**, *18*, 1181. [CrossRef] [PubMed]
45. Murton, J.; Eager, R.; Drury, B. Comparison of flywheel versus traditional resistance training in elite academy male Rugby union players. *Res. Sports Med.* **2021**, *31*, 214–227. [CrossRef]
46. Canos, J.; Corbi, F.; Colomar, J.; Cirer-Sastre, R.; Baiget, E. Effects of isoinertial or machine-based strength training on performance in tennis players. *Biol. Sport* **2022**, *39*, 505–513. [CrossRef] [PubMed]
47. Weng, Y.; Liu, H.; Ruan, T.; Yang, W.; Wei, H.; Cui, Y.; Ho, I.M.K.; Li, Q. Effects of flywheel resistance training on the running economy of young male well-trained distance runners. *Front. Physiol.* **2022**, *13*, 1060640. [CrossRef]
48. Madruga-Parera, M.; Bishop, C.; Fort-Vanmeerhaeghe, A.; Beato, M.; Gonzalo-Skok, O.; Romero-Rodriguez, D. Effects of 8 Weeks of Isoinertial vs. Cable-Resistance Training on Motor Skills Performance and Interlimb Asymmetries. *J. Strength. Cond. Res.* **2022**, *36*, 1200–1208. [CrossRef]
49. Puustinen, J.; Venojarvi, M.; Haverinen, M.; Lundberg, T.R. Effects of Flywheel vs. Traditional Resistance Training on Neuromuscular Performance of Elite Ice Hockey Players. *J. Strength. Cond. Res.* **2023**, *37*, 136–140. [CrossRef]
50. Banks, N.F.; Rogers, E.M.; Berry, A.C.; Jenkins, N.D.M. Progressive iso-inertial resistance exercise promotes more favorable cardiovascular adaptations than traditional resistance exercise in young adults. *Am. J. Physiol. Heart Circ. Physiol.* **2024**, *326*, H32–H43. [CrossRef]
51. Walker, E.; Hernandez, A.V.; Kattan, M.W. Meta-analysis: Its strengths and limitations. *Clevel. Clin. J. Med.* **2008**, *75*, 431–439. [CrossRef]
52. Maroto-Izquierdo, S.; Garcia-Lopez, D.; Fernandez-Gonzalo, R.; Moreira, O.C.; Gonzalez-Gallego, J.; de Paz, J.A. Skeletal muscle functional and structural adaptations after eccentric overload flywheel resistance training: A systematic review and meta-analysis. *J. Sci. Med. Sport* **2017**, *20*, 943–951. [CrossRef]
53. de Keijzer, K.L.; Raya Gonzalez, J.; Beato, M. The effect of flywheel training on strength and physical capacities in sporting and healthy populations: An umbrella review. *PLoS ONE* **2022**, *17*, e0264375. [CrossRef] [PubMed]

54. Sanchez, F.J.N.; de Villarreal, E.S. Does Flywheel Paradigm Training Improve Muscle Volume and Force? A Meta-Analysis. *J. Strength. Cond. Res.* **2017**, *31*, 3177–3186. [CrossRef]
55. Hody, S.; Croisier, J.L.; Bury, T.; Rogister, B.; Leprince, P. Eccentric Muscle Contractions: Risks and Benefits. *Front. Physiol.* **2019**, *10*, 442082. [CrossRef] [PubMed]
56. Ide, B.N.; Nunes, L.A.S.; Brenzikofer, R.; Macedo, D.V. Time Course of Muscle Damage and Inflammatory Responses to Resistance Training with Eccentric Overload in Trained Individuals. *Mediat. Inflamm.* **2013**, *2013*, 204942. [CrossRef]
57. Fernandez-Gonzalo, R.; Lundberg, T.R.; Alvarez-Alvarez, L.; de Paz, J.A. Muscle damage responses and adaptations to eccentric-overload resistance exercise in men and women. *Eur. J. Appl. Physiol.* **2014**, *114*, 1075–1084. [CrossRef] [PubMed]
58. Annibalini, G.; Contarelli, S.; Lucertini, F.; Guescini, M.; Maggio, S.; Ceccaroli, P.; Gervasi, M.; Ferri Marini, C.; Fardetti, F.; Grassi, E.; et al. Muscle and Systemic Molecular Responses to a Single Flywheel Based Iso-Inertial Training Session in Resistance-Trained Men. *Front. Physiol.* **2019**, *10*, 554. [CrossRef] [PubMed]
59. Bishop, P.A.; Jones, E.; Woods, A.K. Recovery from Training: A Brief Review. *J. Strength. Cond. Res.* **2008**, *22*, 1015–1024. [CrossRef] [PubMed]
60. Burt, D.G.; Lamb, K.; Nicholas, C.; Twist, C. Effects of exercise-induced muscle damage on resting metabolic rate, sub-maximal running and post-exercise oxygen consumption. *Eur. J. Sport Sci.* **2014**, *14*, 337–344. [CrossRef]
61. Javier Núñez, F.; Suarez-Arrones, L.J.; Cater, P.; Mendez-Villanueva, A. The high-pull exercise: A comparison between a versapulley flywheel device and the free weight. *Int. J. Sports Physiol. Perform.* **2017**, *12*, 527–532. [CrossRef]
62. Raeder, C.; Wiewelhove, T.; Westphal-Martinez, M.P.; Fernandez-Fernandez, J.; de Paula Simola, R.A.; Kellmann, M.; Meyer, T.; Pfeiffer, M.; Ferrauti, A. Neuromuscular Fatigue and Physiological Responses After Five Dynamic Squat Exercise Protocols. *J. Strength. Cond. Res.* **2016**, *30*, 953–965. [CrossRef]
63. Suarez-Arrones, L.; Saez de Villarreal, E.; Nunez, F.J.; Di Salvo, V.; Petri, C.; Buccolini, A.; Maldonado, R.A.; Torreno, N.; Mendez-Villanueva, A. In-season eccentric-overload training in elite soccer players: Effects on body composition, strength and sprint performance. *PLoS ONE* **2018**, *13*, e0205332. [CrossRef]
64. Sáez-Michea, E.; Alarcón-Rivera, M.; Valdés-Badilla, P.; Guzmán-Muñoz, E. Efectos de seis semanas de entrenamiento isoinercial sobre la capacidad de salto, velocidad de carrera y equilibrio postural dinámico./Effects of six weeks of isoinertial training on vertical jump performance, running velocity, and dynamic postural balance. *Retos Nuevas Perspect. Educ. Física Deporte Recreación* **2023**, *48*, 291–297. [CrossRef]
65. Presland, J.D.; Timmins, R.G.; Bourne, M.N.; Williams, M.D.; Opar, D.A. The effect of Nordic hamstring exercise training volume on biceps femoris long head architectural adaptation. *Scand. J. Med. Sci. Sports* **2018**, *28*, 1775–1783. [CrossRef]
66. Allen, W.J.C.; De Keijzer, K.L.; Raya-Gonzalez, J.; Castillo, D.; Coratella, G.; Beato, M. Chronic effects of flywheel training on physical capacities in soccer players: A systematic review. *Res. Sports Med.* **2023**, *31*, 228–248. [CrossRef]
67. Tous-Fajardo, J.; Maldonado, R.A.; Quintana, J.M.; Pozzo, M.; Tesch, P.A. The Flywheel Leg-Curl Machine: Offering Eccentric Overload for Hamstring Development. *Int. J. Sports Physiol. Perform.* **2006**, *1*, 293–298. [CrossRef] [PubMed]
68. Raya-Gonzalez, J.; Castillo, D.; de Keijzer, K.L.; Beato, M. Considerations to Optimize Strength and Muscle Mass Gains Through Flywheel Resistance Devices: A Narrative Review. *Strength. Cond. J.* **2023**, *45*, 111–121. [CrossRef]
69. Muñoz-López, A.; Nakamura, F.Y. Measuring and Testing with Flywheel Resistance Training Devices. In *Resistance Training Methods;* Lecture Notes in Bioengineering; Springer: Cham, Switzerland, 2022; pp. 181–194.
70. Sagelv, E.H.; Pedersen, S.; Nilsen, L.P.R.; Casolo, A.; Welde, B.; Randers, M.B.; Pettersen, S.A. Flywheel squats versus free weight high load squats for improving high velocity movements in football. A randomized controlled trial. *BMC Sports Sci. Med. Rehabil.* **2020**, *12*, 61. [CrossRef]
71. Alkner, B.A.; Bring, D.K.I. Muscle Activation During Gravity-Independent Resistance Exercise Compared to Common Exercises. *Aerosp. Med. Hum. Perform.* **2019**, *90*, 506–512. [CrossRef]
72. Suchomel, T.J.; Wagle, J.P.; Douglas, J.; Taber, C.B.; Harden, M.; Haff, G.G.; Stone, M.H. Implementing Eccentric Resistance Training-Part 1: A Brief Review of Existing Methods. *J. Funct. Morphol. Kinesiol.* **2019**, *4*, 38. [CrossRef]
73. Petré, H.; Wernstål, F.; Mattsson, C.M. Effects of Flywheel Training on Strength-Related Variables: A Meta-analysis. *Sports Med. Open* **2018**, *4*, 55. [CrossRef]
74. Cokorilo, N.; Horvatin, M.; Dordevic, D.; Stankovic, M.; Pekas, D. Flywheel Training in Older Adults-A Systematic Review. *Sustainability* **2022**, *14*, 4137. [CrossRef]
75. Maroto-Izquierdo, S.; Fernandez-Gonzalo, R.; Magdi, H.R.; Manzano-Rodriguez, S.; Gonzalez-Gallego, J.; De Paz, J.A. Comparison of the musculoskeletal effects of different iso-inertial resistance training modalities: Flywheel vs. electric-motor. *Eur. J. Sport Sci.* **2019**, *19*, 1184–1194. [CrossRef] [PubMed]
76. Martinez-Aranda, L.M.; Fernandez-Gonzalo, R. Effects of Inertial Setting on Power, Force, Work, and Eccentric Overload During Flywheel Resistance Exercise in Women and Men. *J. Strength. Cond. Res.* **2017**, *31*, 1653–1661. [CrossRef] [PubMed]
77. Sabido, R.; Hernández-Davó, J.L.; Capdepon, L.; Tous-Fajardo, J. How Are Mechanical, Physiological, and Perceptual Variables Affected by the Rest Interval Between Sets During a Flywheel Resistance Session? *Front. Physiol.* **2020**, *11*, 663. [CrossRef]

78. Chiu, L.Z.F.; Salem, G.J. Comparison of joint kinetics during free weight and flywheel resistance exercise. *J. Strength. Cond. Res.* **2006**, *20*, 555–562.
79. Newton, R.U.; Murphy, A.J.; Humphries, B.J.; Wilson, G.J.; Kraemer, W.J.; Häkkinen, K. Influence of load and stretch shortening cycle on the kinematics, kinetics and muscle activation that occurs during explosive upper-body movements. *Eur. J. Appl. Physiol. Occup. Physiol.* **1997**, *75*, 333–342. [CrossRef]

www.ingramcontent.com/pod-product-compliance
Lightning Source LLC
Chambersburg PA
CBHW041401210926
51850CB00013B/3149
* 9 7 8 3 7 2 5 8 3 7 2 1 2 *